Probability and Mathematical Statistics (Continued)

*PARZEN · Modern Probability Theory and Its Applications

PRESS · ...tions PUKELSHEIM ·

Opt...

PURI ...

PURI, ...lied

Stat...

RAO ·

RAO ·

RAO a...

Ap...

ROBE... ...nce

ROGE... and Martingales,

Vol...

ROHA... ...matical Statistics

ROSS...

RUBIN...

RUBIN... ...sitivity Analysis and

Sto...

RUZSA...

SCHE...

SEBE...

SEBE...

SEBE...

SERFL... ...tistics

SHOR... ...lications to Statistics

STAU...

STOYANOV · Counterexamples in Probability

STYAN · The Collected Papers of T.W. Anderson: 1943—1985

WHITTAKER · Graphical Models in Applied Multivariate Statistics

YANG · The Construction Theory of Denumerable Markov Processes

Applied Probability and Statistics

ABRAHAM and LEDOLTER · Statistical Methods for Forecasting

AGRESTI · Analysis of Ordinal Categorical Data

AGRESTI · Categorical Data Analysis

ANDERSON and LOYNES · The Teaching of Practical Statistics

ANDERSON, AUQUIER, HAUCK, OAKES, VANDAELE, and WEISBERG ·
Statistical Methods for Comparative Studies

ASMUSSEN · Applied Probability and Queues

*BAILEY · The Elements of Stochastic Processes with Applications to the Natural
Sciences

BARNETT · Interpreting Multivariate Data

BARNETT and LEWIS · Outliers in Statistical Data, *Third Edition*

BARTHOLOMEW, FORBES, and McLEAN · Statistical Techniques for
Manpower Planning, *Second Edition*

BATES and WATTS · Nonlinear Regression Analysis and Its Applications

BELSLEY · Conditioning Diagnostics: Collinearity and Weak Data in Regression

BELSLEY, KUH, and WELSCH · Regression Diagnostics: Identifying Influential
Data and Sources of Collinearity

BHAT · Elements of Applied Stochastic Processes, *Second Edition*

BHATTACHARYYA and WAYMIRE · Stochastic Processes with Applications

BIEMER, GROVES, LYBERG, MATHIOWETZ, AND SUDMAN · Measurement
Errors in Surveys

BIRKES and DODGE · Alternative Methods of Regression

BLOOMFIELD · Fourier Analysis of Time Series: An Introduction

BOLLEN · Structural Equations with Latent Variables

BOX · R.A. Fisher, the Life of a Scientist

BOX and DRAPER · Empirical Model-Building and Response Surfaces

Continued on back end papers

* Now available in a lower priced paperback edition in the Wiley Classics Library.

Fractals, Random Shapes and Point Fields

Fractals, Random Shapes and Point Fields
Methods of Geometrical Statistics

Dietrich Stoyan and Helga Stoyan
Freiberg University of Mining and Technology
(TU Bergakademie Freiberg),
Germany

JOHN WILEY & SONS
Chichester • New York • Brisbane • Toronto • Singapore

Originally published in the German language by
Akademie Verlag GmbH, Leipziger Strasse 3–4,
0-1086 Berlin, Germany under the title "Fraktale-
Formen-Punktfelder. Methoden der Geometrie-Statistik".
Copyright 1992 by Akademie Verlag GmbH

Other Wiley Editorial Offices

John Wiley & Sons, Inc., 605 Third Avenue,
New York, NY 10158-0012, USA

Jacaranda Wiley Ltd, G.P.O. Box 859, Brisbane,
Queensland 4001, Australia

John Wiley & Sons (Canada) Ltd, 22 Worcester Road,
Rexdale, Ontario M9W 1L1, Canada

John Wiley & Sons (SEA) Pte Ltd, 37 Jalan PemImpin #05-04,
Block B, Union Industrial Building, Singapore 2057

British Library Cataloguing in Publication Data

A catalogue record for this book is available from the British Library

ISBN 0 471 93757 6

Typeset by Laser Words, Madras
Printed and bound in Great Britain by Biddles Ltd, Guildford and Kings Lynn

Contents

Chapter 9   Point Description of Figures   141

Chapter 10   Examples   167

PART III   POINT FIELD STATISTICS   187

Chapter 11   Fundamentals   189

Chapter 12   Finite Point Fields   197

Preface

Geometrical objects and structures are studied in many branches of science and engineering. Often these objects have non-regular random shapes or are randomly scattered in space. If the number of the objects under study is very great then statistical analysis makes sense and is indeed necessary. Many methods exist for such analyses; some of the more recent attempt to consider spatial dependences or other complicated correlations.

The aim of this book is to present some statistical methods in a way that may also be understood by non-mathematicians, in particular by materials scientists, geologists, environmental researchers and biologists. We assume that the reader has a basic knowledge of mathematics and statistics, although some concepts and methods that may be unfamiliar to non-mathematicians are explained in the appendices.

Our aim was to write a clear and popular text that is nevertheless mathematically correct. Although many parts of the book may interest applied mathematicians or statisticians, these readers have to accept that this book does *not* contain proofs — it merely outlines the mathematical ideas.

We treat three different subjects: fractals, random shapes and point fields (processes). In discussing these we always restrict attention to planar structures. From the reaction to the book *Stochastic Geometry and its Applications* by Stoyan, Kendall and Mecke, we know that many applied researchers are deeply interested in the first two topics.

Part I gives an introduction to the theory of fractals. This should familiarise the reader with the methods of measuring fractal dimensions. These are used to describe extremely irregular geometric structures. Furthermore, important mathematical models involving fractals are explained, including some of a stochastic nature. We explain the notion of fractal dimension in more detail than is customary for applied mathematicians. Thus Part I has some difficult passages. However, a reader interested only in applications is led quickly to the measurement techniques.

In Part II we recount important modern methods for the statistical analysis of random shapes. Random shapes are studied in such diverse areas as biology and particle science. Biological shapes result from growth process, so that often the geometries correspond to life functions; typically such objects have points on their contours or in their interiors that play specific biological roles. Such points do not usually occur in particles resulting from geological or technological processes.

We consider three approaches: (1) describing objects by their contours and using methods for the analysis of functions; (2) considering them as random compact sets; and (3) describing them as k-tuples of points. These points, frequently called landmarks, are characteristic points on the boundaries or in the interiors of the objects. The mathematics behind these methods is not simple, and hence some expositions are given in outline only. Those interested should consult the relevant references. On the other hand, we hope that our treatment will encourage non-mathematicians to use these statistical methods.

Finally, Part III presents an introduction to the statistical theory of point fields, with and without marks. An important area of application is the analysis of homogeneous systems of particles where the 'points' are particle centres and the 'marks' particle characteristics such as size or orientation. We try to give the theory in an elementary form, with emphasis on the aspects of analysis currently of greatest interest. We also discuss some important classes of point field models, in particular Gibbsian processes. Here some quite modern results are presented. Furthermore, Part III contains, in condensed form, an exposition of the theory of correlations of marked point fields and their statistics, which has been developed during the last decade.

All three parts can be read independently. However, in contrast to the German edition, there is only one list of references, and some German references have been eliminated. Also the chapter and figure numbering is from 1 to N throughout the book. Because of the varied topics considered, it was inevitable that often one symbol will stand for different things in different contexts. This should not, however, lead to any confusion.

We wish to thank many colleagues for their help and support. In particular, we are grateful to the late Ulrich Zähle, who, even in the year of his early death (1989), read Part I of this book and helped with many useful comments. A. Enoch generated the fractal images and A. Bandt supported us with hints on the theory of fractals. Several colleagues sent us their papers, sometimes at precisely the most opportune moment. We mention in particular F. L. Bookstein, C. D. Cutler, I. L. Dryden, W. Gille, K. V. Mardia and H. Ziezold, with whom we had very useful personal discussions. D. G. Kendall helped us to understand the theory of the landmark method. L. Muche and W. Nagel read sections of Part II with care. In the analysis of sand grains we have had the support of R. Schuberth. A. Schwandtke and P. Grabarnik made some calculations for Part III. We thank N. Bamber and R. B. Johnson for translating the book into English. Last, but not least, we thank H. M. Clarke for his meticulous copy-editing and for his improvements to the text.

D. and H. Stoyan
Freiberg, May 1994

List of Symbols

$A(B)$	Area, or Lebesgue measure of the set B
a_k	kth Fourier coefficient
A_k	kth Fourier amplitude
b_k	kth Fourier coefficient
B_x	the set B, shifted by x, $B_x = B + x$
$\mathcal{B}^d$	Borel σ-algebra of R^d
$b(x, r)$	closed disc with centre x and radius r
conv X	convex hull of the set X
$\text{cov}(X, Y)$	covariance of the random variables X and Y, $= \mathsf{E}(X - \mathsf{E}X)(Y - \mathsf{E}Y)$
D	fractal dimension
$D(r)$	distribution function of the distance between a typical point and its nearest neighbour
$D(\varphi)$	deterministic contour function
$D_\mu(x)$	local Hausdorff dimension
$\mathsf{E}X$	expected value of the random variable X
$e_h(x)$	Epanečnikov kernel
$\mathcal{E}(\varphi)$	disturbance function of a contour
$\bar{f}$	mean value of the function $f(x)$
f_{AU}	area-circumference ratio
f_C	convexity ratio
f_{ell}	ellipse ratio
f_R	roundedness ratio
$g(r)$	pair correlation function
h	band width
$h(A, B)$	Hausdorff metric, distance between sets A and B
$\text{H-dim}(A)$	Hausdorff dimension of the set A
$H_s(r)$	spherical contact distribution function
$\mathcal{K}$	set of all compact subsets of R^2
$\mathfrak{K}$	reduced 2nd order moment
$K(r)$	Ripley's K-function
$k_f(r)$	mark correlation function for the test function f
$\mathcal{L}$	Lebesgue measure

$L(\ell)$	chord length distribution function
$l(\ell)$	density function for $L(\ell)$
$L(r)$	L-function
M	mark distribution function
M	mark distribution
$N(B)$	number of points of the point field N in the set B
o	origin of the co-ordinate system
Pr	probability
$p_X(x)$	covering function
$Q_s(r)$	spherical erosion function
$r_X(\varphi)$	radius vector function
$\bar{r}$	mean radius vector length
$s_X(\varphi)$	support function of the set X
$U(X)$	perimeter of the set X
$\mathrm{var}(X)$	variance of the random variable X
α	chemical activity
α_k	kth phase angle
$\alpha^{(2)}$	2nd order factorial moment measure
$\bar{\gamma}_Y(r)$	isotropized set covariance function of the set X
$\Delta(P, Q)$	distance between the landmark configurations P and Q
$\chi_f(\psi)$	contour covariance function of the function f
$\chi_F(\psi)$	contour covariance function of the random contour function F
λ	intensity
$\lambda(x)$	intensity function
$\Lambda(B)$	intensity measure of the set B
$\varrho(P, Q)$	dilation factor for the landmark configurations P and Q
$\varrho^{(2)}$	2nd order product density
$\sigma^2(f)$	variance of the contour function f
$\sigma^2(F)$	variance of the random contour function F
$\phi(r)$	pair potential
$\phi_X^*(r)$	tangent angle function
ω_α	volume of the unit ball in R^α
$\omega(P, Q)$	angle of rotation between the landmark configurations P and Q
$1_B(x)$	indicator function of the set B
$\oplus$	Minkowski addition
$\ominus$	Minkowski subtraction
$\|X\|$	norm of the set X
$\|x\|$	distance of the point x from the origin o in the Euclidean metric
$\log(x)$	$\ln(x)$

PART I

Fractals and Methods for the Determination of Fractal Dimensions

CHAPTER 1

Introduction

Fractals are mathematical models for very irregular, very detailed sets such as given by Figs. 3, 4, 6, 8, 11, 12, 13 and 17. The degree of their 'wildness' or 'roughness' can be characterized by the so-called fractal dimension. In the case of planar sets (only such sets are considered in this book) this lies between 0 and 2 for sets of discrete points and between 1 and 2 for sets of curves. The rougher and wilder a structure is, the greater its fractal dimension. The mathematical theory of fractals is rather complicated — even the sketch given in this book has difficult passages. On the other hand, the methods of measurement of fractal dimension are easily understood, and they also give the user a feeling of the underlying meaning. Therefore non-mathematicians may wish to read up to the end of this chapter and then start again at Chapter 5.

Fractal dimensions

Usually, systems of points have dimension zero and curves dimension one. These values are the topological dimensions. In the case of infinite, very dense systems of points or systems of curves of infinite lengths in bounded areas, it makes sense to make more subtle distinctions. For this purpose fractal dimensions may serve.

The 'denser' the points or the 'longer' (rougher) the curves are, the greater the fractal dimension. The images of fractals shown in this book give an impression of the possibilities that the fractal dimension approach holds for characterizing irregularity and roughness. Of course, there are quite different fractals with the same dimensions: one could not expect that one single number could be sufficient for a unique description of irregularity. (A further parameter, describing the 'rippedness' of fractals, is the lacularity; see Mandelbrot (1982).)

The reader should note that there are several mathematical definitions of fractal dimension, which may lead to different values of dimension for some sets. Nevertheless, many work on the assumption that the different definitions should produce the same results for a given object; frequently several variants of measurement, which are connected with several definitions, are used for the same object with the aim of measuring the same quantity in different ways. If significant differences appear then this may be considered as a sign of a particularly high degree of complexity of the structure studied. There is also a local form of dimension definition, which can be used to describe fluctuations in roughness.

Scale invariance and self-similarity

In nature there are many processes forming structures in which, at decreasing scales, similar principles of formation persist. For example, in biological growth, certain branching processes happen step by step in increasing detail; for example in corals or the cauliflower. Many solid body structures are formed by agglomeration of particles, which may be big initially, but later only smaller ones can be included. Nevertheless, the principles of construction remain the same throughout. Then structures as in Figs. 6, 8, 11 and 12 arise, where small parts have a form similar to that of the whole. In practical measurement one not infrequently observes that under a microscope or pocket-lens parts of the object have similar properties (e.g. random variability) as the whole or those parts visible to the eye.

In such situations one speaks about scale invariance and self-similarity. The theory contains mathematical definitions of self-similarity for both deterministic and random structures. In the case of self-similarity the fractal dimension can be determined by formulae. Self-similar sets are excellent mathematical models, which may help to understand many real-life structures.

Dusts

Sometimes in applications point sets are studied that consist of 'extremely many', 'very small' subsets, e.g. systems of very small pores. In mathematical idealization one could say that such a set consists of uncountably many points but does not contain any piece of a curve. Such sets are here called 'dusts'. A classical mathematical example is the Cantor set or Cantor dust.

Example 1: *Cantor Dust C.* The set C is a subset of the real axis (x-axis). It is obtained by deleting step by step open sub-intervals of $[0,1]$. The intervals deleted are

Step 1: $\left(\frac{1}{3}, \frac{2}{3}\right)$

Step 2: $\left(\frac{1}{9}, \frac{2}{9}\right), \left(\frac{4}{9}, \frac{5}{9}\right), \left(\frac{7}{9}, \frac{8}{9}\right)$

$\vdots$

Step n: $((3k - 2)/3^n, (3k - 1)/3^n), k = 1, \ldots, 3^{n-1}$

(see Fig. 1). (In this process some subintervals are deleted several times, but this does not influence the final result.)

The set C can be written as follows:

$$C = [0, 1] \setminus \bigcup_{n=1}^{\infty} \bigcup_{k=1}^{3^{n-1}} \left(\frac{3k - 2}{3^n}, \frac{3k - 1}{3^n}\right).$$

It is possible to show that C consists of uncountably many points, but the length of C (more precisely, the one-dimensional Lebesgue measure on the x-axis) is zero.

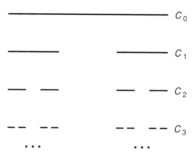

Figure 1   Schematic description of the Cantor dust; see the text for explanation.

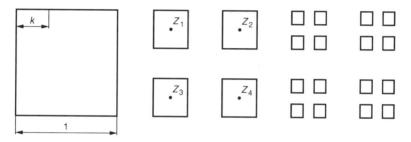

Figure 2   Schematic description of a square dust. At every stage of refinement every square is replaced by four smaller ones. The dimension of the resulting fractal is $\log 4/\log 3 = 1.26$ (p. 25).

Example 2: *Square Dust Q.* The set Q is a planar set. It is obtained by taking from the unit square step by step cross-shaped subsets, as shown in Fig. 2. This set also contains uncountably many disconnected points.

Rough curves and boundaries

Figures 3 and 4 show the fracture lines and the boundary of a planar section of a particle. These curves are very rough. It is plausible to suppose that one could observe similar spikes and cusps at smaller scales if the printing were finer or a pocket-lens were used. In this case length measurement is not easy.

   If one considered the smallest spikes as well then the time for measurement would be rather long and the resulting length gigantic, which would not be very easy to deal with. In a mathematical idealization one could imagine that the spikes occur again and again at smaller scales and that the curve length becomes infinite. A classical example is the von Koch snow flake.

Example. *von Koch Snow Flake S* The von Koch snow flake S is constructed iteratively as shown by Fig. 5. In every step all line segments of the figure are replaced by suitably diminished generators, where the vertices always point outwards. Figure 6 shows a computer-generated snow flake.

Figure 3 Two fractal lines in ceramics of different structures. Such curves can be considered to some approximations as fractals.

Figure 4 Planar section through a graphite particle in cast iron. The structure is so irregular that a fractal model seems to be suitable. The fractal dimension of the contour has been estimated at 1.4 (p. 49).

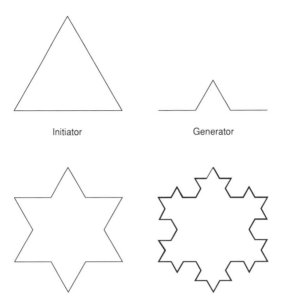

Initiator   Generator

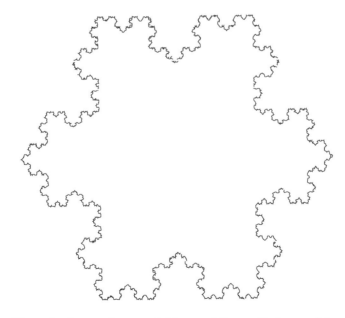

Figure 5 Schematic description of the construction of the von Koch snow flake. At every stage every edge is replaced by a suitably diminished copy of the generator. The fractal dimension of the final structure is $\log 4 / \log 3$ (p. 25).

Figure 6 A computer-generated image of the von Koch snow flake.

Measures of roughness

Fractal dimensions are frequently used as measure of roughness of curves or surfaces (Underwood and Banerji, 1986; Dauskardt *et al.*, 1990). The idea is that a large fractal dimension is related to a high degree of roughness. A disadvantage of this approach is that only one number, the fractal dimension, is used for the characterization. It is interesting to note that there are several different, perhaps better roughness measures (see e.g. Banerji, 1988).

For a curve, as in Fig. 3, the roughness R_L is described by the length L of the curve (measured with maximum accuracy) divided by a length L_p of its projections on a horizontal line:

$$R_L = L/L_p.$$

An excellent characterization of roughness is also possible using the length-weighted distribution of angles of approximating polygons. Additionally, the quantities used in Part II for the description of the texture of figures can serve as roughness measures. To characterize the degree of disorder (of 'chaos') in physics, further parameters are used; for example Lyapunov exponents and entropy characteristics.

Mathematical and natural fractals

For a long time mathematicians have studied geometrical structures as shown in Figs. 6, 8 and 13; see also the historical text of Mandelbrot (1982, 1988). Mathematical idealization makes it possible to assume that the processes of refining used for plotting these figures can be continued indefinitely. In nature one often observes structures and phenomena with similar behaviour (Mandelbrot, 1982; Falconer, 1990) . These are the boundaries of very rough and irregular objects, such as fracture surfaces, systems of pores or fractures in artificial or natural materials (Hornbogen, 1987; Pape *et al.*, 1987; Avnir, 1989; Dauskardt *et al.*, 1990; Herman and Roux, 1990; Kaye, 1993; Genske *et al.*, 1992), complicated physical structures such as gels and polymers (Orbach, 1986); or systems of caves (Curl, 1986), geographical structures (Goodchild and Mark, 1987; Goodchild, 1988) ecological systems with many hierarchy levels (Frontier, 1987; Hastings and Sugihara, 1993), irregular biological structures (Rigaut, 1989), rapidly varying processes such as showers of rain (Lovejoy and Mandelbrot, 1985), fluctuations of share prices (Mandelbrot, 1982) or attractors of chaotic motions. Here also one is speaking of fractal phenomena. The book by Turcotte (1992) is a systematic treatment of fractals in geology and geophysics. In particular, fragmentation, seismicity, tectonics, ore distribution and geomorphology are considered. Of course the behaviour of natural phenomena resembles that of mathematical fractals only up to a certain scale. (In particular, infinite refining is impossible in nature; it is limited at least by the size of elementary particles.) But this does not create serious difficulties in many applications because frequently just that scale is of practical interest and accessible for measurement in which the real objects have their fractal-like behaviour. It is one of the great merits of the theory of fractals to suggest a systematic study of the behaviour of measurements under increasing accuracy. One parameter for characterizing this is the fractal dimension.

Warning! One should not be too enthusiastic in applying the theory of fractals. In nature there are no objects that are fractals in the mathematical sense. Power laws, which are related to scale invariance, have been known for a long time, and are not necessarily to be interpreted in fractal-theoretic terms. As well as the fractal dimension, there are other characteristics that describe similarity properties of geometrical structures.

Stochastic models of fractals

There are a range of stochastic models of fractals, in other words, models of stochastic processes, sets and construction principles that generate fractals. While Mandelbrot (1982) and many physicists have tried to model natural phenomena by such means, this book presents stochastic models principally to enable readers to simulate random fractals and to test measurement methods of fractal dimensions before implementation.

Measurement of fractal dimension

Today many methods of measurement are known. They consist in the analysis of geometrical structures by measurements of varying accuracy, where the correlation of the results with accuracy is systematically studied. Mostly it is assumed that the structure studied can be characterized by a unique value of the dimension and does not consist of various parts with various dimensions. There have been only a few theoretical investigations into fractal statistics; an important problem is the particular role of measurement in the case of high accuracy, where measurement errors have a great influence. One also has to struggle with this problem in the use of image analysers. The investigation of boundaries of rough particles is particularly difficult.

CHAPTER 2

Hausdorff Measure and Dimension

2.1 THE HAUSDORFF MEASURE IN R^2

To fully understand the notion of fractal dimension and empirical methods of measurement, it is first necessary to have some knowledge of measure theory in the real plane R^2. The following is only a sketch; for a full treatment see Falconer (1985, 1990) and Rogers (1970). The usual measures in R^2 are the Lebesgue measure and the one-dimensional Hausdorff measure. (See Appendix A, where the fundamentals of measure theory are explained.) The Lebesgue measure $A(B)$ of a 'sufficiently reasonable' set B is equal to its area; for a curve or a countable set of discrete points the Lebesgue measure is zero. The one-dimensional Hausdorff measure yields the lengths of curves. The length of a 'sufficiently smooth' curve is obtained by approximating it by line pieces, adding the corresponding lengths and refining the approximation; the limit of length sums is then the curve length.

Hausdorff suggested in 1918 a measure definition that generalizes the above notions. First an important special case is explained, namely the (spherical) α-dimensional Hausdorff measure $\mathcal{H}^\alpha$, where $\alpha > 0$. It is defined by

$$\mathcal{H}^\alpha_s(B) = \omega_\alpha \liminf_{\delta \downarrow 0} \left\{ \sum_{(i)} r_i^\alpha : B \subset \bigcup_{(i)} b(x_i, r_i), r_i < \delta \right\}. \qquad (2.1)$$

For integer α, w_α denotes the volume of the unit sphere in R^α. In general,

$$\omega_\alpha = \frac{\pi^{\alpha/2}}{\Gamma\left(1 + \frac{1}{2}\alpha\right)}; \quad \omega_0 = 1, \ \omega_1 = 2, \ \omega_2 = \pi$$

(The symbols inf, sup, lim inf and lim sup are explained in Appendix B.) For a regular set B the quantity $\mathcal{H}^\alpha_s(B)$ is equal to the length ($\alpha = 1$) or the area ($\alpha = 2$).

The definition of $\mathcal{H}^\alpha_s$ can be explained as follows. The subset B of R^2 that is to be measured is covered by closed discs $b(x_i, r_i)$,

$$B \subset \bigcup_{(i)} b(x_i, r_i),$$

with radii all smaller than a given positive number δ. Then the sum

$$\omega_\alpha \sum_{(i)} r_i^\alpha$$

is formed. Of course, there are many such coverings; that is, there is freedom in choosing x_i and r_i. Of particular interest is that choice which yields the smallest value of the above sum. That value is denoted by $S_\delta^\alpha(B)$. It is not difficult to see that for elementary B, in the case $\alpha = 1$ and $\alpha = 2$, $S_\delta^\alpha(B)$ yields the length and area of B respectively. Namely, if B is a straight line segment then B can be completely covered by circles that are touching. Consequently, $S_\delta^\alpha(B)$ is then equal to the length of B. If B is a square then it is intuitively clear that B can be covered by disc of radii smaller than δ, with only a small amount of overlap. Therefore $S_\delta^\alpha(B)$ is practically the area of the square.

A large δ may lead to unreasonable values. For example, if B is a circle of radius r and if $\delta > r$ then B is covered by one disc of radius r with the same centre as B. Taking $\alpha = 1$ since a curve is to be measured, this gives $S_\delta^\alpha(B) = 2r$, which is clearly smaller than $2\pi r$. Thus only those coverings that use small discs are of practical interest. Hence it is reasonable to define

$$\mathcal{H}_s^\alpha(B) = \lim_{\delta \downarrow 0} S_\delta^\alpha(B), \qquad (2.2)$$

which is the same as (2.1).

It is possible to show that $\mathcal{H}_s^\alpha$ is a Borel measure on R^2. Furthermore, for all Borel sets B

$$\mathcal{H}_s^\alpha(B) = A(B), \qquad (2.3)$$

for every smooth curve C

$$\mathcal{H}_s^1(C) = \text{ length of } C, \qquad (2.4)$$

and for every set D of n points

$$\mathcal{H}_s^0(D) = n. \qquad (2.5)$$

Of course, $A(C) = A(D) = 0$. The topological dimensions of C and D are $\alpha = 1$ and $\alpha = 0$, and $\mathcal{H}_s^1$ and $\mathcal{H}_s^0$ respectively give sensible values.

The above coverings may be modified by replacing the discs $b(x_i, r_i)$ by arbitrary compact subsets B_i of R^2, so that each B_i is contained in some discs of radius less than δ. Let rad(B_i) denote half of the maximum distance of two points in B_i. Then the α-dimensional Hausdorff measure $\mathcal{H}^\alpha$ is given by

$$\mathcal{H}^\alpha(B) = \omega_\alpha \liminf_{\delta \downarrow 0} \left\{ \sum_{(i)} \text{rad}(B_i)^\alpha : B \subset \bigcup_{(i)} B_i, \ \text{rad}(B_i) < \delta \right\}. \qquad (2.6)$$

$\mathcal{H}^\alpha$ is also a Borel measure. Since the infimum corresponding to $\mathcal{H}^\alpha$ is obtained by considering a larger set of coverings than for $\mathcal{H}^\alpha_s$,

$$\mathcal{H}^\alpha(B) \le \mathcal{H}^\alpha_s(B). \tag{2.7}$$

Federer (1969, §§2.10.35 and 3.2.26) gives sufficient conditions for equality. In general,

$$c\mathcal{H}^\alpha_s(B) \le \mathcal{H}^\alpha_s(B), \quad \text{with } c = \left(\tfrac{3}{4}\right)^{\alpha/2} \tag{2.8}$$

and

$$\mathcal{H}^2 = \mathcal{H}^2_s = A. \tag{2.9}$$

There are many other definitions of measure with similar properties. For example, instead of $r_i \to r_i^\alpha$ or $\text{rad}(B_i) \to \text{rad}(B_i)^\alpha$, other functions can be considered, or coverings with discs of equal size can be used (Taylor, 1986b). Furthermore, so-called packing measures are used, for example the following counterpart of $\mathcal{H}^\alpha_s$:

$$\mathcal{P}^\alpha_s(B) = \limsup_{\delta \downarrow 0} \left\{ \sum_{(i)} r_i^\alpha : b(x_i, r_i) \text{ disjoint}, x_i \in B, 0 < r_i < \delta \right\} \tag{2.10}$$

(Taylor, 1986a,b). (Here systems of non-overlapping discs with centres in B are considered.)

Particularly important with respect to statistical methods is the Minkowski content $\mathcal{M}^\alpha$. It is based on the following intuitive conception. Let B be a smooth curve of length $l(B)$. The outer parallel set $B_r = B \oplus b(o, r)$ is formed, i.e. B is blown up to a sausage ('Minkowski sausage') of thickness $2r$. For small r the area of B is to a good approximation equal to $2rl(B)$. Consequently, $l(B)$ can be obtained by

$$l(B) = \lim_{r \downarrow 0} \frac{A(B_r)}{2r}$$

In the case of a 'rough' curve this limit does not necessarily exist. Therefore corresponding infima and suprema are considered, which yield for B 'lower' and 'upper' lengths

$$l_*(B) = \liminf_{r \downarrow 0} \frac{A(B_r)}{2r}$$

and

$$l^*(B) = \limsup_{r \downarrow 0} \frac{A(B_r)}{2r}.$$

Once again a positive real number α $(0 < \alpha \le 2)$ may be introduced and used to define the *lower* and *upper* α-dimensional Minkowski contents

$$\mathcal{M}^\alpha_*(B) = \liminf_{r \downarrow 0} \frac{A(B_r)}{r^{2-\alpha} \omega_{2-\alpha}} \tag{2.11}$$

and

$$\mathcal{M}^{*\alpha}(B) = \limsup_{r \downarrow 0} \frac{A(B_r)}{r^{2-\alpha}\omega_{2-\alpha}}. \tag{2.12}$$

(Note that $\mathcal{M}^{\alpha}_{*}$ and $\mathcal{M}^{*\alpha}$ are not measures.) For $\alpha = 0$, 1 and 2 this gives number, length and area as in the case of the Hausdorff measure. If $\mathcal{M}^{\alpha}_{*}(B) = \mathcal{M}^{*\alpha}(B)$ then the notation $\mathcal{M}^{\alpha}(B)$ is used.

2.2　FRACTAL DIMENSION

Now the notion of fractal dimension can be explained. In contrast to the topological dimension, which is 0 for sets of discrete points, 1 for curves and 2 for areas, the fractal dimension can also take non-integer values.

For a given set B the function

$$f(\alpha) = \mathcal{H}^{\alpha}_{s}(B) \quad (0 \le \alpha < \infty)$$

is considered. It has quite a simple form. Namely, there is a number D between 0 and 2 with

$$f(\alpha) = \begin{cases} +\infty & (\alpha < D), \\ 0 & (\alpha > D). \end{cases}$$

The value of $f(D)$ may be zero, infinite or a positive real number. This behaviour of $f(\alpha)$ is a straightforward consequence of the inequality

$$\mathcal{S}^{\alpha}_{\delta}(B) \ge \delta^{\alpha-\beta}\mathcal{S}^{\beta}_{\delta}(B) \quad (\alpha < \beta)$$

and the fact that for any δ with $0 < \delta < 1$ the function $\mathcal{S}^{\alpha}_{\delta}(\mathcal{B})$ is non-increasing in α.

Proof of monotonicity and inequality. Let $\alpha < \beta$. Choose a covering of B that yields (in the sense of p. 12) a value close to $\mathcal{S}^{\alpha}_{\delta}(B)$. Because

$$\sum_{(i)} r_i^{\alpha} \ge \sum_{(i)} r_i^{\beta}$$

for $r_i < 1$, the infimum of the sums $\sum_{(i)} r_i^{\beta}$ taken over all coverings of B cannot exceed $\mathcal{S}^{\alpha}_{\delta}(B)$. Furthermore, because $\alpha - \beta < 0$,

$$\sum_{(i)} r_i^{\alpha} = \sum_{(i)} r_i^{\beta} r_i^{\alpha-\beta} \ge \sum_{(i)} r_i^{\beta} \delta^{\alpha-\beta},$$

which yields the inequality.

It is obvious that the value D plays a special role. For sets of positive area it is equal to 2, for smooth curves it is equal to 1, and for countable sets of points it is 0; in these cases D is equal to the topological dimension. Therefore it is natural to define

$$\text{H-dim}(B) = \inf\left\{\alpha : \alpha > 0, \mathcal{H}_s^\alpha(B) = 0\right\} = \sup\left\{\alpha : \alpha > 0, \mathcal{H}_s^\alpha(B) = \infty\right\}$$
(2.13)

The value H-dim(B) is called the *Hausdorff–Besicovich dimension* of B. Because of (2.6) and (2.7), the same value is obtained if $\mathcal{H}_s^\alpha$ is replaced by $\mathcal{H}^\alpha$.

Analogously, packing dimensions can be defined, starting from $\mathcal{P}_s^\alpha$ (Taylor, 1986a,b). If the starting point of an analogous definition is the lower (upper) α-dimensional Minkowski content then the *lower (upper) Minkowski–Bouligand dimension* M$_*$-dim (M*-dim) is obtained:

$$\text{M}_*\text{-dim}(B) = \inf\{\alpha : \alpha > 0, \mathcal{M}_*^\alpha(B) = 0\} = \sup\{\alpha : \alpha > 0, \mathcal{M}_*^\alpha(B) = \infty\}.$$
(2.14)

The following formulae can be used for calculating this dimension:

$$\text{M}_*\text{-dim}(B) = 2 - \limsup_{r\downarrow 0} \frac{\log A(B_r)}{\log r}$$
(2.15)

or

$$\text{M}_*\text{-dim}(B) = \liminf_{r\downarrow 0} \frac{\log\left[A(B_r)/r^2\right]}{\log(1/r)}.$$
(2.16)

For the determination of M*-dim, sup and inf have to be interchanged.

The determination of fractal dimension is very complicated if particular properties of the sets considered cannot be used, for example self-similarity (§3.2.2). Falconer (1990) describes systematic methods for determining fractal dimensions. It is still more complicated to determine the value of the corresponding measure, for example the Hausdorff measure.

Since in §3.2.2 the formulae for the self-similar case will be given without proof, an example of dimension calculation is given here.

Determination of the fractal dimension and of the Hausdorff measure of the Cantor dust C

For the Cantor dust

$$\text{H-dim}(C) = \frac{\log 2}{\log 3} = 0.6309$$
(2.17)

and

$$\mathcal{H}_s^D(C) = 2^{-D}\omega_D = 1.035$$
(2.18)

for $D = \log 2/\log 3$.

Proof: Let $D = \log 2/\log 3$. In the jth step of the construction described on p. 4 a subset C_j of the x-axis is obtained that is contained in C and that consists of 2^j closed intervals of lengths 3^{-j}. The intervals can be considered to be diameter lines of discs that cover C. For $\delta = \frac{1}{2}3^{-j}$ one obtains

$$\mathcal{S}^{\alpha}_{\delta}(C) \leq 2^j \left(\tfrac{1}{2}3^{-j}\right)^{\alpha} \omega_{\alpha}.$$

In particular, for $\alpha = D$, since $3^D = 2$,

$$\mathcal{S}^{\alpha}_{\delta}(C) \leq 2^{-D}\omega_D.$$

This implies

$$\mathcal{H}^D_s(C) \leq 2^{-D}\omega_D.$$

To prove the converse inequality, an arbitrary covering of C by discs is considered. By slightly enlarging the discs in this covering and using the compactness of C, one can change to a finite covering. Obviously, it is economical to set the disc centres on the x-axis, and thus in the following only finite coverings of this kind are considered. For these, the discs are chosen in such a way that any covering disc $b(x_i, r_i)$ has an intersection of the form $J \cup K \cup J'$ with the x-axis, where K is in the complement of C, and J and J' are intervals contained in C_j and $C_{j'}$, with suitable indices j and j'.

By construction, the sum of the lengths of J, J' and K is not smaller than $\frac{3}{2}$ times the sum of the lengths of J and J'. Thus if $l(X)$ is the length (the one-dimensional Lebesgue measure) of the subset X of the x-axis, the radius r_i of a covering disc satisfies

$$
\begin{aligned}
(2r_i)^D &= [l(J) + l(J') + l(K)]^D \\
&\geq \left\{\tfrac{3}{2}[l(J) + l(J')]\right\}^D \\
&= 2\left[\tfrac{1}{2}l(J) + \tfrac{1}{2}l(J')\right]^D \\
&\geq l(J)^D + l(J')^D.
\end{aligned}
$$

Here the inequality $3^D = 2$ and the concavity of the function $f(x) = x^D$ (since $D < 1$) have been used. Thus if $b(x_i, r_i)$ is replaced by two discs of diameters $l(J)$ and $l(J')$, the sum of the r^D-values is not increased. Using an analogous argument, one can replace the original covering by a covering for which all discs have diameter 3^{-k} for a suitably chosen k and the sum of r^D-values does not exceed the original sum. Since for every k

$$2^k \left(\tfrac{1}{2}3^{-k}\right)^D = 2^{-D},$$

one obtains

$$\mathcal{H}^D_s(C) \geq 2^{-D}\omega_D.$$

This and the inequality above yield

$$\mathcal{H}^D_s(C) = 2^{-D}\omega_D,$$

and consequently the Hausdorff dimension of C is D.

It is not difficult to show that

$$\mathcal{H}_s^D(C) = \mathcal{H}^D(C).$$

Furthermore, it is possible to prove that

$$M_*\text{-dim}(C) = M^*\text{-dim}(C) = \text{H-dim}(C),$$

but the values of $\mathcal{M}_*(C)$ and $\mathcal{M}^*(C)$ differ from $\mathcal{H}_s^D(C)$.

The foregoing proof is typical in as far as it is often easier to give upper bounds for the Hausdorff dimension than lower ones. As for the Cantor dust, it can be shown for the square dust Q of Fig. 2 that its H-dimension is equal to $-\log 4/\log k$. In this case, k takes any value in the range $0 < k < \frac{1}{2}$; thus the dimension can take any value in the interval $[0, 2)$.

There are a number of other definitions of dimension and similar characteristics for irregular sets (Tricot, 1982; Falconer, 1985, 1990; Taylor, 1986b; Cutler, 1991; Jensen, 1993). These show the following basic properties:

monotonicity

$$B \subset B' \Rightarrow \dim(B) \leq \dim(B'); \qquad (2.19)$$

invariance with respect to motions and dilations

$$\dim(\lambda T(B)) = \dim(B)$$

for all Euclidean motions T and all positive dilatation factors λ;

σ-stability

$$\dim\left(\bigcup_{(i)} B_i\right) = \sup_{(i)} \dim(B_i). \qquad (2.20)$$

Fortunately, for many sets (if they are sufficiently 'regular') some, if not all, definitions of dimension lead to the same values (Tricot, 1982; Taylor, 1986b; Falconer, 1988, 1990). A general inequality is $M_*\text{-dim}(B) \geq \text{H-dim}(B)$.

Methods of determining M^*-dim and M_*-dim can be obtained from (2.15) and (2.16). For small r the area of $B \oplus b(0, r)$ is determined and then the limit as $r \downarrow 0$ is estimated. In contrast, a measure in the spirit of the Hausdorff dimension is hardly possible, because it would be necessary to determine the infimum over all coverings. An important starting point for developing measurement methods is the following result of Tricot (1982): for any bounded set B

$$M_*\text{-dim}(B) = -\limsup_{r \downarrow 0} \frac{\log N_r(B)}{\log r}, \qquad (2.21)$$

$$M_*\text{-dim}(B) = -\limsup_{r \downarrow 0} \frac{\log M_r(B)}{\log r}, \qquad (2.22)$$

$$M_*\text{-dim}(B) = -\limsup_{r\downarrow 0} \frac{\log Q_r(B)}{\log r}. \tag{2.23}$$

Here $N_r(B)$ is the smallest number of closed discs of radius r that cover B, and $M_r(B)$ is the maximum number of open discs of radius r and centre in B that do not overlap. Finally, $Q_r(B)$ is connected with a division of the plane in quadratic meshes of width r. The number of squares containing points of B is denoted by $Q_r(B)$.

The right-hand limit in (2.21) is sometimes called the 'entropy dimension' (Hawkes, 1974), that in (2.22) the 'metric dimension' (Kolmogorov and Tihomirov, 1961) and that in (2.23) the 'logarithmic density' (Tricot, 1973), 'quadrat count dimension' or 'box dimension'. The latter is closely related to the so-called 'capacity dimension'. The paper by Hunt (1990) contains a detailed analysis of the errors which arise in the practical determination of this dimension. Methods for the empirical determination of dimensions based on these ideas are considered in Chapter 5.

2.3 LOCAL HAUSDORFF DIMENSION AND DIMENSION DISTRIBUTION

Sometimes it is useful to apply the so-called local Hausdorff dimension, see §5.7. This assigns to every point x of the plane a number $D_\mu(x)$, which may be connected with the dimension of a given set. If this set consists of components of different dimension (it is then called a 'multifractal') then $D_\mu(x)$ may help to find and to describe these differences. Let μ be a probability distribution on R^2. Associate μ with a random point X for which the probability that X lies in E is equal to

$$\Pr(X \in E) = \mu(E) \tag{2.24}$$

for any Borel set E. In particular, $\mu(b(x, r))$ denotes the probability that X lies in the disc $b(x, r)$. Then the *local Hausdorff dimension* $D_\mu(x)$ is defined by

$$D_\mu(x) = \liminf_{r\downarrow 0} \frac{\log \mu(b(x, r))}{\log r} \tag{2.25}$$

(Cutler and Dawson, 1989). The term 'pointwise dimension' is also used (Mandelbrot, 1982; Farmer *et al.*, 1983).

Now let the distribution μ of X be such that there is a compact set B of Hausdorff dimension D with

$$\Pr(X \in E) = 1 \text{ and } \Pr(X \in E) = 0$$

for all subsets E of B with H-dim$(E) < D$. Then for μ-almost all $x^{\dagger}$

$$D_{\mu}(x) = D \qquad (2.26)$$

An example of such a distribution is the 'uniform distribution' of B,

$$\mu(E) = \frac{\mathcal{H}^D(B \cap E)}{\mathcal{H}^D(B)}, \quad \text{where } 0 < \mathcal{H}^D(B) < \infty.$$

If B is the Cantor dust C then the distribution is closely related to the Cantor function (or 'Devil's staircase'). The set C is interpreted as a subset of the x-axis, and X is a random point on the x-axis with the uniform distribution corresponding to $\mathcal{H}^D$,

$$\mu(E) = \frac{\mathcal{H}^D(B \cap E)}{\mathcal{H}^D(B)}, \quad D = \frac{\log 2}{\log 3}.$$

The distribution function of X is

$$\Pr(X < x) = \mu([0, x)) = C(x),$$

where $C(x)$ denotes the Cantor function, (Fig. 7), which is defined by

$$C(x) = \begin{cases} \frac{1}{2} \left(\frac{1}{3} < x \le \frac{2}{3} \right), \\ \frac{1}{4} \left(\frac{1}{9} < x \le \frac{2}{9} \right), \\ \frac{3}{4} \left(\frac{7}{9} < x \le \frac{8}{9} \right), \end{cases}$$

and so on (the continuation is analogous to that for the Cantor dust). Interesting physical applications of the Cantor function are discussed in Mandelbrot (1977, 1982) and Bak (1986). More generally, now let B be the union of n pairwise-disjoint compact sets of Hausdorff dimensions D_i $(i = 1, \ldots, n)$. Let

$$\Pr(X \in B_i) = p_i > 0$$

and

$$\Pr(X \in E_i) = 0 \quad (i = 1, \ldots, n)$$

for any subset E_i of B_i with H-dim$(E_i) < D_i$. Then $D_{\mu}(x)$ takes the value D_i for almost all $x \in B_i$. The values $D_{\mu}(x)$ can be interpreted as samples of a random variable $D_{\mu}(X)$. While X denotes a random point in R^2, $D_{\mu}(X)$ is the corresponding random dimension. The distribution of $D_{\mu}(X)$ is given by the probabilities p_i:

$$\Pr(D_{\mu}(X) = D_i) = \Pr(X \in Bi) = p_i \quad (i = 1, \ldots, n).$$

---

†There is a set N with $\Pr(X \in N) = 0$, so that (2.26) holds for all $x \in N^c$.

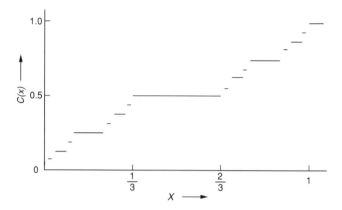

Figure 7 An approximation of the Cantor function (Devil's staircase). Further refining yields more jump points with decreasing jump heights; see text for explanation.

However, note (2.20); the Hausdorff dimension of the set B above satisfies

$$\text{H-dim}(B) = \max_{(i)} \text{H-dim}(B_i).$$

The *dimension distribution function* is given by

$$F_D(\alpha) = \sup\{\mu(E) : E \text{ is a Borel set with H-dim}(E) \le \alpha\}.$$

It satisfies

$$F_D(\alpha) = \Pr(D_\mu(X) < \alpha),$$

(Cutler, 1986, 1990; Cutler and Dawson, 1989).

2.4 FRACTALS

A subset B of R^2 is said to be a fractal if its topological dimension is smaller than H-dim(B). Taylor (1986b) requires additionally that other definitions of dimension give the same value. According to him, nearly all examples in Mandelbrot (1977, 1982) satisfy this condition. Many non-mathematicians seem to believe that all 'natural' fractals have this property. (However, it is known, for example, that for certain attractors different definitions of dimension give different numbers.)

Note that irregular sets with positive area ('islands') are not fractals, since both their topological and fractal dimensions are equal to 2. However, their boundaries (coastlines) may be fractals.

CHAPTER 3

Deterministic Fractals

3.1 GENERAL PROPERTIES OF FRACTALS

Falconer (1985,1990) gives an extensive description of the geometric properties of fractals (in the sense of Hausdorff dimension) in R^d. The following is a short sketch for the case R^2.

Every set B of Hausdorff dimension zero with $0 \leq \mathcal{H}^0(B) < \infty$ is a finite point set. Sets of greater dimension may be unions of sets of different dimension. By (2.20), the subset of maximum dimension determines the dimension of the whole set. Therefore, in the following, only this set is considered. (The Hausdorff measure $\mathcal{H}^D$ of the rest vanishes.)

A set B of dimension 1 with $0 < \mathcal{H}^1(B) < \infty$ consists of two parts: a 'regular' and an 'irregular' one. The regular part is a countable union of rectifiable curves (of finite length), while the irregular part is completely disconnected. Sets of dimension D with $0 < D < 1$ (and $D \neq \frac{1}{2}$) are always totally disconnected. Sets of dimension D with $D > 1$ are always 'irregular' in a certain sense. They do not necessarily consist of curve segments.

There are sets of dimension 2 that only consist of curve segments.

3.2 EXAMPLES OF DETERMINISTIC FRACTALS

3.2.1 Curves of fractal dimension[†]

Let B be the graph of a continuous real function $f(x)$ on $[0, 1]$, i.e.

$$B = \{(x, f(x)), \ 0 \leq x \leq 1\}.$$

If $f(x)$ is sufficiently irregular then B may be a fractal. It is possible to estimate the dimension of B if $f(x)$ satisfies the following Lipschitz condition:

$$\mid f(x + h) - f(x) \mid \leq ch^{2-\alpha} \tag{3.1}$$

[†]Falconer (1985, 1990).

for all x and all h with $0 < h \leq h_0$, where c and h_0 are positive constants. Then

$$\mathcal{H}^{\alpha}(B) < \infty. \tag{3.2}$$

As interesting functions of this kind, consider function series of the form

$$f(x) = \sum_{i=1}^{\infty} a_i f_i(x),$$

where the $f_i(x)$ oscillate increasingly with increasing i. Examples are

$$f_i(x) = \sin(b^i x), \quad b > 1, \tag{3.3}$$

or

$$f_i(x) = g(\lambda_i x) \tag{3.4}$$

with $\lambda_i \to \infty$, where $0 \leq x \leq 4$,

$$g(x) = \begin{cases} x & (0 \leq x < 1), \\ 2 - x & (1 \leq x < 3), \\ x - 4 & (3 \leq x \leq 4), \end{cases}$$

and $g(x)$ is periodically continued outside $[0, 4]$ ('zigzag function').

If (3.3) is used with $a_i = b^{(s-2)i}$ $(1 < s < 2)$, the *Weierstrass function* is obtained. The corresponding fractal dimension is probably s (Falconer, 1985, 1990; Mauldin, 1986; Ledrappier, 1992). In the case of (3.4) the H-dimension can be calculated: let

$$a_i = \lambda_i^{\alpha-2}, \quad 1 < \alpha < 2,$$

and for all i let

$$\lambda_i > 0,$$
$$\lambda_{i+1} > \lambda_i, \quad \text{with } \lambda_{i+1}/\lambda_i \to \infty,$$

but

$$\lim_{i \to \infty} \frac{\log \lambda_{i+1}}{\log \lambda_i} = 1.$$

Then

$$\text{H-dim}(A) = \alpha.$$

3.2.2 Self-similar sets

An important class of fractals are the self-similar sets. These consist of subsets that are geometrically similar to the whole. For self-similar sets it is very easy to calculate the fractal dimension using (3.6). Examples of such sets are the Cantor dust, the square dust Q and the three parts of the von Koch snow flake (see below). The definition of such sets follows Hutchinson (1981).

A mapping Ψ of R^2 into itself is called a *contraction* if for all $x, y \in R^2$

$$\| \Psi(x) - \Psi(y) \| \le c \| x - y \|,$$

with $c \le 1$. The smallest such c is called the *ratio* $c(\Psi)$ of the construction Ψ. In particular there are (contracting) *similarities*; these transform any subset of R^2 to a geometrically similar set. Thus they comprise translations, rotations, reflections and dilatations. The ratio of a similarity is equal to the dilatation factor.

Let $\Psi_1, \ldots, \Psi_m$ be m contractions. The set B is said to be *invariant* with respect to $\Psi_1, \ldots, \Psi_m$ if

$$B = \bigcup_{i=1}^{m} \Psi_i(B). \tag{3.5}$$

If the Ψ_i are similarities and if there exists a positive β with

$$\mathcal{H}^{\beta}(B) > 0 \text{ and } \mathcal{H}^{\beta}(\Psi_i(B) \cap \Psi_j(B)) = 0 \quad (i \ne j)$$

then B is said to be *self-similar*. (The additional assumption ensures that any overlap of the $\Psi_i(B)$ does not produce undesired effects.)

Example 1: *Square.* Any square can be interpreted as a union of four squares of half-side length. Thus here $m = 4$ and $c(\Psi) = \frac{1}{2}$ $(i = 1, \ldots, 4)$.

Example 2: *Cantor dust C* (interpreted as a subset of the real axis)

$$\Psi_1(x) = \tfrac{1}{3}x, \quad \Psi_2(x) = \tfrac{1}{3}(2 + x).$$

The similarity Ψ_1 maps C onto the left part of itself contained in $[0, \frac{1}{3}]$, while Ψ_2 makes the mapping on the right part. Both parts are geometrically similar to C. Thus $m = 2$ and $c(\Psi) = \frac{1}{3}$ $(i = 1, 2)$.

Example 3: *Square dust Q* (Fig. 2). Obviously, Q is geometrically similar to each of the subsets lying in the corners of the original square. Thus the similarities $\Psi_1, \ldots, \Psi_4$ have to be chosen in such a way that they produce a contraction of the ratio k and a shift of the square centre to the centres $Z_1, \ldots, Z_4$ of the subsquares in the first step of the construction of the dust. Here $m = 4$ and $c(\Psi_i) = k$ $(i = 1, \ldots, 4)$.

Example 4: *von Koch snow flake S* (Fig. 6). The set S consists of three congruent subsets, which are each self-similar. Each has $m = 4$ similarities with the ratio $\frac{1}{3}$.

The construction process used for the von Koch snow flake may be generalized, following Mandelbrot (1982). Then one speaks of initiators, substituands and generators. The *substituands* are those parts of the sets that are replaced by another (more complicated) part during the iteration process. In the case of the snow flake the substituands are line segments, while for the square dust of Fig. 2 they are squares.

Those parts that go suitably contracted into the positions of the substituands are called *generators* (Fig. 5 shows the generator of the snow flake). Finally the construction is determined by the starting figure or the *initiator*, which determines the arrangement of the first substituands. In the case of the snow flake the starting figure is an equilateral triangle. Figure 1 shows the generator of the Cantor dust. The initiator here is the interval $[0, 1]$; the line segments are substituands.

The existence of self-similar sets for given similarities and the convergence of iteration procedures is ensured by the following. (Here, even general contractions may be considered.) Let Ψ_i be contractions of R^2 with the factors $c(\Psi_i)$ $(i = 1, \ldots, m)$. With these, a mapping χ is defined by

$$\chi(E) = \bigcup_{i=1}^{m} \Psi_i(E).$$

(This mapping assigns a set to another set; if E is compact then so is $\chi(E)$. Thus χ is interpreted as a mapping of the set $\mathcal{K}$ of all compact subsets of R^2 into $\mathcal{K}$.) The corresponding kth iteration is χ^k:

$$\chi^1(E) = \chi(E) = E_1,$$
$$\vdots$$
$$\chi^k(E) = \chi(\chi^{k-1}(E)) = E_k \quad (k > 1).$$

It is possible to show that there is a unique non-empty compact set B such that

$$B = \chi(B) = \bigcup_{i=1}^{m} \Psi_i(B)$$

if $c(\Psi) < 1$ for all i. For any non-empty compact set K the sequence $\{\chi^k(K)\}$ converges to B with respect to the Hausdorff metric (the proof of this statement uses Banach's fixed point theorem). This makes it possible to plot self-similar sets iteratively (see e.g. Hayashi, 1985; Heesterbeek *et al.*, 1990).

In the following let the Ψ_i $(i = 1, \ldots, m)$ be similarities. The similarity dimension of B is that real number D which satisfies

$$\sum_{i=1}^{m} c_i^D = 1. \tag{3.6}$$

It is positive and uniquely determined.

Example 1: *Square*

$$m = 4, \quad c_1 = \cdots = c_4 = \tfrac{1}{2} : \quad D = 2.$$

Example 2: *Cantor dust C*

$$m = 2, \quad c_1 = c_2 = \tfrac{1}{3} : \quad D = \frac{\log 2}{\log 3}.$$

Example 3: *Square dust Q*

$$m = 4, \quad c_1 = \cdots = c_4 = k: \quad D = \frac{\log 4}{\log k}.$$

Example 4: *von Koch snow flake S*

$$m = 4, \quad c_1 = \cdots = c_4 = \tfrac{1}{3}: \quad D = \frac{\log 4}{\log 3}.$$

Moran (1946) showed that the similarity dimension of the invariant set B is equal to its Hausdorff dimension,

$$\text{H-dim}(B) = D, \tag{3.7}$$

if the following non-overlapping condition is satisfied (Hutchinson, 1981; Falconer, 1985, 1990): there is an open set O with

$$\chi(O) \subset O \text{ and } \Psi_i(O) \cap \Psi_j(O) = \emptyset \quad (i \neq j). \tag{3.8}$$

For the corresponding Hausdorff measure

$$0 < \mathcal{H}^D(B) < \infty, \tag{3.9}$$

and the set B is self-similar, i.e. it satisfies (3.5) and also

$$\mathcal{H}^D(\Psi_i(B) \cap \Psi_j(B)) = 0 \quad (i \neq j)$$

The latter is clear because of (3.9)

$$\sum_{i=1}^{m} \mathcal{H}^D(\Psi_i(B)) = \sum_{i=1}^{m} r_i^D \mathcal{H}^D(B) = \mathcal{H}^D(B),$$

and because $\mathcal{H}^D$ is additive.

It is easy to see that $\mathcal{H}^D(B) < \infty$. For any k, B can be written in the form

$$B = \bigcup_{(i_1,\dots,i_k)} B_{i_1,\dots,i_k}$$

Thus $B_{i_1,\dots,i_k}$ is geometrically similar to B with the ratio $c_{i_1} \cdots c_{i_k} < c < 1$, with $c = \max_{(i)} c_i$. The radii of the smallest covering discs satisfy

$$\sum_{(i_1,\dots,i_k)} \text{rad}(B_{i_1,\dots,i_k})^D = \sum_{(i_1,\dots,i_k)} \text{rad}(B)^D (c_{i_1} \cdots c_{i_k})^D = \text{rad}(B)^D.$$

The second equation follows from

$$\sum_{(i_1,\dots,i_k)} (c_{i_1} \cdots c_{i_k})^D = 1,$$

which follows from the definition of similarity dimension. Choosing $\delta = c^k \, \text{rad}(B)$ in (2.1) yields

$$\mathcal{H}^D(B) \leq \omega_D \, \text{rad}(B)^D.$$

Further very important examples of deterministic fractals are 'strange attractors' (e.g. Julia and Mandelbrot sets) (see e.g. Mandelbrot, 1982; Schuster, 1984; Falconer, 1985, 1990; Peitgen and Richter, 1986; Peitgen and Saupe, 1988; Devaney, 1990; Hastings and Sugihara, 1993; Jensen, 1993).

3.2.3 A program for the generation of generalized von Koch snow flakes

The following is a sketch, by means of an example, of how one could generate approximations of fractal curves by computers. The algorithms used for this may help to understand the architecture of these sets, and the figures obtained can be used to test measurement methods of fractal dimensions. Methods for the generation of figures of fractals are systematically treated in Devaney (1990), Hastings and Sugihara (1993) and Peitgen and Saupe (1988). Our aim here is plotting a generalized von Koch snow flake as in Fig. 8 by computer.

The initiator of the figure is an equilateral triangle of side length l; all line segments are substituands. Figure 9 shows the generator, its parameters are the side length l, the angle α and the ratio a (<0.5). The ratio b is given by

$$b = \frac{\frac{1}{2} - a}{\cos \alpha}.$$

The fractal dimension D of the snow flake is the solution of the equation

$$a^D + b^D = \tfrac{1}{2}, \tag{3.10}$$

which is a particular case of (3.6). Figure 10 shows the dependence of D on a and α. The values above the thick curve are only formal solutions of (3.9); under it the non-overlapping condition (3.8) is satisfied. The open triangle ABC dilated by the factor l plays the role of U in (3.8). A procedure that plots the generator is sketched as follows:

```
Procedure Generator(L);
Begin;
    plot a line segment of length L · a;
        rotate the pen by α degrees to left;
    plot a line segment of length L · b;
        rotate the pen by 2 · α degrees to right;
    plot a line segment of length L · b;
        rotate the pen by α degrees to the left;
    plot a line segment of length L · a;
End;
```

Figure 8 Computer-generated image of a generalized von Koch snow flake. The model parameters are $a = 0.46$ and $\alpha = 85°$. The fractal dimension is 1.78.

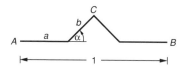

Figure 9 Generator for the generalized von Koch snow-flake. The quantities a and α are model parameters.

Only a few changes are necessary to obtain a procedure that constructs a third of the snow flake. Instead of a line segment, a diminished copy of the generator must be plotted at every stage. The reduction is controlled by the factors a and b. The recursive character of the construction is realized by a recursive procedure. The plotting is halted after some fixed number of steps, called the 'depth'.

```
Procedure Generator(l,depth);
    if depth = 0 then plot line segment of length l
            else Generator(l · a,depth-1);
                    rotate the pen by α degrees to left;
                Generator(l · b,depth-1);
```

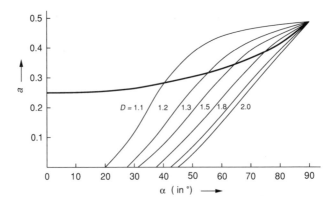

Figure 10   Dependence of the fractal dimension of the generalized von Koch snow flake on the parameters a and α. Only the region below the curve in bold corresponds to fractals, since above the curve the components of the sets generated overlap.

```
            rotate the pen by 2 · α degrees to right;
        Generator(l · b,depth-1);
            rotate the pen by α degrees to left;
        Generator(l · a,depth-1);
    end if
End;
```

The final program plots three figures of this kind so that their base lines form an equilateral triangle.

```
Set pen in start position;
    Generator(l,depth);
    rotate pen by 120 degrees to right;
    Generator(l,depth);
    rotate pen by 120 degrees to right;
    Generator(l,depth).
```

CHAPTER 4

Random Fractals

4.1 RANDOM SELF-SIMILAR SETS

Falconer (1986, 1990), Mauldin and Williams (1986) and Graf (1987) (see also Graf *et al.*, (1988)) have developed a stochastic analogue to the theory of deterministic self-similar sets of Moran and Hutchinson, which was sketched in §3.2.2. Some of its results will be given in this section. But first two examples of compact random self-similar sets are considered.

Example 1: *Random Cantor dust C_L* (Falconer, 1986). The central part of the interval $[0, 1]$ is removed in such a way that the remaining two outer parts have the same length L, where $L < \frac{1}{2}$. Here L is a random variable with density function $f(s)$. From each of these intervals the central part is again removed independently of other removals, where the random length fraction of the remaining outer intervals has the same distribution as L and they are stochastically independent of L. This procedure is continued infinitely, and a random compact set C_L is obtained. It has fractal dimension

$$\text{H-dim}(C_L) = D,$$

where D is the positive solution of the equation

$$-2\mathsf{E}L^D = 2\int s^D f(s)\,\mathrm{d}s = 1; \tag{4.1}$$

see also (4.6). If $f(s)$ is the density function of the uniform distribution on $[0, \frac{1}{2}]$, then $D = 0.457$. By suitable choice of $f(s)$, it is possible to obtain any value between 0 and 1 for H-dim(C_L). If $L \equiv \frac{1}{3}$ then the classical deterministic Cantor dust C is obtained.

Example 2: *Random von Koch snow flake $S_{a,\alpha}$* (Falconer, 1986). As for the deterministic snow flake, the starting figure is an equilateral triangle, but now a randomized form of the construction principle of §3.3 is used. Now a and α are random variables with density functions $f_a(s)$ and $f_\alpha(\delta)$. In each step of the construction in which the suitably diminished figure of Fig. 9 replaces a line segment, a and α are chosen independently of all earlier replacements. This process yields a random compact set $S_{a,\alpha}$, whose distribution depends on $f_a(s)$ and $f_\alpha(\delta)$.

The Hausdorff dimension D is the positive solution of the equation

$$\int \int \left[2s^D + \frac{2(\frac{1}{2} - s)}{\cos^D \delta} \right] f_a(s) f_\alpha(\delta) \, ds \, d\delta = 1. \tag{4.2}$$

In the particular case where a is uniformly distributed on $[\frac{1}{3}, \frac{1}{2}]$ and $\alpha \equiv 60°$, D is the solution of the equation

$$2^{-D} - 3^{-(D+1)} = \frac{1}{6}(D + 1). \tag{4.3}$$

Thus $D = 1.145$. (As Fig. 10 shows, there are no overlappings in this case.)

Figure 11 shows two simulated curves for this model.

The construction of further random self-similar sets is possible similar to the procedure of §3.2.2 (Falconer, 1986, 1990; Graf, 1987; Graf *et al.*, 1988; Patzschke and Zähle, 1990).

Starting with an initiator, in every step substituands are replaced by random generators, which are independent and each identically distributed. The result is a

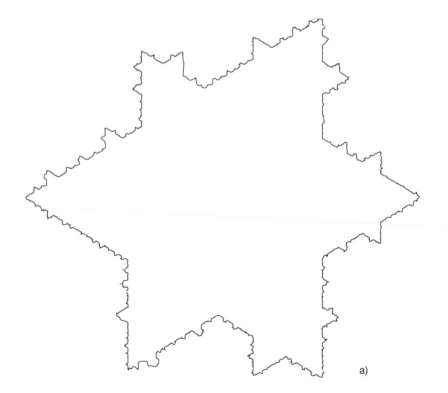

Figure 11   Simulated samples of random Koch snow flakes: (a) $\alpha = 60°$, uniform in $[\frac{1}{3}, \frac{1}{2}]$; (b) $\alpha = 85°$, uniform in $[0.465, 0.5]$.

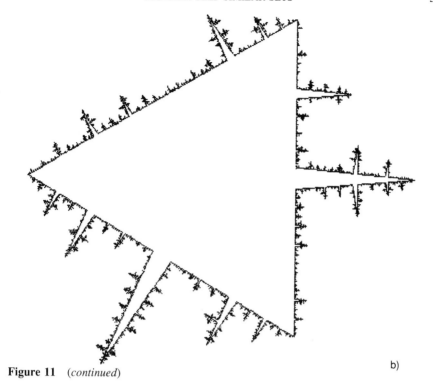

b)

Figure 11   (*continued*)

random compact set B with the following properties:

$$B \stackrel{\mathrm{d}}{=} \bigcup_{j=1}^{m} \Psi_j(B_j),\qquad(4.4)$$

$$B \stackrel{\mathrm{d}}{=} B_j \quad (j = 1, 2, \ldots, m).\qquad(4.5)$$

Here $X \stackrel{\mathrm{d}}{=} Y$ means that X and Y have the same distribution. The B_j are stochastically independent mutually and of $(\Psi_1, \ldots, \Psi_m)$, and the Ψ_j are similarities with the ratios $c_j < 1$. They belong to a set S of similarities, from which they are taken according to a certain distribution. A non-overlap condition analogous to (3.8) must be satisfied (Falconer, 1986, 1990; Graf, 1987).

It can be shown that the Hausdorff dimension of the set B has the value D that is the positive solution of the equation

$$\mathsf{E}\left(\sum_{j=1}^{m} c_j^D\right) = 1\qquad(4.6)$$

The mean has to be calculated according to the distribution of the similarities.

The Hausdorff measures of such sets have been investigated by Graf *et al.* (1988).

The Hausdorff dimension is now calculated for the examples given at the beginning of this section.

Example 1: *Random Cantor dust* C_L. Here $m = 2$, and the set $\mathcal{S}$ of similarities consists of two families of similarities $\{\Psi_{l,s}\}$ and $\{\Psi_{r,s}\}$ with $0 \le s \le \frac{1}{2}$. A point x of the x-axis is mapped by $\{\Psi_{l,s}\}$ and $\{\Psi_{r,s}\}$ onto sx and $1-s+sx$ respectively. The ratio of both mappings is s. Consequently the sum in (4.6) is $2s^D$, and the mean equals

$$2 \int_0^{1/2} s^D f(s) \, \mathrm{d}s.$$

Example 2: *Random von Koch snow flake* $S_{a,\alpha}$. The family $\mathcal{S}$ of similarities now consists of four families $\{\Psi_{1,s}\}$, $\{\Psi_{2,t}\}$, $\{\Psi_{3,t}\}$ and $\{\Psi_{4,s}\}$ with $0 \le s, t \le \frac{1}{2}$. The similarities $\Psi_{1,s}$ and $\Psi_{4,s}$ refine the two base pieces of the generator in the construction of the Koch snow flake; $\Psi_{2,t}$ and $\Psi_{3,t}$ do the same for the sides of the triangle. Their ratios are s and t, with $t = (1 - 2s)/(2\cos\delta)$. The mean in (4.6) is

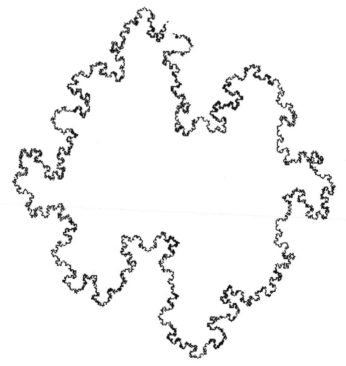

Figure 12   Simulated sample of a random snow flake. At each replacement of a line segment by a generator, the chosen generator is with equal probability 0.5 that of the Koch snow flake or the quadratic snow flake (Mandelbrot, 1982).

equal to

$$\int \int \left[2s^D + \frac{2(\frac{1}{2} - s)}{\cos^D \delta} \right] f_a(s) f_\alpha(\delta) \, ds \, d\delta.$$

Interesting random fractals are obtained if in each step of the construction a random choice is carried out from a set of qualitatively different generator. Figure 12 was obtained in this way.

4.2 MANDELBROT-ZÄHLE CUT-OUTS

Mandelbrot (1972) studied a class of random fractals on the real axis called by him 'random cut-outs'. Zähle (1984b) considered generalizations in R^d. A particular case is the following model. Γ_n is a random union of discs. The random set X is what remains after infinitely many such Γ_n have been deleted from the plane.

The base of the construction is a Boolean model Γ with open convex grains, see Appendix F. The intensity of the corresponding Poisson point field of germs is λ. The mean area and mean perimeter are $\overline{A}$ and $\overline{U}$ respectively, where $0 < \overline{A}$, $\overline{U} < \infty$.

Now imagine infinitely many independent sets $\Gamma^{(n)}$ with the same distribution as Γ $(n = 1, 2, \ldots)$. They are reduced in such a way that new sets Γ_n are obtained:

$$\Gamma_n = r^{-n} \Gamma^{(n)} \quad (n = 1, 2, \ldots),$$

where $r > 1$. Thus Γ_n is again a Boolean model, and its intensity is λr^{2n}, and the mean perimeter of the grains is $\overline{U} r^{-n}$ and the mean area $\overline{A} r^{-2n}$. The cut-out set X is then

$$X = R^2 \backslash \bigcup_{n=1}^{\infty} \Gamma_n.$$

This is a closed set of area zero. Plates 322–325 in Mandelbrot (1982) show simulated samples of such a set. The Hausdorff dimension of this set is

$$\text{H-dim}(X) = 2 - \frac{\lambda \overline{A}}{\log r}$$

if this number is non-negative; otherwise X is empty. Physicists have studied the fractal boundaries of irregular Boolean models; see Hermann (1991).

4.3 RANDOM FRACTALS CONNECTED WITH BROWNIAN MOTION

Brownian motion (or the Wiener process) in R^1 is a stochastic process $\{W_t\}_{t \geq 0}$ with the following properties:

(i)
$$W_0 = 0,$$

which means that the process starts at the origin;

(ii) it has independent increments, i.e. for all n and all $t_1 < \cdots < t_n$ the differences

$$W_{t_n} - W_{t_{n-1}}, \ldots, W_{t_2} - W_{t_1}$$

are stochastically independent;

(iii) the increments are normally distributed and depend only on the time difference:

$$W_{t_2} - W_{t_1}$$

is normal with mean 0 and variance $\sigma^2(t_2 - t_1)$; $\sigma^2 > 0$.

The graph of this process is the curve $\{(t, W_t), 0 \le t \le 1\}$ in the (t, w)-plane. This curve has fractal (Hausdorff) dimension 1.5. A further fractal is the *Brownian bridge*. This is the stochastic process $\{B_t\}_{0 \le t \le 1}$, with

$$B_t = W_t - tW_1 \quad (0 \le t \le 1).$$

It satisfies $B_0 = B_1 = 0$. The graph of $\{B_t\}$, i.e. the curve $\{(t, B_t), 0 \le t \le 1\}$ also has fractal dimension 1.5. The same is true for the contours of smooth star-shaped sets (such as discs or ellipses) that are *noised* by an isotropized Brownian bridge; see Fig. 40 and §7.6.4.

'Still more irregular' curves are given by the *Brownian path*, constructed as follows. Given two stochastically independent Brownian motions

$$\{X_t\}_{0 \le t \le 1} \text{ and } \{Y_t\}_{0 \le t \le 1},$$

assign to each value t the point $P_t = (X_t, Y_t)$ in R^2. The set of all points P_t $(0 \le t \le 1)$ is a connected curve. A simulated piece of it is shown in Fig. 13. The fractal dimension (Hausdorff and packing) is 2, for all σ (Taylor, 1986b). Also, if Wiener processes 'with drift' are considered, where

$$\mathsf{E}W_t = \mu t,$$

then fractal curves are obtained. Their dimensions coincide with those for $\mu = 0$.

Computer simulation of these random fractals is not difficult, the key being the simulation of the Brownian motion $\{W_t\}$ (Ripley, 1987; Saupe, 1988). Most of these simulations are two-step methods.

In the following let $\{S_t\}$ be the simulated process approximating $\{W_t\}$.

Step 1: Simulation of Brownian motion in discrete time steps. Choose a small time length Δt and determine the values $S^{(i)}$:

$$S^{(i)} = S(i\Delta t) \quad \text{with } i = 1, \ldots, n, \text{ where } n\Delta t = 1.$$

Thus

$$S^{(i+1)} = S^{(i)} + Z_i, \quad S^{(0)} = 0.$$

The Z_i form a series of independent normal random numbers with mean 0 and variance $\sigma^2 \Delta t$.

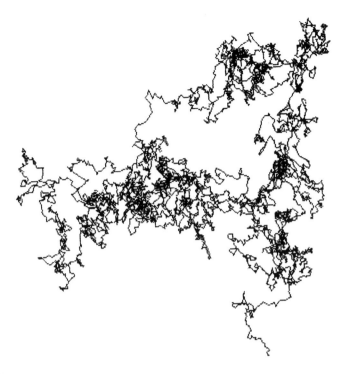

Figure 13   Simulated sample of a Brownian path. Its fractal dimension is 2.

Several methods are known for the generation of normal random numbers (Devroye, 1986; Ripley, 1987). A simple method is the sum-of-12 method based on the central limit theorem. If $u_1, u_2, \ldots$ are uniform random numbers produced by a random number generator (e.g. RND in BASIC) then sums of the form

$$V_1 = \sum_{i=1}^{12} u_i - 6, \quad V_2 = \sum_{i=13}^{24} u_i - 6, \ldots$$

are random numbers with mean 0 and variance 1. Their distribution is close to normal. The V_i obtained can be transformed into the required Z_i by multiplying by $\sigma\sqrt{\Delta t}$.

Step 2: Determination of the intermediate values. Starting from the $S^{(i)}$, the values S_t for t between $i\Delta t$ and $(i+1)\Delta t$ are determined. If Δt is very small then linear interpolation is possible:

$$S_t = S^{(i)} + \frac{S^{(i+1)} - S^{(i)}}{\Delta t}(t - i\Delta t) \quad (i\Delta t \le t \le (i+1)\Delta t).$$

Otherwise random trajectories between $S^{(i)}$ and $S^{(i+1)}$ are generated, using one of several methods. A popular one is the midpoint displacement method suggested by Levy. It is based on the following property of Brownian motion $\{W_t\}$. If t_1 and t_2 are two arbitrary instants between 0 and 1 and if t is between t_1 and t_2 then

$$W_t = \mu(t) + \sigma(t)Z_t,$$

where

$$\mu(t) = \frac{(t_2 - t)W_{t1} + (t - t_1)W_{t2}}{t_2 - t_1},$$

$$\sigma^2(t) = \frac{\sigma^2(t - t_1)(t_2 - t)}{t_2 - t_1}.$$

Here Z_t is a normal random variable with mean 0 and variance 1 that is independent of all W_s for $0 < s < t_1$ and $t_2 < s < 1$. This suggests the following method to obtain $S^{(i)}_{1/2}$, which corresponds to $W_{i\Delta t + \Delta t/2}$ using $S^{(i)}$ and $S^{(i+1)}$: Generate a normal random variable V with mean 0 and variance 1 and set

$$S^{(i)}_{1/2} = \tfrac{1}{2}[(S^{(i)} + S^{(i+1)}) + \sigma\sqrt{\Delta t}\,V].$$

Then one can either linearly interpolate between $S^{(i)}$, $S^{(i)}_{1/2}$ and $S^{(i+1)}$ or make further refinements using $S^{(i)}_{1/4}$ and $S^{(i)}_{3/4}$, etc.; for programs see Saupe (1988). Figure 13 shows a simulated Brownian path generated in a 630×350 raster.

4.4 SELF-SIMILAR STOCHASTIC PROCESSES

Certain important classes of random fractals are closely connected with so-called self-similar processes. A real-valued stochastic process $\{Y_t\}_{0 \leq t < \infty}$ is called *h-self-similar* (or 'scaling at the origin', Mandelbrot and van Ness, 1968) if for all $c > 0$ and all $t \geq 0$

$$Y_{ct} \stackrel{\mathrm{d}}{=} c^h Y_t; \tag{4.7}$$

that is, if Y_{ct} and $c^h Y_t$ have the same distribution.

Example: *Brownian motion.* In this case $Y_t = W_t$ (cf. §4.3) is normal with mean 0 and variance $\sigma^2 t$, where σ^2 is a model parameter. Obviously, for any $c > 0$ the random variable Y_{ct} has a normal distribution with mean 0 and variance $c\sigma^2 t$. Consequently, $\sqrt{c}Y_t$ has the same distribution as Y_{ct}. The Brownian motion is thus $\tfrac{1}{2}$-self-similar.

The situation is different in the case of Brownian motion with drift, where

$$Y_t \sim N(\mu t, \sigma^2 t).$$

Here Y_{ct} is normal with mean ct and variance $c\sigma^2 t$, and there is no h such that (4.7) is true if $\mu \neq 0$.

Sometimes the notion of self-similarity is defined more strictly. It is supposed that for all $c > 0$, for all n and for all $t_1, \ldots, t_n > 0$ the random vectors $(Y_{ct_1}, \ldots, Y_{ct_n})$ and $(c^h Y_{t_1}, \ldots, c^h Y_{t_n})$ have the same distribution. Brownian motion is self-similar in this sense.

There is a vast literature on self-similar processes, see e.g. Sinai (1976), Verwaat (1985) and Taqqu (1986, 1988).

Those values of h that are interesting for applications lie between 0 and 1. The connection to fractals is given by the following. Assuming regularity, the graph of $\{Y_t\}$ (the curve given by $\{Y_t\}$ in the (y, t)-plane) has fractal dimension

$$D = \max\{1, 2 - h\} \tag{4.8}$$

(Kôno, 1986; U. Zähle, 1988). Important moment formulae follow from (4.7):

$$\mathsf{E} Y_t = t^h \mathsf{E} Y_1,$$

$$\mathsf{E} |Y_t| = t^h \mathsf{E} Y_1,$$

$$\mathrm{var} Y_t = t^{2h} \mathrm{var} Y_1.$$

They are the starting points of statistical methods (Mandelbrot and Wallis, 1969; Fairfield-Smith, 1983; Künsch, 1986, 1987). An important class of self-similar processes comprises the h-self-similar processes with stationary increments and $0 < h \leq 1$. These are important stochastic models for phenomena with extensive dependency. Such a process $\{Y_t\}$ is h-self-similar, and the distributions of its increments

$$Y_{t+1} - Y_t$$

are independent of t. It is $h = 1$. In the case $h = 1$

$$Y_t = t Y_1$$

An example of a $\frac{1}{2}$-self-similar process with stationary increments is Brownian motion, which is a famous example of a process with independent increments. With the exception of $h = 1$,

$$\mathsf{E} Y_t = 0$$

always holds. The covariances satisfy

$$\mathsf{E} Y_s Y_t = \tfrac{1}{2} \sigma^2 (|t|^{2h} + |s|^{2h} - |t - s|^{2h}),$$

where $\sigma^2 = \mathrm{var}\, Y_1$

Important examples of such processes are so-called fractional Gaussian noise and fractional Brownian motion (Sinai, 1976; Mandelbrot, 1977, 1982; Graf, 1983; Taqqu, 1986, 1988). Saupe (1988) describes their simulation. Many authors have studied statistical methods (Mandelbrot, 1975; Mandelbrot and Taqqu, 1979; Mohr, 1981; Graf, 1983; Fox and Taqqu, 1986; Beran, 1986, 1991; Künsch, 1987; Taqqu, 1987; Samarov and Taqqu, 1988). Self-similar random measures have also been studied (Daley and Vere-Jones, 1988; U. Zähle, 1988). The support sets of such measures are fractals.

CHAPTER 5

Methods for the Empirical Determination of Fractal Dimension

5.1 INTRODUCTION

There is no generally applicable empirical method of determining fractal dimensions. Rather, there is a range of different methods, which are applied according to the data and the nature of the phenomenon investigated. The methods are closely connected with certain definitions of dimension. It is commonly held among fractal statisticians that for real geometrical structures all definitions of dimension should yield the same value; that is, that the real structures are 'regular' in some sense. Sometimes different methods are used for the same structure, and (nearly) equal values are then taken as a 'proof' of the correctness of the results.

For nearly all natural fractals difficulties appear owing to the limited accuracy of measurement and rounding errors at small scales. Consequently the results do not always behave as for mathematical fractals. Frequently one observes fractal-like behaviour only in a certain region of scales. However, it is reasonable to assume that 'fractal dimensions' that are suitable at these scales are useful as irregularity and roughness parameters. In passing, it should be noted that in the production and analysis of mathematical (i.e. 'true') fractals on image analysers similar phenomena are observed; that is, the behaviour is not fractal-like at very small scales. Starting from this observation, Rigaut (1989) has introduced so-called asymptotic fractals and asymptotic fractal dimensions. The latter are essentially the fractal dimensions which are what one would estimate if one was using with care the results for large and medium scales.

Complicated problems appear on using image analysers. The high accuracy of measurement makes it necessary to consider the geometry of the digital raster on the monitor. The structures investigated are digitized and thus simplified, and for small interpixel distances only rough approximations, e.g. of discs (see Fig. 19), are possible. A particularly difficult case is that where the set to be studied is the very irregular boundary of a particle. It is non-trivial to define the boundary of a set in terms of the digital raster and to measure its length with sufficient precision.

Until now, there have been only a few theoretical investigations into the problem of statistical estimation of dimensions. Self-similar stochastic processes (§4.4) and local dimensions (§5.5) have been the most studied. Surveys have been given by Berliner (1992), Chatterjee and Yilmaz (1992), and Isham (1993).

For the systematic empirical measurement of dimension, pilot studies are useful, which may gauge the measurement method by analysing mathematical fractals. These should be similar to the natural objects to be analysed, and should be presented in the same form (pictures on paper or patterns of pixels on the monitor of an image analyser).

Note that there are also physical methods of dimension measurement, e.g. small-angle X-ray scattering (Schmidt, 1989).

The methods described in this chapter are designed with dusts, systems of curves and particle boundaries particularly in mind.

5.2 DIVIDER STEPPING METHOD

The divider stepping method or yardstick method is a proven and precise method for manual determination of fractal dimensions of planar curves on paper. These curves should be orientated or be capable of being orientated (e.g. the boundary of a particle or a coast line).

The measurement goes as follows. First a starting point S (an endpoint or an arbitrary point on a closed curve is selected on the curve B). Then the dividers are adjusted to width r_1. One point is placed at S, and then the point of intersection of B and the circle centred at S of radius r_1 is determined. This new point is the next point at which these dividers are placed. (If there is more than one point of intersection then the point nearest S when travelling along the curve B should be chosen; see Fig. 14.) The process is repeated as necessary. Let $N(r_1)$ be the number of divider widths that can be placed in B. This procedure is repeated for

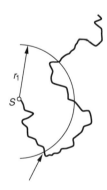

Figure 14 Determination of intersection points for the divider stepping method. If the curve is orientated from left to right then the next intersection point after S is the point indicated by the arrow.

the widths $r_2, r_3, \ldots$ with $r_1 > r_2 > r_3 > \cdots$. Then one obtains approximations $l(r_1), l(r_2), \ldots$ for the length of B:

$$l(r_i) = r_i N(r_i) \quad (i = 1, 2, \ldots).$$

If B is a fractal then one expects

$$l(r) = cr^{1-D}, \tag{5.1}$$

or

$$\log l(r) = \log c + (1 - D) \log r, \tag{5.2}$$

where D is the fractal dimension. Readers who are unhappy with the dimensionality of (5.1) can follow Underwood to rewrite this equation as

$$l_r = l_0 \left(\frac{r}{r_0} \right)^{1-D},$$

where r_0 is an arbitrary constant with dimensions of length, and l_0 is a further length constant.

An estimate of D can be obtained by plotting the points $(r_1, L(r_i))$ on double logarithmic paper, thus obtaining a regression line. The slope is an estimate of D. Since all $L(r_i)$ come from the same picture, there are complicated relationships between the $L(r_i)$. Thus, for the sake of precision, a generalization of the least-squares method should be used (Cressie and Laslett, 1987; Cutler, 1991; Taylor and Taylor, 1991). It is clear that for any r_i the discs corresponding to the circles drawn cover B completely. Thus $N(r_i)$ can be interpreted as an approximation of the disc number in (2.22). Consequently, the divider stepping method is closely related to the Minkowski dimension.

5.3   BOX-COUNTING METHOD

The box-counting method can be used for arbitrary planar structures. It is well suited to image analysers, where the structures are processed as pixel patterns (here it is assumed that there is a square raster). The pixel patterns may be systems of isolated pixels ('dust') or curve-like patterns. Particle boundaries can also be treated using this method, as explained below. First the case of dust or curves is considered.

As a first step all pixels belonging to the structure are counted. The screen is then divided into squares, 2 pixels by 2 pixels, and the number of squares intersecting the structure is counted (Fig. 15). The same is then done for all 3×3 and 4×4 pixel squares, etc. This yields the numbers

$$q_1, q_2, \ldots .$$

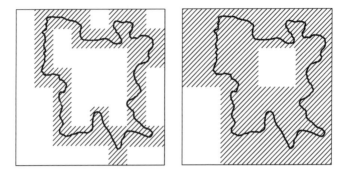

Figure 15   An irregular curve and those pixels and groups of four pixels that are in contact with it. The numbers of these pixels and pixel groups are used in the box-counting method.

Using (2.23), the fractal dimension can now be estimated. For this purpose the points $(1, q_1)$, $(2, q_2)$, ..., are plotted on double logarithmic paper, and a regression line $y = a + bx$ is determined. Its slope b ($b < 0$) yields as an estimate of the fractal dimension as

$$\hat{D} = -b. \tag{5.3}$$

This method can be justified as follows. The formula (2.23),

$$\limsup_{r \downarrow 0} \frac{\log Q_r(B)}{\log r} = -D,$$

corresponds to

$$Q_r(B) = r^{-D}$$

for small r. Using the notation above, this is

$$q_n = (cn)^{-D},$$

where c is the scaling factor. Taking logarithms gives

$$\log q_n = a - D \log n.$$

Tests carried out on mathematical fractals on an image analyser lead to the suggestion that only the points (k, q_k), ..., (N, q_N) rather than $(1, q_1)$, ..., (N, q_N) should be used; for example with $k = 4$ and $N = 10$. These tests show also that the sausage method (§5.4) is more precise than the box-counting method. Taylor and Taylor (1991) have shown that the accuracy of the box-counting method can be increased by filtering the image to be analysed before making the measurements.

Liebovitch and Toth (1980) and Block *et al.* (1990) discuss the problem of an effective numerical computation of the box-counting dimension and generalizations.

If the fractal is the boundary of an irregular particle as on Fig. 4 then first a boundary set should be generated by means of image analysis and then the method described above can be used.

In image analysis there are several definitions of boundary (Kovalevski, 1989). From comparisons of a series of mathematical fractals, it turns out that relatively good measurement results can be obtained using the so-called inner 8-boundary. (Experience shows that this overestimates slightly.) The corresponding definition

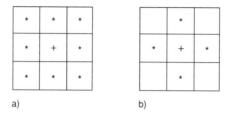

a)   b)

Figure 16 Definition of the inner 8-boundary and the inner 4-boundary. A (+)-pixel of a set of pixels is a boundary pixel if at least one of the (*)-pixels does not belong to the set. Obviously, the inner 8-boundary consists of more pixels.

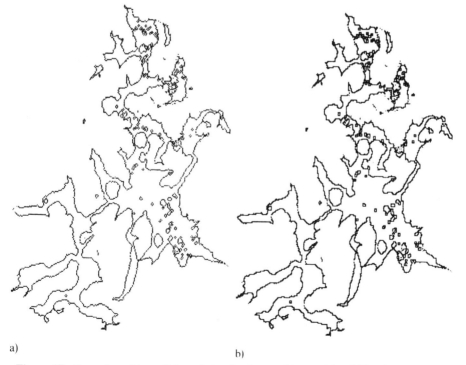

a)   b)

Figure 17 Inner 4- and inner 8-boundaries for the graphite particle of Fig. 4 (a) and (b). The inner 8-boundary is slightly longer and more irregular.

Figure 18   Inner pixels that touch pixels in contact with the curve shown in Fig. 15.

say that a (+)-pixel of the particle is a boundary pixel if one of the (*)-pixels as in Fig. 16 does not belong to the particle. Figure 17 shows the inner 8-boundary of the graphite particle of Fig. 4; for comparison, the so-called inner 4-boundary is also presented, which, as experience shows, underestimates slightly.

Tricot (1986) (see also Tricot *et al.* 1987) suggested the determination of 'inner' and 'outer' dimensions for closed particle boundaries. The inner dimension is obtained by counting the number of squares of side length k ($k = 1, 2, \ldots$) that touch squares of the same side length containing points of the particle (Fig 18). These numbers $q_1^{(1)}, q_2^{(1)}, \ldots$ are used to determine, using (5.3), an inner dimension number D_i. Analogously, an outer dimension D_o is determined based on the squares outside the particle. The larger of the two numbers D_i and D_o is then taken as an estimate of D. As in the case of the divider stepping method, b in (5.3) is obtained from the regression line — possibly using a generalization of the least-squares method (p. 41).

5.4   SAUSAGE METHOD

The sausage or boundary dilation method is still more closely connected to the use of image analysers than the box-counting method (Flook, 1978). The structure analysed is a pixel pattern. It is enlarged by dilation with discs, whose radii increase at each stage. That is, each point of the pattern is taken as the centre of a disc, and the union of all disc pixels is then the dilated set. In every step the number n_r of pixels in the dilated set is noted, and the points (r, n_r) are plotted on double logarithmic paper. Let b be the slope of the resulting regression line; then the dimension should be

$$\hat{D} = 2 - b. \tag{5.4}$$

The theoretical basis of this method is (2.15) (thus it is related to the Minkowski dimension); the rest of the argument applies similarly to that for the box-counting method. Since digital raster discs cannot be presented exactly, it may be helpful here to say something about disc approximations. It is reasonable to use those disc approximations that are suggested by the image analyser. An example is the series shown in Fig. 19. It is important to use the corresponding appropriate 'radii'. These are given by

$$r = \left(\frac{A}{\pi}\right)^{1/2}, \tag{5.5}$$

where A is the area of the corresponding pixel figure. For the figures of Fig. 19 the radii are $(1/\pi)^{1/2}$, $(4/\pi)^{1/2}$, $(9/\pi)^{1/2}$ and $(12/\pi)^{1/2}$, expressed as multiples of the pixel distance.

For systems of curves, Flook (1982b) suggested a correction that considers extensions at the curve ends (Fig. 20). For each curve end the value $\alpha r \left(\frac{1}{4}\pi\right)^{1-\alpha}$ has to be subtracted from the area of the structure, when dilated to radius r. Here α is an apriori estimate of the unknown fractal dimension.

If the set investigated consists of particle boundaries then the inner 8-boundary should be constructed and analysed.

Tests on mathematical fractals, given in a quadratic raster, have shown that acceptable results can be obtained if 5–10 sizes of disc are used. Theoretical dimensions greater than 1.5 are often underestimated. Better results can be obtained if the method given in §5.5 is used and the most frequent value, the mode, of the distribution is taken as the estimate. Creutzburg *et al.* (1992) discuss a fast algorithm for computing the dimension of binary images by the sausage method.

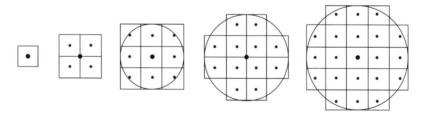

Figure 19   First members of a series of discrete 'discs' in the square raster.

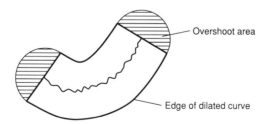

Figure 20   Schematic description of extensions of curve ends. They are considered in the sausage method.

5.5 ESTIMATION OF LOCAL DIMENSION

Several widely used estimation methods are based on the notion of local Hausdorff dimension (p. 18). This includes nearest-neighbour methods or methods for estimating the 'correlation dimension' (Grassberger and Procaccia, 1983; Guckenheimer, 1984; Badii and Politi, 1985; Holzfuss and Mayer-Kress, 1986). Cutler and Dawson (1989) suggested a statistical method for the determination of the local Hausdorff dimension $D_\mu(x)$. Using this method, it is possible to obtain information on the variability and distribution of $D_\mu(x)$. If for a given set B it is possible to find a distribution μ with

$$D_\mu(x) = \text{H-dim}(B) \text{ for } \mu\text{-almost all } x$$

then the dimension of B can also be estimated using the same method. This idea is used below to estimate fractal dimensions of pixel patterns using image analysers.

To apply this method of Cutler and Dawson, it is necessary to generate a series of independent points $X_1, X_2, \ldots$ that have the distribution μ. Let

$$\varrho_n(x) = \min_{i=1,\ldots,n} \| x - X_i \|$$

be the minimum distance between x and the X_i. If x belongs to the support of μ then $\varrho_n(x)$ converges to zero with probability one. Furthermore, the sequence $\{D_n^{-1}(x)\}$ defined by

$$D_n^{-1}(x) = \frac{\log[a\varrho_n(x)]}{\log(b/n)} \quad (n = 1, 2, \ldots) \tag{5.6}$$

almost certainly converges towards $[D_\mu(x)]^{-1}$. This gives a consistent estimate of $D_\mu(x)$. Here a and b are positive numbers determining the speed of convergence. For a distribution concentrated in a disc of radius 1 Cutler and Dawson (1989) characterized the asymptotic behaviour of $D_\mu(x)$. Cutler and Dawson (1990) considered asymptotic normality of related quantities.

Some practical methods can be interpreted as applications of the above method, where μ is the uniform distribution (in the sense of the Hausdorff measure corresponding to the required fractal dimension) on the set R under investigation. It is approximated by the uniform distribution on the pixels of B. To estimate the dimension of B, m test points $t_1, \ldots, t_m$ are used that are chosen according to the uniform distribution on the pixels of B. For every test point t_i the number $s_i(r)$ of pixels of B in the disc $b(t_i, r)$ of radius r centred at t is determined. As in §5.4, the discs are approximated by pixel figures.

For test points near the boundary of the monitor a boundary correction is useful. This can be obtained by multiplying $s_i(r)$ by the factor $f(t_i)$, where

$$f(t_i) = \frac{\pi r^2}{\text{area of } b(t_i, r) \cap W}$$

Then the averages $s(r)$

$$s(r) = \frac{1}{m} \sum_{i=1}^{m} s_i(r)$$

are found for various values r_k of r. The points $(\log r_k, \log s(r_k))$ are then plotted to give a regression line, whose slope is assumed to be an estimate of D. Practical experience suggests that often modal values instead of averages yield better results; see also p. 48.

This method is closely related to a way of estimating the so-called 'correlation dimension'

$$\tilde{D} = \lim_{r \downarrow 0} \frac{\log N(r)}{\log r}, \tag{5.7}$$

where $N(r)$ is the number of pixel pairs with a distance smaller than r. Statistical properties of this method, which is called the Grassberger–Procaccia procedure, are discussed in Wolff (1990) and Smith (1992).

If one has to assume that the structure investigated consists of components of different fractal dimension (a multifractal) then, one may try to determine local dimensions and a dimension distribution. For each of the test points t_i introduced above an individual dimension value D_i is determined. This is done analogously to the above procedure by determining a regression line for the points $(\log r_k, \log s_i(r_k))$ with a series of radii r_k. A frequency distribution of the D_i values gives some impression of the variability of the irregularity and roughness of the geometrical structure under investigation. Another way to characterize multifractals is using a generalized box-counting method and the corresponding generalized dimension (Grassberger, 1983). Here the numbers of pixels in boxes are considered.

5.6 FURTHER METHODS

There are further methods for the determination of dimensions and related characteristics; see e.g. the surveys by Tricot *et al.* (1987), Cutler (1991) and Isham (1993). For curves the following method is useful. The given rough curve is approximated by a smooth curve, and the intersection points of the original curve and the approximating curve form a dust, whose dimension is more easily determined. The dimension of the original curve is obtained by

$$D = 1 + D_{\text{dust}}, \tag{5.8}$$

where D_{dust} is the dimension of the dust (Mattila, 1975, 1981).

It is also possible to measure the dimension of sets with so-called 'variable grey tone' — that is structures that vary in intensity between black and white (Rigaut, 1988). These methods are used, for example, for considering the roughness of surfaces. Still other methods use correlation functions or variograms.

If the fractal considered is a planar section of a three-dimensional fractal then the fractal dimension D_3 of the latter can be obtained as

$$D_3 = 1 + D \tag{5.9}$$

where D is the fractal dimension of the section figure.

Ogata and Katsura (1991) suggested point field statistical methods for determining the fractal dimension of some 'dusts' (epicentres of shallow earthquakes). The pair correlation function $g(r)$ is here proportional to $r^{-(2-D)}$ for small r. (Note that there are non-fractal point fields whose pair correlation function has the same property (see p. 257).

5.7   ESTIMATING THE FRACTAL DIMENSION OF THE BOUNDARY OF A GRAPHITE PARTICLE

For the sake of comparison, some of the methods described have been applied to the boundary of the graphite particle in Fig. 4. The results given are those obtained for the inner 8-boundary and the inner 4-boundary in a 630×350 raster. The box-counting method yields

    1.28 for the 4-boundary,
    1.31 for the 8-boundary.

For the sausage method five disc sizes were used, beginning with those of Fig. 19. Here the measurement results were

    1.34 for the 4-boundary,
    1.39 for the 8-boundary.

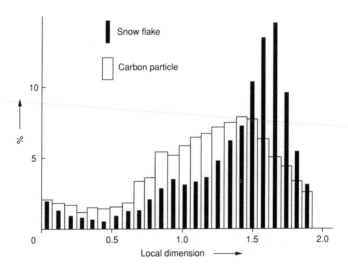

Figure 21   Histogram of the 'local dimension' of the inner 4-boundary of a graphite particle and a von Koch snow flake. Theoretically, the local dimension of the latter should be constant.

Measurements of mathematical fractals suggest the estimate $D = 1.39$. A honest statistician should not end up with such a result in a practical application : 1.4 could be a realistic value.

Finally, to investigate the variability in the wildness of the particle boundary, the measurement method suggested in §5.5 was employed. There the same discs as for the sausage method were used, and all points of the 8-boundary were used as test points, leading to the histogram of Fig. 21. To avoid illusions, this figure also shows the histogram for the generalized von Koch snow flake of Fig. 8 obtained by the same method. The dimension of the snow flake is 1.78. However, since the distribution for the graphite particle is broader than that for the snow flake, this can be taken as a hint of some variability of local dimension for the graphite particle boundary, while for the snow flake the local dimension is constant, 1.78. Note that for the snow flake the modal value is closer than the mean to the theoretical value.

PART II
The Statistics of Shapes and Forms

CHAPTER 6

Fundamental Concepts

In many statistical problems samples of geometrical objects must be studied, for example particles, patterns or living things (or parts of thereof). The aim of the investigations is to obtain information on the form (size and shape) of these objects, because of the influence on technological or biological processes.

- It is hoped to be able to classify the objects with respect to form, as in taxonomic problems. Either the given objects have to be separated into meaningful classes, or single objects or groups of objects have to be assigned to previously given classes.
- It is desirable to describe variations in form statistics.
- Finally, systematic trends in shape and size development have to be detected and described, as they appear in biological or technological processes. Frequently, changes of the form indicate changes in the process; knowledge of form fluctuation may thus help to control processes.

Usually, the form of an object is characterized by a series of real numbers: so-called form parameters. These parameter sets are analysed by means of multivariate statistics.

Figures

In this book only two-dimensional objects or two-dimensional images of three-dimensional objects (projections or sections) are considered. It is assumed that the two-dimensional information is sufficient for a reasonable characterization. (Compare the interesting discussion in Dowdeswell (1982), which tries to justify the use of planar images for sand grains.)

It is always assumed that the objects are sets (in the sense of mathematical set theory), but not grey tone functions (for which see Serra, 1988). It is assumed that the sets are compact and closed in the topological sense. In the following such sets are called figures.

Size and shape

There are close connections between the size and shape of geometrical objects. Statements about size depend heavily on ideas about the shape of the objects considered, since they determine what is measured. Conversely, quantitative shape parameters are based on measurement. Frequently, size and shape are closely connected — in the way that, e.g., small objects are smooth and round but big ones are rough and angular. Therefore it is often necessary to investigate size and shape simultaneously, i.e. to investigate form. The choice of form parameters depends essentially on the aim of the investigation. It makes a big difference whether a parameter is needed that characterizes packing properties or one that is connected with local roughness. This is important, for example, in chemical processes. It is frequently desirable to have parameters that can be interpreted physically or biologically.

In most cases one cannot expect that form parameters uniquely determine figures in the sense that a reconstruction is possible.

Shape parameters

The shape of geometrical objects is often a difficult concept. An essential property is the independence of the shape of a figure X of its position or orientation in the plane. Likewise, scale changes do not change shape. These properties can be formulated mathematically as follows. A shape parameter y is a set function

$$y = f(X).$$

For any X and all translations T and rotations R we have

$$f(X) = f(TX) \tag{6.1}$$

and

$$f(X) = f(RX). \tag{6.2}$$

Furthermore,

$$f(X) = f(\lambda X) \quad \text{for all} \quad \lambda > 0. \tag{6.3}$$

Example. Let $f_1(X)$ be the area of X. This function $f_1(X)$ satisfies (6.1) and (6.2), but not (6.3). Thus, area is not a shape parameter. In contrast, the following quantity

$$f_2(X) = \frac{\text{area of } X}{(\text{perimeter of } X)^2}$$

is a shape parameter.

Tools of form statistics

Taking form statistical measurements is frequently expensive and, when involving manual work, lengthy. Sometimes it is sufficient to measure some distances by ruler or a measuring tape, as in the case of larger biological objects. However, manual work cannot usually be recommended for large samples or for routine work. Particularly difficult are cases of very small objects (investigated by microscope) and very irregular shapes, when even the manual determination of area and perimeter is very difficult. All these observations suggest the use of image analysers and computers. The first step is very important — saving images in computer memory. Once this has been done, simple image analysis is straightforward; in principle, any personal computer with graphics suffices.

In form analysis it is usual to analyse binary images, i.e. images consisting only of black and white points (pixels). Such analysis is done by means of specialized hardware or by suitable software; in the latter case good graphics is required.

Mathematical approaches

This book presents three approaches to form statistics: the objects investigated are treated as subsets of R^2, or sequences of discrete points are assigned to them, or they are described using functions.

Set description. The geometrical object to be analysed is considered as a compact subset of R^2, that is as a figure. Ideas of set geometry and mathematical morphology in the sense of Matheron and Serra (Serra, 1992) are used in the analysis. The corresponding elementary characteristics are area and perimeter. Many form parameters are based on these and similar characteristics. A typical method of this approach is 'deviation from convexity'. This approach is widely used; certainly it is well suited to description of particles but not so well suited to biological objects.

Point description. A series of typical points are assigned to the geometrical objects. Usually, these 'landmarks' (cf. Bookstein, 1978) have a particular meaning; the tips of the nose and chin are natural landmarks of skull profiles. Sometimes, landmarks are defined geometrically as points of extreme curvature or of particular tangent directions. These points or secondary characteristics (e.g. distances between certain landmarks or angles between connecting lines) are then used for statistical analysis. This method is very appropriate for biological objects.

Function description. The contour of the geometrical object is described by a function. There are two main variants.

(a) The contour function may be periodic. An example is the radius-vector function; here the distances of the contour points from a suitably chosen central reference point are considered as a function of the angle that this line makes.

(b) If the object is symmetric with respect to a line then one may take the orthogonal distance of the contour point from the line as a function of position on the line (Fig. 24).

In particular, variant (a) often leads to the use of Fourier series. Then Fourier coefficients serve as form parameters.

This method, which could be called contour parametrization, is used both for biological and technological problems. It is rather expensive, particularly in measuring the contour; thus image analysers should be used for this.

There are close connections between these three descriptive approaches. In different ways they make it possible to use a method which can be formulated abstractly as follows. The figures analysed are considered as samples of a random compact set X. The distribution of X has to be determined. For simplicity, X is transformed into another random variable, whose values lie in a space mathematically simpler than the space $\mathcal{K}$ of all compact subsets of R^2. Such spaces of particular interest are R^1, R^2 and certain function spaces.

Very simple random variables assigned to X are its area $A(X)$ and its perimeter $U(X)$. More complicated random variables are obtained by, for example, describing the contour by a function. For instance, pairs of random variables are obtained when the contours are considered as ellipses and are described by the lengths of their semi-axes. Finally, contour functions can be interpreted as elements of function spaces.

Another important approach in form statistics is the theory of Grenander (Grenander 1976, 1978, 1981, 1989; Grenander and Keenan 1989; Grenander *et al.*, 1991). This considers configurations obtained by forming random geometrical objects from combinations of random generators. Markov models are of particular interest.

Biological objects and particles

In this book two types of geometrical objects are considered. In the first case particular points on the contour or in the interior of the object are selected with a certain meaning ('landmarks'). Clearly these play an important role in quantitative analyses — such a point can be taken as a starting point for a parametric description of the contour. Since this case is typical for objects resulting from growth processes, the term *biological object* will be used in the sequel. This is a slight abuse of language, and it should not be taken to mean that the corresponding methods are of no interest to non-biologists.

Objects are frequently of a quite different nature. They may have convexities and concavities, but neither plays a role comparable to the landmarks of biological objects. (Of course, particular points could be selected here as well, e.g. points of large curvature or very large deviation from the centre of gravity. However, such points are clearly different from, say, nose tips.) As an example consider the sand grains in Fig. 65. When describing their contours by functions, the starting points are chosen arbitrarily (or one is not quite certain of the right choice, if, say, a particular extremum has been chosen as the starting point). In such cases the term *particles* will be used. Again, this name should not lead non-engineers to conclude that the corresponding methods are of no interest.

Form problems in pattern recognition

Further problems appear in connection with problems of pattern recognition and artificial intelligence, when automata have to recognize and classify figures. Patterns of points or dots are a special case. Here the first problem is to find, extract or filter out structures and relations; 'landmarks' are not given a priori and have to be defined. For typical methods for the analysis of point patterns by graphs (like the stick-man representation) and drawings see Radke (1988) and Toussaint (1988b). One may also determine 'contours' or 'hulls' (shape hulls) for point patterns, perhaps by taking the union of discs of chosen radii centred at appropriate points (Edelsbrunner *et al.*, 1983).

Stochastic models
Stochastic models may help is gaining better understanding of shape fluctuations. They are mathematical models of random sets, functions or point systems. Practicable stochastic models are based on an intuitive principle of construction containing easily interpreted components depending on only a few parameters. These parameters are statistically estimated, and the goodness-of-fit of the models tested.

Form and growth processes
Form changes during both biological growth processes, and technological processes, (e.g. by baking or erosion). The methods presented in this book may help to demonstrate these differences statistically. The mathematical descriptions of the change processes are considered only marginally — in the literature there are few approaches in this direction, see §8.5, Firey (1974), Grenander (1976) and Bookstein (1978).

Normalization
Many methods of shape statistics require normalization of the figures in order to eliminate size differences. All figures must be of 'equal size' in some sense; so, for example, the scale may be changed, making length, area or perhaps the distance from the nose-tip to the chin equal for all objects.

Homologization
Besides normalization, so-called *homologization* is often useful, particularly for biological objects. The aim is to inter-relate equal landmarks. For example, a coordinate system may be chosen in such a way that a particular landmark h as the same (x, y)-coordinates (e.g. nose tip at the origin) for all figures. By suitable normalization it is possible to have a further landmark that has the same coordinates (e.g. $x = 1$, $y = 0$) for all figures.

   Homologization may also consist in orientating the figures so that the line connecting two particular landmarks is the same for all figures (e.g. the x-axis).

   Consequently the scaling of the variate t may vary between the different landmarks.

Image analysers for form statistics
These are, in effect, the main ways of acquiring an image:

- 'direct' imagery, e.g. by camera, video or microscopy,
- 'reconstructed' imagery, e.g. by using radar, ultrasound or X-rays.

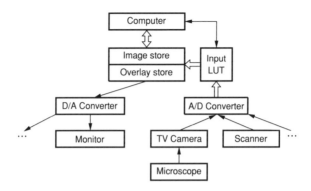

Figure 22   Schematic description of an image analysis system.

Reconstructed imagery is distinguished by the fact that the incoming data are 'signals' which are not directly visualisable; reconstruction involves solving an inverse problem.

Figure 22 shows schematically a possible configuration of devices for the acquisition phase of image analysis. In commercial systems, often some of the elements (given in Fig. 22) are united in one device or are omitted. The input of images is made by a scanner, TV camera, or a digitizing board. If it is not already in a digital form it is necessary to transform it into the language of the computer. This is done by analogue–digital converter. The digitized images are shown on the monitor, where they can be observed and perhaps corrected by means of a light pen or other instrument. The grey-scale image can be reduced to a binary image by operations such as thresholding, in which grey values lighter than a chosen threshold are set to white, and others to black. Operations of mathematical morphology, such as opening or closing (Appendix D) may be used to computer-enhance the images yet further. The computer then analyses the images geometrically, statistically, etc. An efficient image memory is useful.

This book does not consider problems of computational geometry, although they play an important role in form analysis on computers. The reader is referred to Toussaint (1988a) and references therein.

Shape, roundness and boundary structure

In the analysis of particles, form is sometimes described by three (somewhat vague) aspects (Barrett, 1980). As sketched in Fig. 23, one speaks of

- 'shape' for large-scale variations;
- 'roundness' for the smoothness at vertices, and more generally for larger variations from a given shape;
- surface texture for short-range fluctuations superposed on the shape and roundness fluctuations.

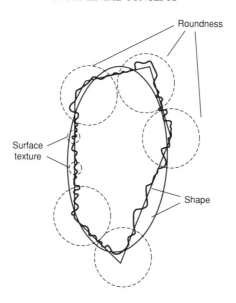

Figure 23 A typical particle and several aspects of its description. The general form can be given approximately by an ellipse or a polygon. The situation on the contour at greater scales may be characterized by roundness, while the notation of boundary texture is connected with fine small-scale deviations.

In the case of elliptical particles variations in the ratio of the semi-axis lengths are considered as roundness variations. If the boundaries of the ellipses are roughened then the corresponding local variations in shape are considered as texture variations.

Non-geometrical methods of form analysis for three-dimensional particles

In engineering several methods of form analysis are used that are based on physical principles (see e.g. Willetts and Rice, 1983; Huller, 1985). The form of particles may also be characterized by means of sieving and weighting (see e.g. Willetts *et al.*, 1982; Nielsen, 1985).

Methods of multivariate statistics in form statistics

After analysing the form, a series of parameters is given for each object. These data are analysed using multivariate statistics. The aim is to find relations between each groups of objects, to classify the objects by means of the parameters and to determine which parameters are closely related to the form of the object. The following are used as parameters:

- distances between landmarks and other lengths of characteristic parts of the objects;
- further geometrical characteristics (perimeter, area, appropriate angles);
- results of shape analysis (e.g. normalized Fourier coefficients).

Before statistical analysis, it may be useful to transform the data to obtain data of the same order and dimension; or it may be useful to take logarithms. It is also useful to try a pilot analysis with relatively many parameters, so that experience can be obtained and uninteresting parameters eliminated.

The following are particularly important methods for application in form statistics.

Discriminant analysis helps to separate samples of objects into one or more groups. The division is made by a so-called discriminant function which is based on the form parameters. An example is given on p. 172.

Principal component analysis helps to reduce a large number of parameters describing the analysed object to a smaller set of new parameters. The new parameters should reflect the variety of the object. It may be that every object is originally given by three parameters and thus can be considered as a point in R^3. If the object points form a cloud shaped like an ellipsoid then it is natural to use a new coordinate system corresponding to the axes of the ellipsoid. The variable corresponding to the longest axis may be a useful new parameter.

Experience in statistics shows that the variate representing maximum variability is often closely related to the size of the object (§10.1).

Cluster analysis helps to classify sets of objects, that is, to divide them into groups or clusters. Objects come into the same cluster that are in a certain sense 'close' together. An application is given in §10.2.

Reyment *et al.*, (1984) have described in detail the application of multivariate statistics in morphometry.

CHAPTER 7

Representation of Contours

7.1 INTRODUCTION

An important method of characterizing a figure is the representation of its contour by a function, which might be

- the cross-section function (for symmetric figures);
- the radius-vector function (for star-shaped figures);
- the support function (mainly for convex figures); or
- the tangent-angle function.

Large data sets should be processed automatically using image analysers. Early applications are described by Flook (1982b), Lohmann (1983) and Huller (1985). The first author used the QTM 720 and the second a Digital Graphics Systems CAT-100/C real-time Video Digitizer. Many other image analysers can be used in a similar way. Frequently, a smoothing of the contours before measurement is useful (§7.3).

The automatic processing of contours is usually done in the following steps:

(1) loading and saving the figure;

(2) smoothing;

(3) normalization and homologization;

(4) choosing the interpolation nodes of the contour function;

(5) measurement of the function values;

(6) transformation of the measurement results into a series of equidistant interpolation nodes.

The approximating functions are then chosen as in §7.5 and the corresponding coefficients are determined numerically. The latter are frequently used as form parameters. Finally, one may look for suitable stochastic models. To orientate the reader, the following are some advantages and disadvantages of the function description method. Advantages are

- effective data reduction: frequently only a few coefficients of approximating functions are needed for rather precise form description;

- a convenient description of complicated forms understandable by anybody accustomed to working with functions;

- an intuitive characterization of many form properties, such as small form fluctuations around a 'mean' figure.

  Disadvantages of the function description method are

- the necessary choice of a reference point taken to be the origin (e.g. the centre of gravity or a certain contour point, which often appears to be arbitrary);

- the complicated formulae for functions and characteristics/even for simple and common figures such as ellipses.

7.2 DEFINITION AND MEASUREMENT OF CONTOUR FUNCTIONS

7.2.1 Cross-section functions for symmetric figures

Symmetric figures as in Fig. 24 are particularly considered in biological investigations. It is natural to describe the form by the *cross-section function* $q(x)$, the half-breadth at x.

Problems occur if there is more than one contour point on the line through x orthogonal to the symmetry axis (x-axis), as is the case for x' in Fig. 24. This ambiguity can be removed by choosing the mean distance as the function value. Thus for the value x' in Fig. 24 one sets

$$y' = q(x') = \tfrac{1}{3}(y'_1 + y'_2 + y'_3).$$

Another possibility is choosing the outer contour point, namely $y' = y'_3$.

Such difficulties can be avoided by smoothing the contours before the statistical analysis (§7.3). Of course, these simplifications may destroy essential features, and so the cross-section function may not be a good basis for form analysis. In this case the tangent-angle function should be used.

If a figure is given for which the half-breadth can be uniquely determined for every value of x then measurement yields a sequence of function values

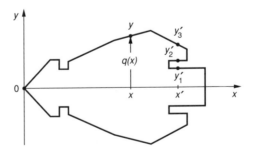

Figure 24 Definition of the cross-section function for a symmetric figure. At x' the function value may be $q(x') = \tfrac{1}{3}(y'_1 + y'_2 + y'_3).$

$$x_1, q(x_1), \ldots, x_n, q(x_n).$$

The x_i are usually chosen equidistantly. The number n should be in a reasonable proportion to the number p of parameters of the approximating function; something like $n > 2p$ is desirable.

In the case of small samples and no microscopic objects the values $q(x_i)$ can be determined manually. Otherwise image analysers are necessary, with their possibilities for measurement of chord lengths.

7.2.2 Radius-vector functions

Frequently, the contour of a figure X is described by the radius-vector function. For this, it is necessary to choose a reference point in the interior of X. This may be, for example, the centre of gravity, the centre of the smallest disc that completely contains X or a biologically important point. Then the figure is translated such that this point lies at the origin o. For simplicity of notation, the translated figure X will in the following also be denoted X.

It is necessary that X be star-shaped with respect to o. That is, for any contour point x of X the whole line segment from o to x lies within X. Figure 25 shows a star-shaped and a non-star-shaped set.

If the star-shapedness is violated only by small irregularities in the contour, one may try to recover it by pre-smoothing.

The *radius-vector function* $r_X(\varphi)$ depends on the angle φ made by a ray emanating from o with the x-axis (Fig. 25a). The quantity $r_X(\varphi)$ is equal to the length of the line segment from o to the contour point x in which the φ-ray intersects the boundary. The radius vector function precisely characterizes X: if $r_X(\varphi)$ is given, then X can be uniquely reconstructed. Clearly,

$$r_{\lambda X}(\varphi) = \lambda r_X(\varphi), \quad X \subset Y \Rightarrow r_X(\varphi) \leq r_Y(\varphi).$$

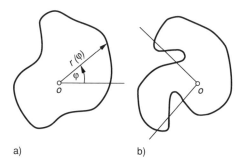

a)   b)

Figure 25 (a) A star-shaped figure. Each ray starting at o forms only one line segment in it. The length of the line segment for a ray of direction φ ($0 \leq \varphi \leq 2\pi$) is denoted by $r_X(\varphi)$. This is the radius-vector function. (b) A non-star-shaped figure. There are rays starting at o that cross the boundary more than once.

The map $X \to r_X$ transforms figures into elements of a function space. If the figures have the property that the radius-vector function is continuous then the Banach space $C[0, 2\pi]$ is suitable, subject to the acceptability of the supremum norm.

Formulae for $r_X(\varphi)$ can be given for some geometrical figures.

Example. Let X be an ellipse with principal semi-axis lengths a and b, $(a \geq b)$, and $\varphi = 0$ in direction of the longer semi-axis; then

$$r_X(\varphi) = \left(\frac{\cos^2 \varphi}{a^2} + \frac{\sin^2 \varphi}{b^2} \right)^{-1/2}.$$

Integrating $r_X(\varphi)$ yields the perimeter and area of X:

$$U(X) = \int_0^{2\pi} \left[r^2(\varphi) + r'^2(\varphi) \right]^{1/2} d\varphi, \tag{7.1}$$

$$A(X) = \frac{1}{2} \int_0^{2\pi} r^2(\varphi) \, d\varphi. \tag{7.2}$$

A further geometrical parameter (depending on the choice of the reference point) is the *mean radius-vector length*

$$\bar{r} = \frac{1}{2\pi} \int_0^{2\pi} r(\varphi) \, d\varphi.$$

The formulae for $r_X(\varphi)$ are often rather complicated, and it seems natural to consider figures X for which $r_X(\varphi)$ has a simple analytical form. An example is the piecewise-linear continuous radius-vector function

$$r_X(\varphi) = \frac{\varphi - \phi_i}{\Delta_i} R_{i+1} + \left(1 - \frac{\varphi - \phi_i}{\Delta_i} \right) R_i$$
$$(\phi_i \leq \varphi \leq \phi_{i+1}; i = 0, 1, \ldots, n - 1),$$
$$\Delta_i = \phi_{i+1} - \phi_i, \quad R_0 = R_n, \quad \phi_0 = 0, \quad \phi_n = 2\pi$$

Figure 26 shows the case $n = 4$ with $\phi_i = \frac{1}{2}i\pi$. In particular, there are 'radial rhombi' as in §7.4.2. The contours of figures with piecewise-linear radius-vector functions are curved outwards. The points corresponding to $\varphi = \phi_i$ are cusp points, which may also be pulled into the interior; thus these figures are not necessarily convex. It is easy to give a necessary and sufficient condition on $r_X(\varphi)$ for the convexity of X. Given any φ_1 and φ_2 with $0 < \varphi_2 - \varphi_1 < \pi$, for any φ with $\varphi_1 < \varphi < \varphi_2$ the value $r_X(\varphi)$ has to satisfy

$$r_X(\varphi) \geq \frac{r_X(\varphi_1) r_X(\varphi_2) \sin(\varphi_2 - \varphi_1)}{r_X(\varphi_1) \sin(\varphi - \varphi_1) + r_X(\varphi_2) \sin(\varphi_2 - \varphi)}.$$

If equality holds then the corresponding contour point lies on the line connecting $(\varphi_1, r_X(\varphi_1))$ and $(\varphi_2, r_X(\varphi_2))$.

Figure 26   Figure with piecewise-linear radius-vector function $r_X(\varphi)$ with four interpolation nodes at $\varphi = \frac{1}{2}i\pi$; $(i = 1, \ldots, 4)$.

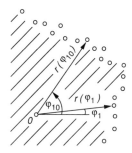

Figure 27   Determination of the radius-vector function by an image analyser. Referring to the estimated centre of gravity ○ (which need not be a pixel), the distances to all boundary pixels of the figure are measured.

The measurement of radius-vector functions can be carried out manually for small samples and non-microscopic objects. The radius-vector lengths are measured for an equidistant series of angles of φ_i. However, in the case of large samples or smallish objects image analysers are again necessary.

The centre of gravity can be determined from the pixels corresponding to the object, and is generally not a pixel. Then the image analyser determines all contour points $p_1, \ldots, p_m$ of X. (In solving this problem for regular objects the form of the definition of the boundary does not have such heavy consequences as when determining fractal dimensions as in Part I.) It is easy to calculate for every point p_i the angle φ and the radius-vector length $r_X(\varphi_i)$ (Fig. 27). Thus the φ_i are in general not equidistant. Frequently, these data are modified by interpolation to obtain equidistant interpolation nodes. If the raster is fine enough, the errors connected with this manipulation may be neglected.

7.2.3   Support functions

Support functions are frequently used for the description of convex sets. For a compact subset X of R^2 the *support function* is defined as follows. Let g_φ be the orientated line through the origin with direction φ $(0 < \varphi < 2\pi)$. Furthermore, let $g_\varphi^\perp$ be the line orthogonal to g_φ with the property that the set X lies completely in that half-plane determined by $g_\varphi^\perp$ with $g_\varphi^\perp \cap X \neq \emptyset$, which is opposite to the direction of g_φ, (Fig. 28).

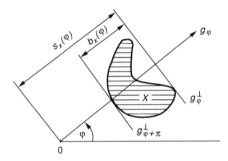

Figure 28 Definition of the support function of a figure.

The absolute value of the support function is equal to the distance from o to $g_\varphi^\perp$, and $s_X(\varphi)$ is negative if X lies behind $g_\varphi^\perp$ as seen from the origin; that is,

$$s_X(\varphi) = \max_{x \in X}\{e_\varphi \cdot x\} = \max_{x \in X}\{x_1 \cos \varphi + x_2 \sin \varphi\},$$

where $e_\varphi = (\cos \varphi, \sin \varphi)$ is the unit vector along g_φ and $x = (x_1, x_2)$.

If o is an element of X then $s_X(\varphi) \geq 0$ for all φ. It is obvious that the support function of X coincides with that of the corresponding convex hull conv X. Thus there exist different non-convex figures with the same support function. In contrast, convex figures are uniquely determined by their support functions. (For non-convex X another definition of the support function is also used, an additive extension of the convex form to the convex ring, the class of finite unions of convex compact sets.)

The support functions of convex compact sets are dealt with in detail in Matheron (1975) and Gruber and Wills (1993). Some fundamental properties are

for $Y = \lambda X$	$s_Y(\varphi) = \lambda s_X(\varphi)$,
for $X \subset Y$	$s_X(\varphi) \leq s_Y(\varphi)$,
for $Z = X \oplus Y$	$s_Z(\varphi) = s_X(\varphi) + s_Y(\varphi)$.

For any X the support function is continuous in φ. The distance between two convex compact sets X and Y with respect to the Hausdorff metric satisfies

$$h(X, Y) = \sup_{\varphi \in [0, 2\pi]} |s_X(\varphi) - s_Y(\varphi)|.$$

Consequently, the convex compact sets can be embedded isometrically in the Banach space $C[0, 2\pi]$ of all continuous functions on $[0, 2\pi]$. For simple geometrical figures the support function can be easily given.

Example 1: For a disc $b(x, r)$ with centre $x = (x_1, x_2)$ and radius r,

$$s_X(\varphi) = x_1 \cos \varphi + x_2 \sin \varphi + r \quad (0 \leq \varphi \leq 2\pi).$$

Example 2: For square of side length a centred at o with one edge parallel to the direction given by $\varphi = 0$,

$$s_X(\varphi) = \begin{cases} \frac{1}{2}\sqrt{2}\,a\cos\left(\frac{1}{4}\pi - \varphi\right) & \left(0 \le \varphi \le \frac{1}{2}\pi\right), \\ s_X\left(\varphi - \frac{1}{2}\pi\right) & \left(\varphi \ge \frac{1}{2}\pi\right). \end{cases}$$

If the convex figure X has a smooth boundary, then the support function $s_X(\varphi)$ determines the curvature. If $\rho_X(\varphi)$ is curvature radius corresponding to φ, then

$$\rho_X(\varphi) = s_X(\varphi) + s_X''(\varphi) \qquad (0 \le \varphi < 2\pi),$$

see Kallay (1974, 1975).

The perimeter $U(X)$ and the area $A(X)$ satisfy

$$U(X) = \int_0^{2\pi} s_X(\varphi)\,\mathrm{d}\varphi$$

and

$$A(X) = \frac{1}{2}\int_0^{2\pi} s_X(\varphi)[s_X(\varphi) + s_X''(\varphi)]\mathrm{d}\varphi.$$

Closely connected with the support function is the *width function* $w_X(\varphi)$:

$$w_X(\varphi) = s_X(\varphi) + s_X(\varphi + \pi) \quad (0 \le \varphi \le \pi).$$

Obviously, $w_X(\varphi)$ is the breadth of X in the direction φ (Fig. 28): it is the support function of $X \oplus \check{X}$, $\check{X} = -X$. For form analysis the width function has the advantage that for all x

$$w_{X+x}(\varphi) = w_X(\varphi),$$

i.e. it is invariant with respect to translation. In contrast, rotations do change the width function (with the exception of rotations through an angle π). If X is symmetric then

$$s_X(\varphi) = s_X(\varphi + \pi) \quad (0 \le \varphi \le \pi).$$

In this case the width function determines the figure X uniquely. In general, the form of X is not uniquely given by the function $w_X(\varphi)$. Figure 29 shows a set that has, as the disc, a constant width function (it is called the Reuleaux triangle; Santaló, 1976). Mathematicians study sets of constant width systematically (Chakerian and Groemer, 1983; Gruber and Wills, 1993).

The width function yields the perimeter $U(X)$ by

$$U(X) = \int_0^{2\pi} w_X(\varphi)\mathrm{d}\varphi.$$

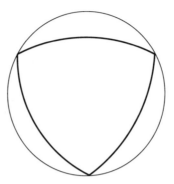

Figure 29   Reuleaux triangle. This figure has the same width in all directions.

However, the area $A(X)$ of X is not uniquely determined by $w_X(\varphi)$. It is

$$A_{\min} \leq A(X) \leq \tfrac{1}{8} \int_0^{2\pi} w_X(\varphi)[w_X(\varphi) + w_X''(\varphi)]\mathrm{d}\varphi = A_{\max},$$

with $A_{\min} \geq \tfrac{2}{3} A_{\max}$, see Amit and McCullagh (1993).

This paper also gives a formula for the boundary curve of the maximal set and a construction of the minimal set. The latter has, if X is smooth, three corners, similarly as the Reuleaux triangle. Kallay (1974, 1975) characterizes the set of all convex sets with the same width function.

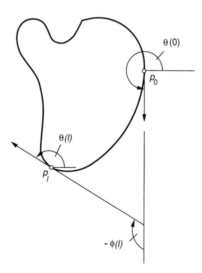

Figure 30   Explanation of the tangent angles $\theta(0)$ and $\theta(l)$, and the corresponding value $\phi(l)$ of the tangent-angle function. The latter is equal to the negative angle shown in the right lower corner of the figure. $\phi(0)$ is equal to zero.

7.2.4 Tangent-angle functions

Sometimes the tangent angle of the contour is used for the description of figures (Zahn and Roskies, 1972). Of course, it must be assumed that the contour of the figure X considered is piecewise-smooth so that a tangent is non-existent at only a finite number of points.

Let the perimeter of the figure X be $U(X) = L$. Every point p_l on the contour of X can thus be identified with a number l, with $0 \leq l \leq L$, run through clockwise. A pointer is placed at p_0 so that its zero position coincides with the tangent direction. If the position of the pointer moves on the contour then the pointer changes its direction in such a way that it is always in the direction of the tangent, where its orientation is given by the direction of movement.

The angle given by the pointer direction in p_l is denoted $\phi(l)$, where $-2\pi \leq \phi(l) \leq 2\pi$, $\phi(0) = 0$ and $\phi(l) = -2\pi$ (Fig. 30). The function $\phi(l)$ is called the *tangent-angle function* or angular-bend function. Figure 31 gives an example.

For shape analysis it is helpful to normalize the function $\phi(l)$. A useful variant has been suggested by Zahn and Roskies (1972):

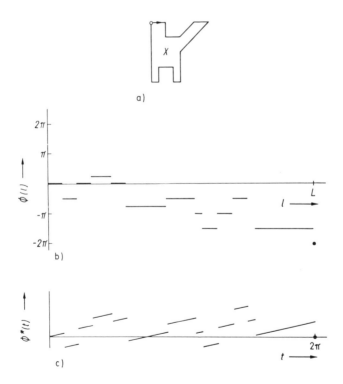

Figure 31 The tangent-angle function of a polygon X. (a) The polygon; the left upper vertex is the starting point $(l = 0)$. (b) The tangent-angle function $\phi_X(l)$ of X. (c) Normalized tangent-angle function $\phi_X^*(t)$, satisfying $\phi_X^*(t) = \phi_X(Lt/2\pi) + t$.

$$\phi_X^*(t) = \phi\left(\frac{Lt}{2\pi}\right) + t \quad (0 \le t \le 2\pi).$$
$$\phi_X^*(0) = \phi_X^*(2\pi) = 0.$$

The perimeter length is eliminated by the normalization, and the function is then defined on $[0, 2\pi]$, like the radius-vector and the support function. Of course, the variable t plays a role quite different to ϕ in the other two functions.

For a disc $\phi_X^*(t)$ vanishes identically. The figures X and λX have the same tangent-angle functions for all positive λ.

The manual determination of the tangent-angle function is quite difficult, and the use of an image analyser is suggested. Using it, the contour is approximated by a polygon. With the vertex coordinates of the polygon, it is easy to determine $\phi(l_i)$ for an arbitrary series of interpolation nodes l_i between 0 and L.

7.2.5 Comparison of the three variants of contour functions

The three functions described in §§7.2.2–7.2.4,

the radius-vector function	$r_X(\varphi)$,
the support function	$s_X(\varphi)$,
the tangent-angle function	$\phi_X^*(t)$,

have certain advantages and disadvantages. The definition of $r_X(\psi)$ is easier to understand than that of $s_X(\varphi)$ or $\phi_X^*(t)$. Similarly, $r_X(\varphi)$ is preferable with respect to robustness from digitization errors and to convenience of use in automatic image analysis.

The field of application for $\phi_X^*(t)$ is the largest; it is 'only' required that the contour of the figure X be smooth. In contrast, the radius-vector function requires the figure to be star-shaped, while for the support function convexity is desirable.

All three functions depend on the choice of a reference point. Its influence is strong in the case of the radius-vector function; the same figure may have quite different radius-vector functions for different reference points. In the case of the support function the translation by (a, b) of the reference point (origin of the coordinate system) generates an additional term $a\cos\varphi + b\sin\varphi$. Also the form of $\phi_X^*(t)$ depends on the choice of starting point on the contour; but it is quite easy to transform the function corresponding to a certain starting point into that starting at another point. Scale changes leave $\phi_X^*(t)$ invariant, while $r_X(\varphi)$ and $s_X(\varphi)$ are multiplied by the scaling factor.

7.2.6 Smoothing of contours

When describing contours as functions, rough boundaries lead to problems. These problems can be reduced if the contours are smoothed prior to the analysis. Two methods are suggested here. They should be tested before routine application to ensure that the smoothing process does not destroy essential information.

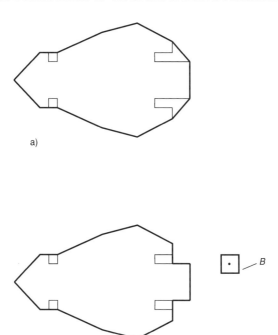

a)

b)

Figure 32 Two ways to smooth the contour of the figure in Fig. 24: (a) convex hull; (b) closing with the given square.

Forming the convex hull

Instead of the set X, its convex hull $\text{conv} X$ is analysed. Its boundary is often smoother than that of X. Figure 32(a) shows the convex hull for the figure in Fig. 24. The cross-section function for the convex hull can be determined without difficulty.

Closing

Instead of X, its closure X_B with a suitably chosen structuring element B is analysed (Appendix D). The operation of closing smooths small inlets and closes gaps. A good choice of the structuring element is a disc or a regular polygon (according to the type of image analyser). The size of B (the radius) should be chosen after some tests; the larger B is, the smoother X_B is. If B is too big then information can be lost. Figure 32(b) shows the figure of Fig. 24 after closing with a square.

7.3 INVARIANT PARAMETERS OF CONTOUR FUNCTIONS FOR PARTICLES

The contour functions $f(x)$ discussed in §7.2 all depend in some manner on the choice of a reference point or a coordinate system. In the case of particles without

natural landmarks this choice is arbitrary. Therefore quantities that are connected
with the contour function used but are independent of the choice of the reference
point are of particular interest. In the following some simple quantities of this
kind are discussed. They are of use if the radius-vector function, the breadth
function or the tangent-angle function are used. They may be used as starting
points for forming form parameters. It is always assumed that $f(x)$ is continuous
and continuously differentiable, with the expectation of a finite number of cusp
points.

The *mean value of the function* is given by

$$\bar{f} = \frac{1}{2\pi} \int_0^{2\pi} f(x)\,dx. \tag{7.3}$$

The *variance of the function* is given by

$$\sigma^2(f) = \frac{1}{2\pi} \int_0^{2\pi} [f(x) - \bar{f}]^2\,dx. \tag{7.4}$$

Clearly,

$$\sigma^2(f) = \frac{1}{2\pi} \int_0^{2\pi} f^2(x)\,dx - \bar{f}^2.$$

If $f(x)$ is the radius-vector function then (7.2) yields

$$\sigma^2(r) = \frac{A}{\pi} - \bar{r}^2, \tag{7.5}$$

where A is the area of the figure.

Contour distribution function. The quantities $\bar{f}$ and $\sigma^2(f)$ can be interpreted
as mean and variance corresponding to a distribution function F_f, defined as
follows:

$$F_f = \frac{1}{2\pi} \mathcal{L}(\{\varphi : f(\varphi) \leq x; 0 \leq \varphi \leq 2\pi\}) \quad (x \geq 0).$$

Here $\mathcal{L}$ denotes the Lebesgue measure of R^1; For a regular set $\mathcal{A}$, $\mathcal{L}(\mathcal{A})$ is its
length. Of course, different contour functions may have the same distribution
function.

The *variance of the derivative* is given by

$$\sigma^2(f') = \frac{1}{2\pi} \int_0^{2\pi} f'(x)^2\,dx, \tag{7.6}$$

and

$$\bar{f}' = \int_0^{2\pi} f'(x)\,dx = f(2\pi) - f(0) = 0. \tag{7.7}$$

Thus the name 'variance' is justified.

The *contour covariance function* is given by

$$\chi_f(\varphi) = \frac{1}{2\pi} \int_0^{2\pi} [f(x) - \bar{f}][f(x + \varphi) - \bar{f}] \, dx \quad (0 \le \varphi \le 2\pi). \tag{7.8}$$

Here it is $f(2\pi + t) = f(t)$ for all t.

The contour covariance function $\chi_f(\varphi)$ does not determine $f(\chi)$ uniquely; there are different f with the same $\chi_f(\varphi)$ (§7.5.4). We have

$$\chi_f(0) = \sigma^2(f). \tag{7.9}$$

Obviously

$$\chi_f(\varphi) = \chi_f(2\pi - \varphi) \tag{7.10}$$

for $\varphi > \pi$; thus $\chi_f(\varphi)$ is given only for $0 \le \varphi \le \pi$. Also

$$\chi'_f(0) = 0. \tag{7.11}$$

Finally, $\sigma^2(f')$ and $\chi_f(\varphi)$ are related by

$$\sigma^2(f') = -\chi''_f(0), \tag{7.12}$$

which can be shown by integration by parts.

Further invariant form parameters are the Fourier coefficients A_n (§7.5.4).

Which form properties are characterized by the quantities introduced above?

- $\bar{f}$ is a size parameter.

- $\sigma^2(f)$ and $F_f(x)$ are size, shape and roundness parameters.

- $\sigma^2(f')$ describes local properties, i.e. roundness and texture variations of the contours.

- $\chi_f(\varphi)$ describes many aspects of form fluctuations, both local and global ones. For example, a shape property such as n-poleness can be very well diagnosed. The normalized function $\chi_f(\varphi)/\chi_f(0)$ characterizes the shape of the corresponding figure.

These parameters are used in §10.1 for the statistical analysis of three samples of sand grains.

7.4 TWO CLASSES OF FIGURES

7.4.1 Superellipses

A large class of figures is formed by the so-called superellipses[†] (Gardner 1965). These are figures with two symmetry axes as in Fig. 33. The contour in the first

---

[†]This word was coined by the Danish poet Piet Hein, who in his youth was interested in mathematics and physics.

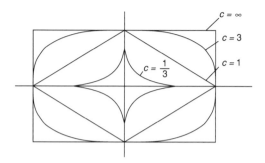

Figure 33   Four superellipses. For $c = 2$ an ellipse is obtained (not shown here), and for $c = 3$ a figure with a shape between an ellipse and a rectangle. In the extreme cases $c = 1$ and $c = \infty$ a rhombus and a rectangle are obtained respectively, while for $c < 1$ non-convex figures are obtained.

quadrant of the (x, y)-coordinate system is given by

$$\left(\frac{x}{a}\right)^c + \left(\frac{y}{b}\right)^c = 1.$$

Here a and b are positive parameters with $a \geq b$, the so-called semi-axis lengths. The positive parameter c determines the shape and roundness of the figure. That is, for $c \geq 1$ the figure is convex, for $c = 1$ a rhombus is obtained and for $c = 2$ an ellipse, while as c tends to infinity the figure approaches a rectangle with the side lengths $2a$ and $2b$.

The calculation of geometrical parameters for superellipses is generally not easy. Its radius-vector function is

$$r(\varphi) = \left(\frac{\cos^c \varphi}{a^c} + \frac{\sin^c \varphi}{b^c}\right)^{-1/c} \quad \left(0 \leq \varphi \leq \tfrac{1}{2}\pi\right). \tag{7.13}$$

Of course,

$$r(\varphi) = \begin{cases} r(\pi - \varphi) & \left(\tfrac{1}{2}\pi \leq \varphi \leq \pi\right), \\ r(\varphi - \pi) & (\varphi \geq \pi). \end{cases}$$

The area is

$$A = \frac{2}{c}\frac{[\Gamma(1/c)]^2}{\Gamma(2/c)} ab, \tag{7.14}$$

where Γ denotes the gamma function. The factor before ab is the area E of the super-unit-disc ($a = b = 1$), obtained by integration as follows:

$$E = 4 \int_0^1 (1 - x^c)^{1/c}\, dx = \frac{4}{c} \int_0^1 (1 - y)^{1/c} y^{(c-1)/c}\, dy = \frac{2}{c}\frac{[\Gamma(1/c)]^2}{\Gamma(2/c)}. \tag{7.15}$$

In the case of an ellipse ($c = 2$)

$$A = \pi ab, \tag{7.16}$$

and for the perimeter U the approximation formula

$$U \approx \pi[1.5(a+b) - \sqrt{ab}] \tag{7.17}$$

is well known. The mean radius-vector length $\bar{r}$ satisfies

$$\bar{r} = \frac{2b}{\pi} N \left(\sqrt{1 - \frac{b^2}{a^2}} \right) \quad (a \geq b), \tag{7.18}$$

where

$$N(k) = \int_0^{\pi/2} \frac{dx}{\sqrt{1 - k^2 \sin^2 x}}.$$

For the calculation of the elliptic integral the series expansion

$$N(k) = \frac{\pi}{2} \left\{ 1 + \left(\frac{1}{2} \right)^2 k^2 + \left(\frac{1 \cdot 3}{4 \cdot 2} \right)^2 k^4 + \cdots + \left[\frac{(2n-1)!!}{2^n \cdot n!} \right]^2 k^{2n} + \cdots \right\}$$

can be used, where $i!! = 1 \cdot 3 \cdot 5 \cdots i$ for odd i. However, its convergence is not very fast.

The contour covariance function $\chi(\varphi)$ can be determined by numerical integration. Since

$$r(\varphi) = r(\varphi + \pi) \text{ and } r(\varphi) = r(\pi - \varphi),$$

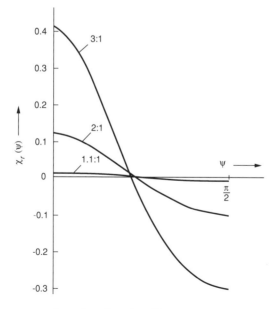

Figure 34   Contour covariance functions for ellipses of different semi-axis length ratios.

the relations

$$\chi_r(\varphi + \pi) = \chi_r(\varphi) \tag{7.19}$$

and

$$\chi_r(\varphi) = \chi_r(\varphi - \pi) \tag{7.20}$$

hold for all φ. Thus it is sufficient to calculate $\chi_r(\varphi)$ for $0 \le \varphi \le \frac{1}{2}\pi$.

Figure 34 shows contour covariance functions for ellipses. There are large differences for different ratios of axis lengths. In contrast, the normalized functions $\chi_r(\varphi)/\chi_r(0)$ are nearly independent of this ratio.

7.4.2 Radial-rhombi

Radial-rhombi are figures with a particularly simple radius-vector function. They form a class of figures that at first glance look somewhat strange (Fig. 35). They have two axes of symmetry, and their radius-vector function in the first quadrant of the (x, y)-coordinate system is

$$r(\varphi) = a - 2\frac{a - b}{\pi}\varphi \quad \left(0 \le \varphi \le \frac{1}{2}\pi\right). \tag{7.21}$$

Here a and b are positive parameters with $a \ge b$, which could be called 'semi-axis lengths'. A radial-rhombus is convex only if $a = b$; then it is a disc of radius a. Otherwise, it has a cusp at $\varphi = 0$ (on the x-axis) and a slot at $\varphi = \frac{1}{2}\pi$ (on the y-axis). The area is

$$A = \frac{1}{3}\pi(a^2 + ab + b^2) \tag{7.22}$$

and the mean radius-vector length

$$\bar{r} = \frac{1}{2}(a + b) \tag{7.23}$$

If $a \ne b$ then the formula for the perimeter is rather complicated:

$$U = 2\left(\frac{as - bt}{x} + x\log\frac{a + s}{b + t}\right) \tag{7.24}$$

with

$$x = \frac{2}{\pi}(a - b), \quad s = (a^2 + x^2)^{1/2}, \quad t = (b^2 + x^2)^{1/2}.$$

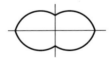

Figure 35 A radial-rhombus with a semi-axis ratio 2 : 1. The radius-vector function is piecewise-constant, but, in contrast to the figure in Fig. 26, this figure has two axes of symmetry.

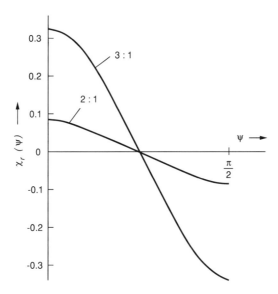

Figure 36 Contour covariance function for radial-rhombi of different axis-length ratios.

Since the radius-vector lengths are uniformly distributed on $[b, a]$, the variance of radius-vector length, $\sigma^2(r)$, is

$$\sigma^2(r) = \tfrac{1}{12}(a - b)^2, \tag{7.25}$$

and that of the derivative of the radius-vector function is

$$\sigma^2(r') = \frac{4(a - b)^2}{\pi^2}. \tag{7.26}$$

The contour distribution function $F_r(x)$ is the distribution function of a uniform random variable on $[b, a]$.

The contour covariance function $\chi_r(\varphi)$ satisfies, as in the case of ellipses, the relations (7.19) and (7.20), and for $0 \leq \varphi \leq \tfrac{1}{2}\pi$ is given by

$$\chi_r(\varphi) = \tfrac{1}{12}(a - b)^2[1 - 6x^2 + 4x^3], \quad x = 2\varphi/\pi \tag{7.27}$$

Figure 36 shows the contour covariance functions of radial-rhombi.

7.5 DETERMINATION OF APPROXIMATING CONTOUR FUNCTIONS

7.5.1 Introduction

Using the methods given in §7.2, contour functions can be determined for figures. These empirical functions (often given only in the form of lists) are mostly too complicated for statistical analysis. Therefore they are frequently approximated by simpler functions, and these are then analysed. In many cases only a few form parameters are needed. Then rough approximations, reflecting only the main aspects, are sufficient, even if local details are lost.

Approximations by Fourier series are particularly popular. (There are also in the literature approximations by polynomials, Walsh or Hadamard functions (Flook, 1982b, 1987; Meloy, 1977).) As form factors, the first coefficients of the function series are used. Their intuitive interpretation is often difficult, in particular for polynomial approximations or Fourier approximations if no periodicity is visible in the contours.

Let there be given a sequence of interpolation nodes x_i and function values y_i $(i = 1, \ldots, n)$:

$$x_1, y_1$$
$$\vdots$$
$$x_n, y_n$$

A suitable approximation by mathematically simple functions has to be determined. Thus a function $F(x; c_1, \ldots, c_p)$ is chosen with parameters $c_1, \ldots, c_p$, which are to be determined such that the sum of squares of deviations,

$$\sum_{i=1}^{n} [y_i - F(x_i; c_1, \ldots, c_p)]^2,$$

is minimized. The Marquardt procedure is an effective numerical method for determination of the c_i (Ortega and Rheinhold, 1970; Dennis and Schnabel, 1983). A popular form is

$$F(x; c_1, \ldots, c_p) = \sum_{j=1}^{p} c_j g_j(x),$$

where the $g_j(x)$ are certain functions, for example

$$g_j(x) = x^j \quad (\text{'polynomial approximation'})$$

or

$$g_j(x) = \sin k_j x \text{ and } \cos k_j x \quad (\text{'Fourier series approximation'})$$

In addition to these two variants, there is the possibility of choosing particular functions appropriate for a given problem. An important method of this kind is that of *eigenfunctions* (Lohmann, 1983; Full and Ehrlich, 1986; Rohlf, 1986; Rice and Silverman, 1991); cf. p. 94.

7.5.2 Approximation by ellipses and radial-rhombi

There are various methods for estimating the principal axis lengths of approximating ellipses or radial-rhombi. The choice is determined by the possibilities for measurement and the deviations of the figures from the ideal. Five possibilities are as follows.

(1) The D_{max} method. This involves measurement of area A and determination of the line g that has the maximum intersection with X. If d_{max} is the length of $g \cap X$,

$$2a = d_{max}, \tag{7.28}$$

$$b = \frac{2A}{\pi d_{max}}. \tag{7.29}$$

(2) The $A-U$ method. Measurement of U and A leads to a and b using the formulae for U and A in §7.4. In particular, for ellipses

$$a = \alpha + \left(\alpha^2 - \frac{A}{\pi} \right)^{1/2}, \tag{7.30}$$

where

$$\alpha = \frac{1}{3} \left[\left(\frac{A}{\pi} \right)^{1/2} + \frac{U}{\pi} \right], \quad b = \frac{A}{\pi a}. \tag{7.31}$$

Similar methods for three-dimensional figures have been suggested by Riss (1988) and Riss and Grohier (1986).

(3) The R method. This involves the measurement of $\bar{r}$ and A and determination of a and b by means of the formulae in §7.4. In particular, for radial-rhombi

$$a = \bar{r} + \sqrt{3}\sigma(r), \tag{7.32}$$

$$b = \bar{r} - \sqrt{3}\sigma(r), \tag{7.33}$$

with

$$\sigma^2(r) = \frac{A}{\pi} - \bar{r}^2. \tag{7.34}$$

(4) The covariance method. This involves the determination of $\chi_r(\varphi)$ and the choice of a and b such that $\chi_r(\varphi)$ and the theoretical counterpart (dependent on a and b) are close in the sense of the least-squares method.

(5) The radius-vector function approximation. Approximating ellipses or radial-rhombi are determined by the least-squares method, using $r_X(\varphi)$. The centre of gravity of the figure is chosen as the origin of the coordinate system; by the optimization procedure, not only a and b but also the directions of the axes are determined.

The $A-U$ method is somewhat unstable in the case of a rough contour or for large deviations of the figure from an ellipse or a radial-rhombus. The fourth and fifth methods are rather complicated. All methods can be stabilized by replacing X by conv cX, where c is chosen such that $A(\text{conv } cX) = a(X)$.

In §7.7.3 the first three methods are compared for a stochastic model of fluctuations of X around an ellipse.

7.5.3 Fourier analysis

Introduction

Contour functions are frequently approximated by Fourier series:

$$f(x) \approx a_0 + \sum_{k=1}^{p} (a_k \cos kx + b_k \sin kx) \tag{7.35}$$

or

$$f(x) \approx A_0 + \sum_{k=1}^{p} A_k \cos(kx + \alpha_k), \tag{7.36}$$

with

$$A_0 = a_0, \tag{7.37}$$

$$A_k^2 = a_k^2 + b_k^2, \tag{7.38}$$

$$\alpha_k = \begin{cases} \arctan(-b_k/a_k) & (a_k > 0), \\ \arctan(-b_k/a_k) + \pi & (a_k < 0), \end{cases} \tag{7.39}$$

with $k = 1, 2, \ldots, p$. The coefficient A_k is called the *kth* harmonic amplitude, and α_k the *kth* phase angle. The numbers a_k/A_0, and b_k/A_0, or A_k/A_0 and α_k, serve as shape factors.

The approximation of a function $f(x)$ by a Fourier series may be recommended for the following reasons:

- there exist good numerical methods and programs for calculation of the Fourier coefficients a_k and b_k;
- frequently a non-mathematical interpretation of the a_k and b_k, or the A_k and α_k, is possible; this is the case if shape properties are discussed that are connected with periodicity (e.g. *n*-poleness).

Fourier approximation is discussed in detail in Schwarcz and Shane (1969), Beddow and Philip (1975), Beddow (1980, 1984), Beddow and Meloy (1980), and Huller (1985). Frequently, the approximation by Fourier series has a formal character only. In particular, this is the case for biological objects, which do not show any periodicity, for example human skull profiles. The a_k, b_k, A_k and α_k are then formal form parameters only. The critical evaluation of the Fourier approximation by Bookstein *et al.* (1982) makes worthwhile reading.

Some facts on Fourier series

The function $f(x)$ defined for $0 \le x \le 2\pi$ has to be approximated by a function $g_p(x)$:

$$g_p(x) = a_0 + \sum_{k=1}^{p} (a_k \cos kx + b_k \sin kx).$$

For this the least-squares method is used, i.e. the a_k and b_k are determined so that the integral of squared deviations is minimized.

$$S = \int_0^{2\pi} [f(x) - g_p(x)]^2 \, dx \to \text{Min.}$$

The integral can be interpreted as a function of the $a_0, \ldots, a_p$ and $b_1, \ldots, b_p$. The optimal values of $a_0, \ldots, b_p$ are obtained by taking the derivatives of S with respect to $a_0, \ldots, b_p$ and setting them equal to zero. This method leads to the following formulae:

$$a_0 = \frac{1}{2\pi} \int_0^{2\pi} f(x) \, dx, \qquad (7.40)$$

$$a_k = \frac{1}{\pi} \int_0^{2\pi} f(x) \cos kx \, dx, \qquad (7.41)$$

$$b_k = \frac{1}{\pi} \int_0^{2\pi} f(x) \sin kx \, dx \quad (k = 1, \ldots, p). \qquad (7.42)$$

Thus the Fourier coefficients can be determined independently, and they do not depend on p. On increasing the number of terms from p to $p + p'$, the first p coefficients remain the same, and only the new coefficients $a_{p+1}, \ldots, a_{p+p'}$ and $b_{p+1}, \ldots, b_{p+p'}$ have to be computed.

Sometimes also the infinite series

$$g(x) = a_0 + \sum_{k=1}^{\infty} (a_k \cos kx + b_k \cos kx)$$

is considered, with the a_k and b_k being defined by (7.40)–(7.42), under the condition that

$$\int_0^{2\pi} f(x)^2 \, dx < \infty.$$

The functions $f(x)$ and $g(x)$ almost coincide if the so-called Dirichlet condition is satisfied, i.e. if

(1) $f(x)$ is bounded in $[0, 2\pi]$, and
(2) $f(x)$ has in $[0, 2\pi]$ only a finite number of maxima, minima and discontinuities.

Then $f(x) = g(x)$ holds at all points of continuity, x, while at points of discontinuity, x,

$$g(x) = \tfrac{1}{2}[f(x - 0) + f(x + 0)].$$

As $n \to \infty$, the coefficients a_k, b_k and A_k tend to zero because

$$2a_0^2 + \sum_{k=1}^{\infty}(a_k^2 + b_k^2)^2 = 2A_0^2 + \sum_{k=1}^{\infty} A_k^2 = \frac{1}{\pi} \int_0^{2\pi} [f(x)]^2 \, dx. \tag{7.43}$$

The speed of convergence is higher for a 'smooth' function $f(x)$ than for a 'rough' one with discontinuities or cusps.

If $f(x)$ is given by a formula then formulae can frequently be given for the Fourier coefficients (Lighthill, 1958).

Example. Let $f(x)$ be the radius-vector function of a square of unit area (with origin at the centre, and sides parallel to the coordinate axes). Then

$$r(x) = A_0 + A_4 \cos(4x + \pi) + A_8 \cos 8x + \cdots, \tag{7.44}$$

with

$$A_0 = (2/\pi)c, \tag{7.45}$$

$$a_{4l} = \frac{4}{\pi} \left[c + 2\sqrt{2} \sum_{i=1}^{l} \frac{(-1)^i}{(4i-1)(4i-3)} \right], \tag{7.46}$$

$$c - \log(1 + \sqrt{2}), \quad l = 1, 2, \ldots$$

(Bandemer et al., 1985). All other Fourier coefficients vanish. The first three coeffi cients are

$$A_0 = 0.561, \quad A_4 = 0.078, \quad A_8 = 0.025.$$

A good approximation of the square is already obtained with these values (Fig. 37).

Properties and interpretation of Fourier coefficients

(a) The coefficient a_0 is equal to the mean $\bar{f}$ of $f(x)$ on the interval $[0, 2\pi]$. If $f(x)$ is the radius-vector function then a_0 is equal to the mean radius of the figure considered. If $f(x)$ is the support function of a convex figure X then

$$a_0 = \frac{U(X)}{2\pi}. \tag{7.47}$$

(b) If $f(x)$ is the radius-vector function, then

$$\tfrac{1}{2}\pi(2A_0^2 + A_1^2 + A_2^2 + \cdots)$$

is equal to the area of the figure.

(c) The Fourier coefficients depend in different ways on the form of the object. The a_k, b_k and A_k with small indices k (up to $k = 5$) tend mainly to describe the global characteristics, while the coefficients for large k describe roughness (boundary texture), see Fig. 38 and §10.1.

Beddow *et al.* (1977) suggested for the case of the radius-vector function the following form parameters:

parameters of global structure

$$L_{n_1} = \sum_{k=1}^{n_1} A_k^2;$$

parameters of roughness:

$$R_{n_2}^{n_3} = \sum_{k=n_2}^{n_3} A_k^2.$$

Here n_1, n_2 and n_3 are suitable natural numbers chosen with the figures to be analysed in mind (e.g. $n_1 = n_2 = 5$ and $n_3 = 20$). Zahn and Roskies (1972) considered relations between shape and Fourier coefficients for the case of the tangent-angle function.

(d) Let the function $f(x)$ have a smaller period than 2π, namely $2\pi/l$, where l is a natural number greater than one. Then if a_0 is excluded, only the coefficients a_k and b_k with

$$k = lm \quad (m = 1, 2, \ldots)$$

differ from zero. The same is true for the A_k. Thus symmetry properties of figures are reflected by the Fourier coefficients.

Example 1. For the unit square $l = 4$, so that only the coefficients $A_4, A_8, \ldots$ are different from zero; see (7.47).

Example 2. Schuberth (1987) calculated the coefficients A_k corresponding to the radius-vector function for some figures. Table 1 shows some of the results (see also Rösler *et al.*, 1987). For these figures the A_k decrease continuously with increasing approximation to a disc. (For a disc all A_k vanish for $k > 1$.) A similar result holds for ellipses.

(e) The operation that assigns to a function $f(x)$ its Fourier coefficients a_k and b_k is linear. That is, let $f(x)$ be given in the form

$$f(x) = \sum_{i=1}^{m} \gamma_i f_i(x)$$

and let $a_k^{(i)}$ and $b_k^{(i)}$ be the Fourier coefficients of the functions $f_i(x)$. Then the Fourier coefficients of $f(x)$ satisfy

$$a_k = \sum_{i=1}^{m} \gamma_i a_k^{(i)} \tag{7.48}$$

$$b_k = \sum_{i=1}^{m} \gamma_i b_k^{(i)} \quad (k = 1, 2, \ldots). \tag{7.49}$$

Table 1 Fourier coefficients $A'_k = (A_k/A_0) \times 1000$ for a regular n-gon and a rectangle.
(a) Regular n-gon.

$$A_0 = \frac{nr\cos(\pi/n)}{2\pi} \log\left[\frac{1+\sin(\pi/n)}{1-\sin(\pi/n)}\right],$$

and r is the distance of the corner from the centre.

								A'_k								
n \ k	1	2	3	4	5	6	7	8	9	10	11	12	13	14	15	16
3	—	—	275.1	—	—	104.3	—	—	53.7	—	—	32.3	—	—	21.5	—
4	—	—	—	139.4	—	—	—	44.0	—	—	—	20.9	—	—	—	12.1
5	—	—	—	—	85.6	—	—	—	—	24.8	—	—	—	—	11.4	—
6	—	—	—	—	—	58.2	—	—	—	—	—	16.1	—	—	—	—
7	—	—	—	—	—	—	42.2	—	—	—	—	—	—	11.4	—	—
8	—	—	—	—	—	—	—	32.0	—	—	—	—	—	—	—	8.5
9	—	—	—	—	—	—	—	—	25.2	—	—	—	—	—	—	—
10	—	—	—	—	—	—	—	—	—	20.3	—	—	—	—	—	—

(b) Rectangle:

$$A_0 = \frac{b}{\pi} \log\left(\frac{a+\sqrt{\alpha^2+b^2}}{b}\right) - \frac{a}{\pi} \log\left(\frac{\sqrt{\alpha^2+b^2}-b}{a}\right),$$

and a, b are the side lengths $(a > b)$, $c = a/b$.

								A'_k								
c \ k	2	4	6	8	10	12	14	16	18	20	22	24	26	28	30	32
1	—	139.4	—	44.0	—	20.9	—	12.1	—	7.8	—	5.5	—	4.1	—	3.1
2	400.0	17.4	75.8	57.4	12.4	17.5	21.9	9.6	4.6	10.7	7.4	0.1	5.4	5.5	1.6	2.5
3	597.1	193.5	19.0	48.1	58.1	41.1	16.5	4.2	15.6	17.4	12.2	4.0	3.4	7.6	8.1	5.4
4	716.4	322.6	124.3	18.6	32.8	50.4	47.8	34.8	18.3	3.1	8.1	14.2	15.4	12.8	8.0	2.4
10	1000.7	672.1	480.1	348.5	251.7	177.9	120.8	76.3	41.9	15.6	4.0	18.2	27.8	33.8	36.7	37.1

An analogous relation does not hold for A_k and α_k.

(f) The A_k are invariant with respect to shifts of the variable x. That is, instead of the function $f(x)$, the function $f_a(x)$ is considered where

$$f_a(x) = \begin{cases} f(x+a) & (x+a \le 2\pi), \\ f(x+a-2\pi) & \text{otherwise.} \end{cases}$$

Here a is a fixed number between 0 and 2π. The Fourier series corresponding to $f_a(x)$ is obviously

$$f_a(x) = A_0 + \sum_{k=1}^{\infty} A_k \cos[k(x+a)+\alpha_k] = A_0 + \sum_{k=1}^{\infty} A_k \cos(kx + ka + \alpha_k).$$

Thus the function $f_a(x)$ has the same coefficients A_k as $f(x)$. This property is very important for form description by means of the radius-vector, support and tangent-angle functions. The choice of the starting point ($\varphi = x = 0$) does not influence the A_k. These quantities are thus invariant function parameters in the sense of §7.3. If figures have to be analysed for which the choice of the starting point is arbitrary (e.g. for particles) then the A_k/A_0 are reasonable shape parameters. In the case of biological objects the situation is different. Here the starting point is a certain landmark, and it is frequently not sufficient to consider the A_k alone.

Example 1. The first three non-vanishing terms of the Fourier series corresponding to the radius-vector function of a unit square yield the function $g(x)$,

$$g(x) = A_0 + A_4 \cos(4x + \pi) + A_8 \cos 8x$$

(p. 82). Figure 37 shows the corresponding figure. It is quite similar to a square. If instead of the correct phase angles $\alpha_4 = \pi$, $\alpha_8 = 0$ the values $\alpha_4 = \frac{1}{2}\pi$ and $\alpha_8 = \frac{1}{6}\pi$ are used then the dotted figure in Fig. 37 is obtained, which differs more from a square.

Example 2. Figure 38 shows a simplified fish profile and the approximating radius-vector contours corresponding to 19 and 36 Fourier coefficients A_k. (The direction $\varphi = x = 0$ corresponds to a line from the centre to the middle of the tail. 72 interpolation nodes were used in the calculation.) If all phase angles are set equal to zero then the fourth curve is obtained, which only vaguely resembles a fish. Corresponding to the dipole character of the fish profile, the coefficient a_2 is the largest after a_0. Those values a_k/a_0 that are absolutely greater than 0.05 are

$$\begin{aligned}
a_2/a_0 &= 0.335, & a_7/a_0 &= 0.058, \\
a_4/a_0 &= 0.128, & a_8/a_0 &= 0.056, \\
a_6/a_0 &= 0.088, & b_3/a_0 &= -0.124.
\end{aligned}$$

Davies and Hawkins (1979) also tried to transform the α_k into parameters that are independent of the choice of the starting point ($x = 0$).

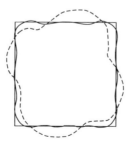

Figure 37 Approximation of a square by a figure that corresponds to a Fourier series using the first three A_k different from zero ($k = 0, 4, 8$). The dotted contour has been obtained by replacing the right phase angles $\alpha_4 = \pi$ and $\alpha_8 = 0$ by $\alpha_4 = \frac{1}{2}\pi$ and $\alpha_8 = \frac{1}{6}\pi$.

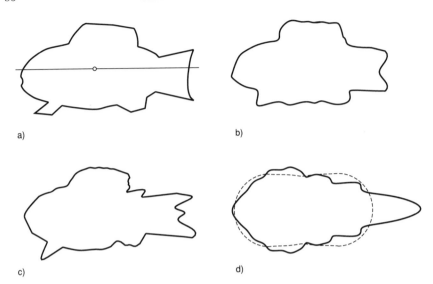

Figure 38 (a) Fish profile and approximations to (b) 19 and (c) 36 Fourier coefficients A_k. (d) The figures with phase angle set to zero — the unbroken curve being generated by nineteen A_k, and the broken by only two.

Calculation of Fourier coefficients

Suppose we are given the function values $y_j = f(x_j)$ for n interpolation nodes, $j = 0, 1, \ldots, n - 1$. Then the Fourier coefficients can be calculated using (7.40)–(7.42) by numerical integration. The particular form of the integrals and properties of the trigonometric functions enables essential simplifications. Consequently, there are elegant algorithms, for example the well-known 'fast Fourier transform' (Chatfield, 1980; Nussbaumer, 1981; Elliot and Rao, 1982). Here particular formulae are given. Let n be even and let the interpolation nodes be equidistant, i.e.

$$x_j = j\Delta, \quad j = 0, 1, \ldots, n - 1, \text{ with } \Delta = \frac{2\pi}{n}.$$

Then altogether n Fourier coefficients can be determined, namely $a_0, a_1, \ldots, a_{n/2}$, $b_1, \ldots, b_{n/2-1}$. They are calculated using the formulae

$$a_0 = \frac{1}{n} \sum_{j=0}^{n-1} y_j,$$

$$a_k = \frac{2}{n} \sum_{j=0}^{n-1} y_j \cos k x_j \quad \left(k = 1, \ldots, \tfrac{1}{2}n - 1\right),$$

$$a_{n/2} = \frac{1}{n} \sum_{j=0}^{n-1} (-1)^j y_j$$

$$b_k = \frac{2}{n} \sum_{j=0}^{n-1} y_j \sin k x_j \quad \left(k = 1, \ldots, \tfrac{1}{2}n - 1\right).$$

(7.50)

The corresponding Fourier series is then

$$g_{n/2}(x) = a_0 + \sum_{k=0}^{n/2} a_k \cos kx + \sum_{k=0}^{n/2-1} b_k \sin kx$$

It has the property that for all j

$$g_{n/2}(x_j) = y_j = f(x_j);$$

thus $g_{n/2}(x)$ is an interpolation function.

In passing, the formulae above are interpretable as the result of averaging of two numerical integration procedures by the so-called Simpson method; the error is of order Δ^5 (Zurmühl, 1963).

Zahn and Roskies (1972) give particular formulae for the case when $f(x)$ is a tangent-angle function.

Relations between the contour covariance function and Fourier coefficients

As in the theory of stochastic processes, there is a close connection between the contour covariance function $\chi_f(\varphi)$ and the Fourier coefficients A_k of a given function $f(x)$. That is,

$$\chi_f(\varphi) = \tfrac{1}{2} \sum_{k=1}^{\infty} A_k^2 \cos k\varphi \tag{7.51}$$

and

$$A_k^2 = \frac{2}{\pi} \int_0^{2\pi} \chi_f(\varphi) \cos k\varphi \, d\varphi \quad (k = 1, 2, \ldots), \tag{7.52}$$

with $A_0 = \bar{f}$. The formula (7.51) shows that the contour covariance function does not uniquely determine the contour function $f(x)$, because the Fourier coefficients A_k (without the α_k) do not determine $f(x)$ uniquely. The validity of (7.51) follows from $f(x) - \bar{f} = \sum_{k=1}^{\infty} A_k \cos(kx + \alpha_k)$, using the definition of $\chi_f(\varphi)$:

$$\chi_f(\varphi) = \frac{1}{2\pi} \int_0^{2\pi} \sum_{k=1}^{\infty} A_k \cos(k\varphi + \alpha_k) \sum_{n=1}^{\infty} A_n \cos(n\varphi + n\varphi + \alpha_n) \, d\varphi$$

$$= \frac{1}{2\pi} \sum_{k=1}^{\infty} A_k^2 \int_0^{2\pi} \cos(k\varphi + \alpha_k) \cos(k\varphi + k\varphi + \alpha_n) \, d\varphi$$

$$= \frac{1}{2\pi} \sum_{k=1}^{\infty} A_k^2 \int_0^{2\pi} \cos^2 k\varphi \cos k\varphi \, d\varphi$$

$$= \frac{1}{2\pi} \sum_{k=1}^{\infty} A_k^2 \pi \cos k\varphi = \tfrac{1}{2} \sum_{n=1}^{\infty} A_k^2 \cos k\varphi.$$

The formula (7.52) is a consequence of (7.51), since, as in the calculation above,

$$\int_0^{2\pi} \cos n\varphi \cos m\varphi \, d\varphi = \begin{cases} \pi & (n = m), \\ 0 & \text{otherwise.} \end{cases}$$

7.6   STOCHASTIC MODELS FOR THE CONTOUR FUNCTION APPROACH

7.6.1   Invariant contour function parameters of random figures

Let X be a random figure with random contour function $F(x)$. Then the parameters introduced in §7.3, $\bar{f}$, $\sigma^2(f)$ etc., become random variables, denoted by $\bar{F}$, $\sigma^2(F)$ etc.

It is useful to introduce in addition to $\sigma^2(F)$ and $\chi_F(\varphi)$ the quantities $\sigma^2(F)$ and $\chi_F(\varphi)$:

$$\sigma^2(F) = \frac{1}{2\pi} \int_0^{2\pi} [F(x) - \mathsf{E}\bar{F}]^2 \, dx, \qquad (7.53)$$

$$\chi_f(\varphi) = \frac{1}{2\pi} \int_0^{2\pi} [F(x) - \mathsf{E}\bar{F}][F(x + \varphi) - \mathsf{E}\bar{F}] \, dx, \qquad (7.54)$$

where $\sigma(F)$ is the standard deviation of F away from the *expected* mean, rather than its own mean. They satisfy

$$\sigma^2(F) = \sigma^2(F) + (\bar{F} - \mathsf{E}\bar{F})^2 \qquad (7.55)$$

and

$$\chi_F(\varphi) = \chi_F(\varphi) + (\bar{F} - \mathsf{E}\bar{F})^2. \qquad (7.56)$$

The random Fourier coefficients A_n of F are related to $\chi_F(\varphi)$ by

$$A_n^2 = \frac{2}{\pi} \int_0^{2\pi} \chi_F(\varphi) \cos n\varphi \, d\varphi \quad (n = 1, 2, \ldots) \qquad (7.57)$$

(in (7.57) it makes no difference whether $\chi_F(\varphi)$ or χ_F is used.) Now let F be the sum of two stochastically independent components:

$$F = F_1 + F_2.$$

This makes sense for the radius-vector or support function, but not for the tangent-angle function. Then

$$\bar{F} = \bar{F}_1 + \bar{F}_2$$

and

$$\mathsf{E}\sigma^2(F_1 + F_2) = \mathsf{E}\sigma^2(F_1) + \mathsf{E}\sigma^2(F_2), \qquad (7.58)$$

$$\mathsf{E}\chi_{F_1+F_2}(\varphi) = \mathsf{E}\chi_{F_1}(\varphi) + \mathsf{E}\chi_{F_2}(\varphi) \qquad (7.59)$$

(but not necessarily $E\sigma^2(F_1+F_2) = E\sigma^2(F_1)+E\sigma^2(F_2)$). Analogously, the Fourier coefficients $A_n(F)$ of F satisfy

$$EA_n^2(F) = EA_n^2(F_1) + EA_n^2(F_2) \quad (n = 1, 2, \ldots). \tag{7.60}$$

Let the contour function $F(x)$ now have the same mean $EF(x)$ and the same variance var $F(x)$ for all x (this is true if the figure X is isotropic). Then the expected value of $\bar{F}$, $\sigma^2(F)$ etc. can be expressed by quantities that depend on an arbitrary (but fixed x):

$$\begin{aligned}
E\bar{F} &= EF(x), \\
E\sigma^2(F) &= \mathrm{var}(F(x)), \\
E\sigma^2(F) &= \mathrm{var}(F(x)) - \mathrm{var}\,\bar{F} \quad (0 \le x \le 2\pi).
\end{aligned}$$

7.6.2   Random radial-rhombi

Let X be a radial-rhombus with random semi-axis lengths A and B. Let the corresponding means be m_A and m_B, the corresponding variances σ_A^2 and σ_B^2 and the correlation coefficient ϱ_{AB}. Then the invariant parameters of the radius-vector function $R_x(\varphi)$ satisfy

$$\bar{R} = \tfrac{1}{2}(A + B), \tag{7.61}$$

$$E\bar{R} = \tfrac{1}{2}(m_A + m_B), \tag{7.62}$$

$$\mathrm{var}\,\bar{R} = \tfrac{1}{4}E(A + B - m_A - m_B)^2 \tag{7.63}$$

or

$$\mathrm{var}\,\bar{R} = \tfrac{1}{4}(\sigma_A^2 + \sigma_B^2 + 2\varrho_{AB}\sigma_A\sigma_B). \tag{7.64}$$

Furthermore, by (7.25),

$$\sigma^2(R) = \tfrac{1}{12}(A - B)^2 \tag{7.65}$$

and

$$E\sigma^2(R) = \tfrac{1}{12}D_\rho, \text{ with } D_\rho = (m_A - m_B)^2 + \sigma_A^2 + \sigma_B^2 - 2\sigma_{AB}\sigma_A\sigma_B. \tag{7.66}$$

Because of the symmetry of the radial-rhombus, equations analogous to (7.19) and (7.20) hold for the contour covariance function $\chi_R(\phi)$. Thus $\chi_R(\varphi)$ is needed only for $0 \le \varphi \le \tfrac{1}{2}\pi$. For these φ-values

$$\chi_R(\varphi) = \tfrac{1}{12}(A - B)^2(1 - 6x^2 + 4x^3), \tag{7.67}$$

with $x = 2\varphi/\pi$, or, more briefly,

$$\chi_R(\varphi) = \tfrac{1}{12}(A - B)^2 c(\varphi).$$

This formula can be used to obtain $\chi_R(\varphi)$ by means of (7.56). The expected value $\mathsf{E}\chi_R(\varphi)$ is given by

$$\mathsf{E}\chi_R(\varphi) = \tfrac{1}{4}(\sigma_A^2 + \sigma_B^2 + 2\varrho_{AB}\sigma_A\sigma_B) + \tfrac{1}{12}D_\rho c(\varphi). \qquad (7.68)$$

7.6.3 Randomly disturbed figures

This class of models is described for the case of the radius-vector function, for which it makes most sense (but it is not clear, for example for particles, that the deviations always appear in the radial direction). These figures fluctuate randomly around a mean deterministic figure:

$$R(\varphi) = D(\varphi) + \mathcal{E}(\varphi) \quad (0 \le \varphi \le 2\pi). \qquad (7.69)$$

Here $D(\varphi)$ is the deterministic radius-vector function of the mean figure, and $\mathcal{E}(\varphi)$ is a disturbing radius-vector function.[†] Figure 39 shows two examples, where $D(\varphi)$ corresponds to an ellipse and $\mathcal{E}(\varphi)$ is a disturbing function of the type considered in §7.6.4. Suppose that for all φ

$$\mathsf{E}\mathcal{E}(\varphi) = 0.$$

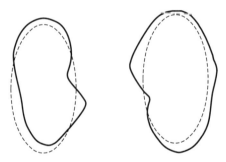

Figure 39 Two randomly disturbed ellipses. A tooth contour was used as the disturbing function, (p. 91).

This implies

$$\mathsf{E}R(\varphi) = D(\varphi), \quad \operatorname{var}R(\varphi) = \operatorname{var}\mathcal{E}(\varphi).$$

Note that $R(\varphi)$ and the corresponding characteristics are related to the central point (perhaps the centre of gravity) of the figure corresponding to $D(\varphi)$.
    Obviously,

$$\mathsf{E}\bar{R} = \mathsf{E}\bar{D}, \qquad (7.70)$$

[†] Similarly Underwood (1980) considered discs disturbed by deterministic sine curves.

$$E\sigma^2(R) = E\sigma^2(D) + E\sigma^2(\mathcal{E}) \tag{7.71}$$

and

$$E\chi_R(\varphi) = \chi_D(\varphi) + E\chi_\mathcal{E}(\varphi). \tag{7.72}$$

The mean area of the figure X corresponding to $R(\varphi)$ is

$$EA(X) = A(D) + \pi E\sigma^2(\mathcal{E}) \tag{7.73}$$

where $A(D)$ is the area of the figure corresponding to $D(\varphi)$. There is no such simple formula for the perimeter (§7.7.3). In particular, for particles it seems to be natural to assume that $\mathcal{E}(\varphi)$ is isotropic.[‡] Under this condition for all φ,

$$\operatorname{var} R(\varphi) = \operatorname{var} \mathcal{E}(\varphi) = E\sigma^2(\mathcal{E}).$$

7.6.4 Three disturbance models

Tooth contour

A fairly simple model for an isotropic disturbance function follows, which will be used in (7.69) and which is a rough model for particles. This model has as parameters a natural number l and a distribution function $F(x)$. Let ζ be a random variable uniformly distributed on $[0, 2\pi]$, and let $Z_1, \ldots, Z_l$ be random variables with distribution function $F(x)$, mean m and variance σ^2, which are independent of ζ ($l \geq 4$). Then $\mathcal{E}(\varphi)$ is the following random function on $[0, 2\pi]$:

$$\mathcal{E}(\varphi) = \lambda Z_i + (1 - \lambda)Z_{i+1}$$

for

$$\varphi = \lambda \varphi_i + (1 - \lambda)\varphi_{i+1} \quad (0 \leq \lambda \leq 1),$$

with

$$\varphi_i = \zeta + i\Delta, \quad i = 1, \ldots, l, \quad \Delta = 2\pi/l;$$

Here values of φ and φ_i greater than 2π are interpreted as $\varphi - 2\pi$ and $\varphi_i - 2\pi$, respectively, and Z_{l+1} is set equal to Z_1.

Figure 39 shows how these functions change the form of ellipses (for details see §7.7.3).

[‡]This means that the functions $\mathcal{E}_a(\varphi)$ defined by

$$\mathcal{E}_a(\varphi) = \begin{cases} \mathcal{E}(\varphi + a) & (x + a \leq 2\pi), \\ \mathcal{E}(\varphi + a - 2\pi) & \text{otherwise} \end{cases}$$

have the same distribution for any a between 0 and 2π.

By construction, $\mathcal{E}(\varphi)$ is isotropic; thus $E\bar{\mathcal{E}} = E\mathcal{E}(\varphi)$ and $E\sigma^2(\mathcal{E}) = \text{var}\,\mathcal{E}(\varphi)$ for all φ. We have independent of l or Δ for all φ

$$E\mathcal{E}(\varphi) = m, \tag{7.74}$$

$$\text{var}\,\mathcal{E}(\varphi) = \tfrac{2}{3}\sigma^2. \tag{7.75}$$

Furthermore,

$$E\chi_\mathcal{E}(\varphi) = E\chi_\mathcal{E}(2\pi - \varphi)$$

and

$$E\chi_\mathcal{E}(\varphi) = \tfrac{1}{6}\sigma^2 \times \begin{cases} (4 - 6x^2 + 3x^3) & (\varphi < \Delta), \\ (2 - x)^3 & (\Delta \leq \varphi \leq 2\Delta), \\ 0 & (k\Delta \leq \varphi \leq (k+1)\Delta, \\ & \text{with } k \leq l - 2), \end{cases} \tag{7.76}$$

with $x = \varphi/\Delta$ and

$$E\sigma^2(\mathcal{E}') = -[E\chi_\mathcal{E}(\varphi)]''_{\varphi=0} = 2\left(\frac{\sigma}{\Delta}\right)^2. \tag{7.77}$$

Isotropized Brownian bridge

The following model will also be used in (7.69) as a disturbance function $\mathcal{E}(\varphi)$, and may be used for particles. In contrast to the tooth contour, it has infinitely many tiny roughnesses. The construction is as follows. Let $\{X(\varphi)\}$ be a Wiener process with parameter σ^2; $X(\varphi)$ is thus normally distributed with mean zero and variance $\varphi\sigma^2$. The corresponding Brownian bridge (on $[0, 2\pi]$) is $\{Y(\varphi)\}$:

$$Y(\varphi) = X(\varphi) - \frac{\varphi}{2\pi}X(2\pi) \quad (0 \leq \varphi \leq 2\pi). \tag{7.78}$$

The random function $Y(\varphi)$ is continuous on $[0, 2\pi]$ and zero for $\varphi = 0$ and $\varphi = 2\pi$. 'Isotropization' of $Y(\varphi)$ yields $\mathcal{E}(\varphi)$:

$$\mathcal{E}(\varphi) = Y(\varphi + \zeta), \tag{7.79}$$

where ζ is a random variable uniformly distributed on $[0, 2\pi]$ and independent of $\{X(\varphi)\}$. If $\varphi + \zeta > 2\pi$ then the Y-value corresponding to $\varphi + \zeta - 2\pi$ should be taken.

Some properties of $\mathcal{E}(\varphi)$ (which is nowhere-differentiable) may be derived quite easily:

$$E\mathcal{E}(\varphi) = 0, \tag{7.80}$$

$$\text{var}\,\mathcal{E}(\varphi) = \tfrac{1}{3}\pi\sigma^2, \tag{7.81}$$

and

$$\mathsf{E}\chi\varepsilon(\varphi) = 2\pi\sigma^2 \left[\tfrac{1}{6} - \tfrac{1}{2}\left(\frac{\varphi}{2\pi}\right) + \tfrac{1}{2}\left(\frac{\varphi}{2\pi}\right)^2 \right] \quad (0 \le \varphi \le \pi). \tag{7.82}$$

Figure 40 shows a curve that has been obtained by superimposing an isotropized Brownian bridge onto a disc. The corresponding theoretical curve has fractal dimension 1.5. For a statistical application of this model see Stoyan and Lippmann (1993).

It is possible as well to perturb a unit circle by means of other random processes. For this, put as in (7.79)

$$Y(\varphi) = \zeta(\varphi) - \frac{\varphi}{2\pi}\zeta(2\pi) \qquad (0 \le \varphi \le 2\pi)$$

where $\zeta(\varphi)$ is a right-continuous random process on $[0, 2\pi]$ such that $\zeta(0) = 0$. (If ζ is not continuous, then the closure of the perturbed set has to be taken.) Then, similar to (7.80), it is possible to isotropize the process Y.

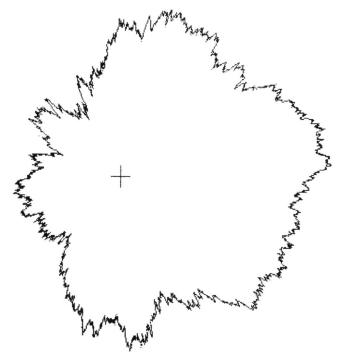

Figure 40 Simulated isotropized Brownian bridge, which was used as a disturbance function to randomly disturb a circle.

Eigenshapes

The following model is frequently used, particularly for biological objects:

$$X(\varphi) = D(\varphi) + \sum_{k=1}^{\infty} C_k f_k(\varphi), \tag{7.83}$$

where the $f_k(\varphi)$ are deterministic functions, and the C_k are independent random variables with

$$EC_k = 0$$

and

$$\text{var } C_k = \gamma_k \quad (k = 1, 2, \ldots).$$

Since isotropy is not assumed, the covariance function of $\mathcal{E}(\varphi)$ has to be used, which is more complicated than $E_X(\psi)$:

$$\kappa(\varphi, \psi) = E\mathcal{E}(\varphi)\mathcal{E}(\psi) \quad (0 \le \varphi, \psi \le 2\pi).$$

It is easily shown that

$$\kappa(\varphi, \psi) = \sum_{k=1}^{\infty} \gamma_k f_k(\varphi) f_k(\psi). \tag{7.84}$$

The γ_k are the eigenvalues of the kernel $\kappa(\varphi, \psi)$, and the $f_k(\varphi)$ are the corresponding eigenfunctions. Frequently they can be interpreted biologically or physically. Usually the $f_k(\varphi)$ corresponding to large γ_k characterize global form variation while the others reflect local irregularities.

7.7 STATISTICAL ANALYSIS FOR THE CONTOUR FUNCTION APPROACH

7.7.1 Values of the invariant parameters of single figures

The determination of the values of $\bar{f}$, $\sigma^2(f)$, $\sigma^2(f')$ and $\chi_f(\varphi)$ for a given function $f(x)$ is a problem of numerical mathematics. Starting from the function values $f_i = f(x_i)$ $(i = 1, \ldots, N)$ corresponding to the nodes of interpolation x_i, integrals such as

$$\frac{1}{2\pi} \int_0^{2\pi} f(x) \, dx \quad \text{or} \quad \int_0^{2\pi} f^2(x) \, dx$$

must be calculated. If there are sufficient nodes and they are equidistant then the integrals may be approximated by the following sums:

$$\bar{f} \approx \frac{1}{N} \sum_{i=1}^{N} f_i,$$

$$\sigma^2(f) \approx \frac{1}{N} \sum_{i=1}^{N} (f_i - \bar{f})^2,$$

$$\chi_f(k\Delta) \approx \frac{1}{N} \sum_{i+1}^{N} (f_i - \bar{f})(f_{i+k} - \bar{f}),$$

where

$$f_{N+1} = f_1, \; \Delta = \frac{2\pi}{N}.$$

There are several ways of calculating $\sigma^2(f')$:

$$\sigma^2(f') \approx \frac{1}{N} \sum_{i=1}^{N} \left(\frac{f_{i-2} - 8f_{i-1} + 8f_{i+1} - f_{i+2}}{12\Delta} \right)^2 \qquad (7.85)$$

or

$$\sigma^2(f') \approx -\frac{\chi_f(\Delta) + 4\chi_f(2\Delta) + 9\chi_f(3\Delta) - 14\chi_f(0)}{49\Delta^2}. \qquad (7.86)$$

The first formula is derived by approximating the derivative $f'(\Delta)$; the second comes from (7.12). There $\chi_f(\varphi)$ is approximated by a parabola of the form

$$\chi_f(\varphi) = \chi_f(0) + a\varphi^2,$$

obtained by the least-squares method. (There is no linear term, because $\chi_f'(0) = 0$.) The values $\chi_f(0)$, $\chi_f(\Delta)$, $\chi_f(2\Delta)$ and $\chi_f(3\Delta)$ are used in the approximation. The optimal coefficient is

$$a = \frac{\chi_f(\Delta) + 4\chi_f(2\Delta) + 9\chi_f(3\Delta) - 14\chi_f(0)}{98\Delta^2}.$$

If Δ is not 'very small' then the results of the two methods differ, and are only rough approximations to $\sigma^2(f')$. Nevertheless, the values can be compared for different figures, and are thus useful parameters for describing the boundary texture if the same formula is always used.

7.7.2 Statistical determination of distributional characteristics of the invariant parameters

Let $X_1, \ldots, X_n$ be a random sample of independently identically distributed figures with contour functions $F_1(x), \ldots, F_n(x)$. Quantities of interest include

$$\mathsf{E}\bar{F}, \; \mathsf{E}\sigma^2(F), \; \mathsf{E}\chi_f(\varphi).$$

On account of the independence assumption, the methods of classical statistics are applicable. Thus this section is very short. An unbiased asymptotically normally distributed estimator of $\mathsf{E}\bar{F}$ is given by

$$\bar{\bar{F}} = \frac{1}{n} \sum_{i=1}^{n} \bar{F}_i. \qquad (7.87)$$

If measurements are given for N nodes of interpolation then one may use

$$\tilde{\bar{F}} = \frac{1}{nN} \sum_{i=1}^{n} \sum_{j=1}^{N} F_{ij}. \qquad (7.88)$$

Here F_{ij} is the function value of the ith figure at the jth node of interpolation. The variance of $\bar{F}$ is estimated by

$$s^2(\bar{F}) = \frac{1}{n-1} \sum_{i=1}^{n} (\bar{F}_i - \bar{\bar{F}})^2. \qquad (7.89)$$

Analogously, all other means and variances may be estimated.

7.7.3 Statistics for randomly disturbed functions

Let the observed radius-vector functions $R_i(\varphi)$ have the form (2.69),

$$R_i(\varphi) = D(\varphi) + \mathcal{E}_i(\varphi),$$

where the random functions $\mathcal{E}_i(\varphi)$ are identically and independently distributed and isotropic with $\mathsf{E}\mathcal{E}_i(\varphi) \equiv 0$. The aim of the statistics is the estimation of the deterministic function $D(\varphi)$ and of the distributional characteristics for $\mathcal{E}(\varphi)$. There are two cases that have to be considered.

(1) The position $\varphi = 0$ corresponds to a landmark, as in the case of biological objects. Then it makes sense to determine means and variances for selected values of φ.

(2) The position $\varphi = 0$ has been chosen arbitrarily on the contour as in the case of particles. In this case invariant contour parameters should be studied.

In the simpler first case $D(\varphi)$ may be estimated as

$$\overline{D(\varphi)} = \frac{1}{n} \sum_{i=1}^{n} R_i(\varphi). \qquad (7.90)$$

A smooth estimate of $D(\varphi)$ can be obtained by the least-squares method. It may be useful to apply a penalized least-squares approach, where smoothing parameters are introduced in order to diminish the roughness of the estimations (Rice and Silverman, 1991).

Now the characteristics of $\mathcal{E}(\varphi)$ will be estimated:

$$\operatorname{var}\mathcal{E}(\varphi) : s^2(\varphi) = \frac{1}{n-1}\sum_{i=1}^{n}[R_i(\varphi) - \overline{D(\varphi)}]^2, \tag{7.91}$$

$$\mathsf{E}\sigma^2(\mathcal{E}) : s^2(\mathcal{E}) = \frac{1}{n}\sum_{i=1}^{n}\sigma^2(R_i - \bar{D}), \tag{7.92}$$

$$\mathsf{E}\chi_{\mathcal{E}}(\varphi) : k_{\mathcal{E}}(\varphi) = \frac{1}{n}\sum_{i=1}^{n}\frac{1}{2\pi}\int_0^{2\pi}[R_i(\psi) - \bar{D}(\psi)]$$
$$\times[R_i(\psi + \varphi) - \bar{D}(\psi + \varphi)]\,d\psi, \tag{7.93}$$

where the integral may be replaced by a sum, as in §7.7.1.

Example. Let $D(\varphi)$ be the radius-vector function corresponding to an ellipse with semi-axis lengths $a = 2$ and $b = 1$. Let $\mathcal{E}(\varphi)$ be a tooth contour function with $l = 8$ and $F(x)$ corresponding to a uniform distribution on $[-0.5, 0.5]$.

By simulation, 10 figures as in Fig. 39 have been generated and then used as a sample. The function $D(\varphi)$ is then estimated by (7.92); Fig. 41 shows the result. The estimate of $\mathsf{E}\sigma^2(\mathcal{E})$ is $s^2(\mathcal{E}) = 0.0561$, while the theoretical value is 0.0555. Table 2 shows the true and estimated values of $\mathsf{E}\chi_{\mathcal{E}}(\varphi)$. The accuracy of estimation for this example is quite good.

If the figures to be analysed belong to a sample of independent identically distributed figures then it is possible to use the methods of classical statistics to evaluate the quality of the estimates. For example, for large n or normally distributed $\mathcal{E}(\varphi)$ the following confidence interval based on the t-distribution can be used:

$$\Pr\left(\bar{D}(\varphi) - \frac{S(\varphi)}{\sqrt{n}}t_{n-1,1-\alpha/2} \le D(\varphi) \le \bar{D}(\varphi) + \frac{S(\varphi)}{\sqrt{n}}t_{n-1,1-\alpha/2}\right) = 1 - \alpha. \tag{7.94}$$

Example (continued). Since the function $\mathcal{E}(\varphi)$ is isotropic,

$$\mathsf{E}\sigma^2(\mathcal{E}) = \operatorname{var}\mathcal{E}(\varphi)$$

for all φ. Thus, $s^2(\mathcal{E})$ is an acceptable estimate of $\operatorname{var}\mathcal{E}(\varphi)$ for any φ.

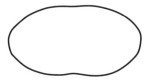

Figure 41 Mean of 10 randomly disturbed ellipses as in Fig. 39. This is an estimate of the ellipse contour function $D(\varphi)$.

Table 2 Estimated and true values of the expectations
$E\chi_\varepsilon(\varphi)$ of the random contour covariance function of
the disturbance function.

φ	$k_\varepsilon(\varphi)$	$E\chi_\varepsilon(\varphi)$
0°	0.0561	0.0555
8°	0.0539	0.0532
16°	0.0483	0.0469
24°	0.0403	0.0382
32°	0.0312	0.0284
40°	0.0221	0.0190
48°	0.0141	0.0113
56°	0.0078	0.0060
64°	0.0031	0.0027
72°	−0.0005	0.0009
80°	−0.0029	0.0002
88°	−0.0046	0.0000

For $\alpha = 0.05$ and $n = 10$ the t-value is

$$t_{9;0.975} = 2.262,$$

so that the following confidence interval for $D(\varphi)$ is obtained:

$$(\bar{D}(\varphi) - 0.040, \bar{D}(\varphi) + 0.040).$$

Statistics for eigenshape analysis

The model parameters γ_k and $f_k(\varphi)$ are statistically estimated using a sample of
independent figures with $X_1(\varphi), \ldots, X_n(\varphi)$. $D(\varphi)$ is estimated and then the discrete
data values x_{ij} are analysed:

$$x_{ij} = X_i(\varphi_j) - \hat{D}(\varphi_j) \quad (i = 1, \ldots, n; \; j = 1, \ldots, s).$$

The φ_j might be equidistant interpolation nodes. Then the covariance matrix Γ
is calculated for the x_{ij}:

$$\Gamma = (\hat{\kappa}(\varphi_k, \varphi_l)),$$

and

$$\hat{\kappa}(\varphi_k, \varphi_l) = \sum_{i=1}^{n} x_{ik} x_{il} \quad (k, l = 1, \ldots, s).$$

The corresponding eigenvalues are estimates for the first γ_k and the corresponding
eigenvectors estimates for the $f_k(\varphi)$. Applications in shape statistics can be found
in Lohmann (1983) and Full and Ehrlich (1986).

Connections with Fourier analysis are discussed by Rohlf (1986), who also
discusses the formal similarity to principal component analysis. Rice and Silverman

(1991) describe a smoothed form of the analysis; however, they do not assume that the $f_k(\varphi)$ are periodic. As an example, they analyse a sample of curves.

Now it is shown by an example how the more complicated second case of p. 96 can be treated, using the methods of §7.5.2. The figures to be analysed are generated by simulation as above. That is, one considers ellipses with semi-axis lengths $a = 2$ and $b = 1$, roughened by a tooth contour with $l = 8$ and deviations uniformly distributed on $[-0.5, 0.5]$. These figures are approximated by ellipses, and the lengths of the corresponding semi-axes have to be estimated.

The methods used will be the D_{max} method, the $A-U$ method and the R method. (the latter is based on measurement of area and mean radius$^\dagger\bar{r}$). The calculation of semi-axis lengths is carried out using (7.16) and (7.18). The complexity of the formula for $\bar{r}$ necessitates the application of the *regula falsi* to calculate b after elimination of a by $A = \pi ab$. By the way it is not possible to determine suitable a and b for all values A and r.

Mean values of a and b for samples of figures yield estimates of the semiaxis lengths. Another possibility is to determine means of A, U and d_{max} for the samples and to then use (7.28)–(7.31).

An exact theoretical investigation of the properties of these three estimates seems to be difficult, even for the simple disturbance model used in the simulations. Note that the means of area and perimeter of the random figures are greater than the area and perimeter of the original ellipses. Namely, by (7.73)

$$\mathsf{E}A = \pi ab + \pi \mathsf{E}\sigma^2(\mathcal{E}).$$

For the values used in the simulation this gives $\mathsf{E}A = 6.4577$ (instead of 6.2832 for the original ellipse). Simulation yielded the value $\mathsf{E}U = 10.09$ (instead of 9.6943 for the ellipse).

By simulation, the properties of the estimators for a and b have been investigated for all three methods. This has been done both by estimating a and b for each figure separately (and then averaging) and also using sample means for A, U and d_{max}. The calculations show the latter procedure to be preferable.

Of the three methods, the D_{max} method turned out to be clearly the best, yielding estimates of a and b closest to 2 and 1. While the $A-U$ method still gives acceptable values, the quality of the estimates from the R method is rather poor. For samples of each 10 figures the values of bias (= mean of estimate − true value) and standard deviation are given in Table 3.

As the variability of $\mathcal{E}(\varphi)$ decreases, the quality of the estimators increases. For an interval of variation $[-0.25, 0.25]$ (instead of the original interval $[-0.5, 0.5]$) all three methods have bias close to zero; even for the R method they turned out to be (a) 0.04 and (b) −0.02.

$^\dagger\bar{r}$ pertains to the radius-vector function of the disturbed figure, whose centre of gravity does not usually coincide with the centre of the original ellipse.

Table 3 Bias and estimated standard deviations for the estimators for a and b using three estimation methods.

	D_{max} method		A–U method		R method	
	a	b	a	b	a	b
Bias	0.01	−0.03	0.14	−0.95	0.41	−0.16
Standard deviation	0.07	0.04	0.08	0.03	0.45	0.14

7.7.4 Statistics for random radial-rhombi and related figures

As in §7.6.2, random radial-rhombi are now considered. The parameters m_A, m_B, σ_A, σB and ϱ_{AB} must be estimated from a sample of independent radius-vector functions $R_1(\varphi), \ldots, R_n(\varphi)$, and the distributions of A and B must be determined.

The mean values m_A and m_B may be estimated by the method of moments using (7.62) and (7.65). This yields the unbiased estimators

$$\hat{m}_A = \bar{\bar{R}} + \sqrt{3}\,\overline{\sigma(R)} \tag{7.95}$$

and

$$\hat{m}_B = 2\,\bar{\bar{R}} - \hat{m}_A. \tag{7.96}$$

Here

$$\bar{\bar{R}} = \frac{1}{n}\sum_{i=1}^{n}\bar{R}_i$$

and

$$\overline{\sigma(R)} = \frac{1}{n}\sum_{i=1}^{n}\sqrt{\sigma^2(R_i)}.$$

Further distributional characteristics can be obtained by determining the parameters a_i and b_i separately for each function $R_i(\varphi)$ using the formulae in §7.4.2 and subsequent statistical analysis:

$$a_i = \bar{R}_i + [3\sigma^2(R_i)]^{1/2}, \quad b_i = 2\bar{R}_i - a_i,$$

with

$$\bar{a} = \frac{1}{n}\sum_{i=1}^{n}a_i.$$

One obtains $\hat{m}_A = \bar{a}$ and an analogous estimator for m_B. An unbiased estimator of σ_A^2 is given by

$$s_a^2 = \frac{1}{n-1}\sum_{n-1}^{n}(a_i - \bar{a})^2.$$

σ_B^2 can be estimated analogously. The empirical correlation coefficient r_{AB} for the series $(a_1, b_1), \ldots (a_n, b_n)$ is an estimator for the parameter ϱ_{AB}.

The empirical distribution function corresponding to the a_i and b_i is an estimator of the distribution function of A and B, respectively.

If these methods are applied to figures that are only approximately radial-rhombi then the quality of the approximation must be characterized. This can be done, for example, using the mean contour covariance function $E\chi_R(\varphi)$. The theoretical values corresponding to radial-rhombi,

$$
\begin{aligned}
E\chi_R(\varphi) = {}& \tfrac{1}{4}(\sigma_A^2 + \sigma_B^2 + 2\varrho_{AB}\sigma_A\sigma_B) \\
& + \tfrac{1}{12}[(m_A - m_B)^2 + \sigma_A^2 + \sigma_B^2 \\
& - 2\varrho_{AB}\sigma_A\sigma_B] \left[1 - 6\left(\tfrac{2\varphi}{\pi}\right)^2 + 4\left(\tfrac{2\varphi}{\pi}\right)^3 \right] \quad (0 \leq \varphi \leq \tfrac{1}{2}\pi) \\
E\chi_R(\varphi) = {}& E\chi_R(\pi - \varphi) \quad (\tfrac{1}{2}\pi \leq \varphi < \pi),
\end{aligned}
$$

can be compared with the statistically obtained values

$$
\overline{\chi_R(\varphi)} = \frac{1}{n} \sum_{i=1}^{n} \int_0^{2\pi} [R_i(x) - \overline{\overline{R}}][R_i(x + \varphi) - \overline{\overline{R}}] \, dx.
$$

The degree of concurrence of both functions characterizes the goodness-of-fit of the radial-rhombus model.

CHAPTER 8

Set Theoretic Analysis

8.1 INTRODUCTION

The methods of form analysis considered in this section are based on analysis of the objects as figures, i.e. as subsets of R^2. For these sets, geometrical characteristics are considered. Since the description of form is the aim, those characteristics are considered that are independent of the position and orientation of the figures. Examples are area and perimeter or the distribution of chord lengths formed by random lines.

Ideas from the theory of random sets are used to obtain means and variances. The results desired determine whether or not the figures are normalized and homologized. The same is true for a smoothing before the analysis.

The use of image analysers is recommended; without such devices even area measurement is barely feasible for large samples of figures.

8.2 SIMPLE GEOMETRICAL SHAPE RATIOS

There are many shape ratios, often describing in a very intuitive manner certain geometrical properties (or complexes of properties). Experience shows that if the aim is discrimination and classification, the use of shape ratio is frequently equivalent to or more effective than more complicated methods. It is useful to consider those quantities which are somehow related to the actual problem; for example, the contour should be considered if relations between particles and their neighbourhood are of interest.

Area-perimeter ratio

$$f_{AU}(X) = \frac{4\pi A(X)}{U(X)^2},$$
(8.1)

where $A(X)$ is the area of the figure X and $U(X)$ its perimeter. This shape ratio characterizes, for example, deviations from circular form. For any disc $f_{AU} = 1$, while for all other figures $f_{AU} < 1$ (this is a consequence of the so-called isoperimetric inequality; Hadwiger, 1958). The smaller f_{AU} is, the greater is the deviation from circular shape. For example, an ellipse with ratio $\alpha = a : b\ (> 1)$ of semi-axis lengths satisfies

$$f_{AU}(X) = \frac{4}{\left(1.5(1+\alpha)/\sqrt{\alpha} - 1\right)^2};$$

f_{AU} decreases monotonical with α.

Schmidt-Kittler (1986) describes similar parameters for objects given only as fragments.

Note that shape ratios that use the perimeter are of dubious value in the case of very rough contours. Sometimes pre-analysis smoothing may be useful. Bandemer *et al.* (1989) have studied a fuzzy version of f_{AU}, which may be useful when only a vague definition of the contour is possible.

Circularity shape ratio (Wadell, 1935)

$$f_K(X) = \frac{\text{diameter of the circle with area } A(X)}{L(X)}, \qquad (8.2)$$

where $L(X)$ is the maximum distance of a pair of contour points of X. For connected X, $L(X)$ is the same as the so-called Feret diameter, the maximum length of the orthogonal projection of X on a line. f_K is also equal to one for a disc, and otherwise is less than one. Similar shape ratios are obtained using in-circle radius or circum-circle radius. (The in-circle radius, the radius of the largest circle in X, can be obtained by means of an image analyser, applying the erosion operation with discs. A method for the determination of the circum-circle radius has been suggested by Jourlin and Laget (1988).)

Symmetry factor of Blaschke

$$f_s(X) = 1 - \frac{A(X)}{A(S(X))}. \qquad (8.3)$$

Here $S(X)$ denotes the so-called 'central symmetrization' of X:

$$S(X) = \tfrac{1}{2}(X \oplus \check{X}),$$

where $\check{X}$ is the set reflected at the origin, $\check{X} = \{x : -x \in X\}$. For convex X, Rademacher's inequality (Burago and Zalgaller, 1988) gives

$$A(X) \le A(S(X)) \le \tfrac{3}{2}A(X),$$

or

$$\tfrac{1}{3} \le f_s(X) \le 1.$$

For symmetric convex sets $f_s = 1$, while for triangles it is $\tfrac{1}{3}$ (Moreau and Rubio, 1987; Jourlin and Laget 1988). These authors describe in detail the determination of $f_s(X)$ by means of image analysers.

Further symmetry characteristics are discussed in Jourlin and Laget (1988). Matheron and Serra (1988) have studied a relation for convex sets that gives an ordering with respect to their degree of asymmetry. They coined the term 'totally symmetric'.

Convexity ratio

$$f_C(X) = \frac{A(X)}{A(\text{conv } X)},\tag{8.4}$$

where $A(\text{conv } X)$ is the area of the convex hull of X. This shape ratio characterizes deviations from convexity. Clearly a convex figure X has the convexity factor $f_C(X) = 1$, while in all other cases $f_C(X) < 1$.

Another shape ratio, also characterizing deviations from convexity, is given by

$$f_{\text{costar}}(X) = \frac{A(X) - A_{\text{st}}(X)}{-A(X)};\tag{8.5}$$

Here $A_{\text{st}}(X)$ is the mean of the area of the 'star' in X seen from a random point in X (Fig. 42). For convex sets $f_{\text{costar}}(X) = 0$; otherwise $f_{\text{costar}}(X) > 0$.

See Wegmann (1980) for more non-convexity characteristics.

Roundness factor (Wadell, 1935)

$$f_R(X) = \frac{n^{-1}\displaystyle\sum_{i=1}^{n} r_i(X)}{\text{radius of the in-circle of } X}.\tag{8.6}$$

It is assumed that the contour of X contains n points of large curvature $P_1, \ldots, P_n$, where the radius of curvature at P_i is $r_i(X)$, (Fig. 43). (Here n is a quantity that depends on the shape of X.) For a disc $f_R(X) = 1$; otherwise $f_R(X) < 1$. This shape ratio, which was for a long time not easy to obtain in practice and which appears somewhat vague, is used successfully in petrology (Pettijohn, 1975). For example, Krumbein (1941) used it to describe the relationship between the degree of abrasion of stones in a tumbling barrel and the duration of the abrasion process (Fig. 49). Pirard (1992, 1994a,b) shows how $f_r(X)$ can be measured by means of an image analyser.

The measurement of curvatures is also discussed in detail in Russ (1989).

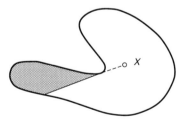

Figure 42 Definition of the 'star' of a figure. The non-shaded part is visible from x and forms the star with respect to x.

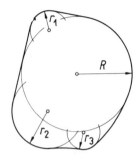

Figure 43 Definition of 'roundness ratio'. For the given figure radii of curvature have to be determined.

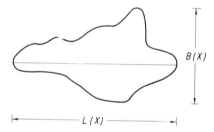

Figure 44 Definition of 'elongation ratio' — maximum Feret diameter $L(X)$ and corresponding breadth $B(X)$.

Length ratios. For many biological objects length ratios are valuable shape parameters. Mainly landmark distances and their ratios are considered (see e.g. Mosimann, 1970). Applications are also given in Schmidt-Kittler (1984) and §10.2.

Elongation factor

$$f_l(X) = \frac{B(X)}{L(X)}, \tag{8.7}$$

where $B(X)$ is the breadth of X measured orthogonally to the line $L(X)$ connecting the two most extreme points on X (Fig. 44).

Further shape ratios characterize figures in comparison to certain geometrical figures.

Ellipse ratio

$$f_{\mathrm{ell}}(X) = \frac{a_e(X)}{b_e(X)}, \tag{8.8}$$

where $a_e(X)$ and $b_e(X)$ are the lengths of the long and short semi-axes respectively of the ellipse of area $A(X)$ and perimeter $U(X)$. They are determined by the $A{-}U$ method, i.e. by (7.30) and (7.31).

Radial-rhombus ratio

$$f_{\text{rr}}(X) = \frac{a_r(X)}{b_r(X)}, \tag{8.9}$$

where $a_r(X)$ and $b_r(X)$ are the lengths of the long and short semi-axes respectively of the radial-rhombus of area $A(X)$ and mean radius vector $\bar{r}(X)$. They are calculated using (7.32) and (7.33).

A further shape ratio is the ratio of side lengths of the smallest rectangle containing the figure; sometimes the sides are taken parallel to the axes of a given coordinate system. Worth reading for general discussions of shape indices are Exner (1987) and Pirard (1994a,b).

8.3 CHARACTERISTICS OF RANDOM COMPACT SETS

8.3.1 Introduction

Under certain conditions, ideas and results of the theory of random sets can be used in form statistics. This is possible if the position of the figures is given. Frequently this situation can be generated in a natural manner by introducing a 'natural' coordinate system connected with the landmarks, (§10.2). This situation also holds with spreading processes. For example, Vorob'ev (1984) has studied those figures X that describe the area during a forest fire that is either burning or in ashes at a given instant. Cressie (1984) has studied the growth of tumours using the starting point of the tumour as the origin of the coordinate system.

Of particular importance for applications are the covering function and various means and medians. Furthermore, sets can be described by functions independently of their position and orientation in the plane. Examples are the set covariance function, the erosion function and the chord length distribution function.

8.3.2 Random compact sets

Let $\mathcal{K}$ be the set of all compact subsets of R^2. On $\mathcal{K}$ the Hausdorff metric h is defined by

$$h(K_1, K_2) = \inf\{\epsilon > 0 : K_1 \subseteq K_2 \oplus b(0, \epsilon), k_2 \subseteq K_1 \oplus b(0, \epsilon)\}$$

(Appendix E). Under the metric h, $\mathcal{K}$ is a complete separable metric space. The corresponding open subsets generate a σ-algebra, the Borel σ-algebra $\mathcal{B}_K$ of $\mathcal{K}$.

A random compact set is a measurable function from a probability space $[\Omega, \mathcal{A}, \text{Pr}]$ into $[K, \mathcal{B}_K]$. Random compact sets in this sense are also random closed sets as in Matheron (1975). Consequently their distribution is given by the probabilities

$$\text{Pr}(X \cap K = \emptyset) \quad (K \in \mathcal{K}).$$

In passing, it should be noted that the distribution of a random compact convex set is also given by the system of all inclusion probabilities $\Pr(X \subset K)$ (Vitale, 1983; Cressie and Laslett, 1987).

For $K = \{x\}$ the probability $\Pr(x \in X)$ is obtained, which satisfies

$$\Pr(x \in X) = 1 - \Pr(x \notin X).$$

Thus the *covering function* $P_X(x)$ is given by

$$p_X(x) = \Pr(x \in X) \quad (x \in R^2). \tag{8.10}$$

Of course, $p_X(x)$ can also be interpreted as the mean of the indicator function $1_X(x)$:

$$p_X(x) = \mathsf{E}1_X(x).$$

The covering function takes values between 0 and 1. The set b_x of all $x \in R^2$ with $p_X(x) > 0$ is called the support of X. The set K_t of all $x \in R^2$ with $p_X(x) = 1$ is called the kernel, the set of fixed points, or essential minimum $e(X)$. If $X_1, X_2, \ldots,$ is a sequence of i.i.d. random compact sets, then almost surely

$$\bigcap_{i=1}^{\infty} X_i = e(X)$$

and $\bigcap_{i=1}^{n} X_i$ converges almost surely to $e(X)$.

8.3.3   Mean value formulae for compact convex sets

Let Y and Z be two independent random compact convex sets with finite mean areas $\mathsf{E}A(Y)$ and $\mathsf{E}A(Z)$ and perimeters $\mathsf{E}U(Y)$ and $\mathsf{E}U(Z)$. Furthermore let Z be isotropic;[†] then for the set $X = Y \oplus Z$ the following relations hold:

$$\mathsf{E}A(X) = \mathsf{E}A(Y) + \frac{1}{2\pi}\mathsf{E}U(Y)\mathsf{E}U(Z) + \mathsf{E}A(Z) \tag{8.11}$$

and

$$\mathsf{E}U(X) = \mathsf{E}U(Y) + \mathsf{E}U(Z). \tag{8.12}$$

These formulae are generalizations of the famous Steiner formula (Matheron, 1975, p. 85).

8.3.4   Means of random compact sets

In the literature one may find several attempts at defining the mean of random compact sets. Here those variants which are particularly important for applications are discussed.

[†]This means that Z and any random set obtained by rotation around the origin have the same distribution.

Aumann's definition. This is closely connected with convexity and support functions, and thus is not very attractive for non-convex sets. A *selection point* ξ of a random compact set X is a random variable on the probability space $[\Omega, \mathcal{A}, \mathrm{Pr}]$ with values in R^2, i.e. a random vector or a random point, for which almost certainly $\xi \in X$.

Example. Let X be a random line segment with endpoints $u(X)$ and $o(X)$. Then $u(X)$, $o(X)$ and the centre of the line segment are selection points.

The *Aumann mean* $\mathsf{E}_A X$ of a random compact set X is the set of means of all selection points:

$$\mathsf{E}_A X = \{\mathsf{E}\xi : \xi \text{ is a selection point and } \mathsf{E}\xi \in R^2\} \qquad (8.13)$$

(Aumann, 1965).

With respect to compactness and convexity of $\mathsf{E}_A X$, the following can be said. The *norm* $\|X\|$ of a random compact set is that random variable which yields the distance of X to the singleton $\{o\}$ with respect to the Hausdorff metric:

$$\|X\| = h(\{o\}, X).$$

The set $\mathsf{E}_A X$ is compact if and only if $\mathsf{E}\|X\| < \infty$.

If the distribution of X is free of atoms (i.e. there is no compact set X_1 with $\mathrm{Pr}(X = X_1) > 0$) then $\mathsf{E}_A X$ is even convex (Richter, 1963). The same is true if the σ-algebra A of the probability space $[\Omega, \mathcal{A}, \mathrm{Pr}]$ is 'sufficiently rich'. This means that for any λ with $0 < \lambda < 1$ and any pair of selection points ξ_1 and ξ_2 there is a random event A_λ with $\mathrm{Pr}(A_\lambda) = \lambda$ that is independent of X, ξ_1 and ξ_2 (Kruse and Meyer, 1987). In this case the following random variables are also selection points

$$\eta_\lambda = \begin{cases} \xi_1(\omega) & (\omega \in A_\lambda) \\ \xi_2(\omega) & (\omega \in A_\lambda^c) \end{cases} \quad (0 < \lambda < 1).$$

Clearly, the mean of η_λ satisfies

$$\mathsf{E}\eta_\lambda = \mathrm{Pr}(A_\lambda)\mathsf{E}\xi_1 + \mathrm{Pr}(A_\lambda^c)\mathsf{E}\xi_2 = \lambda\mathsf{E}\xi_1 + (1 - \lambda)\mathsf{E}\xi_2.$$

Thus for two arbitrary points of $\mathsf{E}_A X$ the whole connecting line lies completely in $\mathsf{E}_A X$.

Because of these properties, only compact convex sets X are considered from now on. It is natural then to work with support functions (§7.23). Since it is possible to embed the set of support functions of convex sets isometrically into the Banach space $C[0, 2\pi]$ of continuous functions on $[0, 2\pi]$, the theory of random variables on Banach spaces and, in particular, the corresponding limit theorems can be used (Araujo and Gine, 1980). The mean may be defined by means of the so-called Bochner integral. The support functions of X and $\mathsf{E}_A X$ satisfy

$$s_{\mathsf{E}_A X}(\varphi) = \mathsf{E}s_X(\varphi) \quad (0 \le \varphi \le 2\pi). \qquad (8.14)$$

This offers a simple method for the determination of $E_A X$. For all φ the mean of $s_X(\varphi)$ is determined, and then $E_A X$ is that convex set whose support function is $E s_X(\varphi)$.

If X is discrete, i.e. if there are convex sets $K_1, K_2, \ldots$ with $\Pr(X = K_i) = p_i$ and $\sum_{(i)} p_i = 1$, then

$$E_A X = p_1 K_1 \oplus p_2 K_2 \oplus \cdots \tag{8.15}$$

Matheron (1975, Chapter 9) has given a more general integral formula.

If X is isotropic then $E_A X$ is a disc centred at o. That is, in the isotropic case the random variable $s_X(\varphi)$ has the same distribution for all φ, and $E s_X(\varphi)$ is constant.

Example. Let X be a line segment of random length $l(X)$ centred at o. Let the orientation of X be uniform. Clearly, X is isotropic. The Aumann mean $E_A X$ is the disc $b(o, r)$ with

$$r = \frac{1}{\pi} E l(X).$$

Certainly many shape statisticians would not like this result. Instead, they would rather measure the lengths of line segments and then formulate their result as

'shape: line segment, size = mean length: $E l(X)$'.

Note that variances connected with the Aumann mean may also yield form information (§8.3.5).

Two geometrical statements about $E_A X$ are possible. If X is convex then the perimeters satisfy

$$U(E_A X) = E U(X) \tag{8.16}$$

In general, the mean areas satisfy the inequality

$$\sqrt{A(E_A X)} \geq E \sqrt{A(X)} \tag{8.17}$$

(Vitale, 1987a).

The first statement is a simple consequence of a well-known theorem of Cauchy (that the perimeter of a planar convex set can be expressed in terms of the mean breadth (Santalo, 1976, p. 3)), the relationship between width and support functions, and (8.14). By the way, it is

$$E_A(a_1 X_1 \oplus \cdots \oplus a_n X_n) = a_1 E_A X_1 \oplus \cdots \oplus a_n E_A X_n$$

for real a_1 and random compact X_i.

Artstein and Vitale (1975) proved a law of large numbers for compact sets. Let $X_1, X_2, \ldots$ be a sequence of independent identically distributed random compact sets with $E\|X_1\| < \infty$. Then, almost certainly,

$$\frac{1}{n}(X_1 \oplus \cdots \oplus X_n) \to E_A(\text{conv } X_1)$$

in the sense of convergence with respect to the Hausdorff metric. The case of dependent X_i was studied by Schürger (1983). Weil (1982a) proved a central limit theorem for random compact sets (convergence to 'Gaussian' random sets).

Vitale (1990) suggested a modification of Aumann's definition, which yields non-convex means as well. Let $E_{red}X$ denote the Vitale mean of a random compact set X. For applications it may be sufficient to say that in the case of a set X with a discrete distribution

$$E_{red}X = p_1 X_1 \oplus \cdots \oplus p_n X_n, \tag{8.18}$$

analogously to (8.15), but the X_i do not need to be convex. In general,

$$E_{red}X \subseteq E_A X \tag{8.19}$$

The definition of $E_{red}X$ is based on selection points as in $E_A X$. However, these points are not defined as maps from $[\Omega, \mathcal{A}, \mathrm{Pr}]$ to R^2. Rather, they are mappings of the 'canonical' probability space $[K, B_K, P_K]$ to R^2. Here P_X is the distribution of X on the Borel σ-algebra B_X. The points are called K-selection points. The definition of the mean is then

$$E_{red}X = \{E\xi : \xi \text{ is a } K\text{-selection point and } E\xi \in R^2\}. \tag{8.20}$$

For a discrete distribution of X, the notion of K-selection point ξ is defined as follows. For every set X_i an element ξ_i is chosen. If $X(\omega) = X_i$ then $\xi(\omega) = \xi_i$.

If the samples of X form a continuous one-parameter family then $E_{red}X$ is given by the so-called Stieltjes–Minkowski integral (Matheron, 1975, Chapter 9).

Another generalization of the Aumann expectation is given in Molchanon (1993a). It is determined by the family of functions on the setting space. The Aumann expectation appears then as a particular case when all these functions are linear.

Radius-vector mean. Let the random compact set be star-shaped and contain the origin with probability one. Then the radius-vector function $r_X(\varphi)$ can be used to describe X. Let $Er_X(\varphi) < C < \infty$ for all φ. The radius-vector mean of X is the compact set $E_r X$ that has the radius-vector function

$$r(\varphi) = Er_x(\varphi) \quad (0 \le \varphi \le 2\pi).$$

Naturally, it is not necessarily convex.

As in the case of the Aumann mean, $E_r X$ is a disc when X is isotropic. In general,

$$E_r X \subseteq E_{red}X \subseteq E_A X. \tag{8.21}$$

The difference between $E_A X$ and $E_r X$ can be quite considerable. The mean area satisfies the inequality

$$EA(X) \ge A(E_r X).$$

Fréchet's definition. Compact sets can be considered as elements of the metric space K with Hausdorff metric h. For particular classes of sets other metrics are also used (§9.4). Therefore it is interesting for form statistics that there is a general theory of means in metric spaces (Ziezold, 1977, 1989).

Let $\mathcal{X}$ be an abstract set and d a metric on $\mathcal{X}$. Furthermore, let X be a random variable with values in $\mathcal{X}$. Suppose that

$$\mathsf{E}d(X, a)^2 < \infty$$

for any element a of $\mathcal{X}$. Then any element m of $\mathcal{X}$ such that

$$\mathsf{E}d(X, m)^2 = \inf_{a \in \mathcal{X}} \mathsf{E}d(X, a)^2 \tag{8.22}$$

is called a *mean value element* of X (Fréchet, 1948). The set $\mathcal{E}_F X$ of all mean value elements of X is called *Fréchet mean* of X. In particular, if $\mathcal{X} = \mathcal{K}$ and $d = h$, each of these elements is a compact set.

In most cases the determination of mean value elements is a rather complicated optimization problem. Unfortunately, this is true also of the case $\mathcal{X} = \mathcal{K}$ and $d = h$.

Fréchet's definition is not satisfactory for shape statistics if the position and orientation of the figures in space is uninteresting and directly congruent figures cannot be distinguished (two sets are directly congruent if they can be transformed into each other by translation and rotation). In such cases a quotient space $\mathcal{K}_C$ with the corresponding quotient metric h_C is considered instead of $\mathcal{K}$. All figures that are directly congruent to a given figure form a class. The set of all these classes is $\mathcal{K}_C$. On $\mathcal{K}_C$ a metric h_C is given by

$$h_C(A_C, B_C) = \inf_{S, T \in \mathcal{M}} \{h(TA, SB)\} \quad (A_C, B_C \in \mathcal{K}_C),$$

where A and B are arbitrary elements of A_C and B_C respectively. Here $\mathcal{M}$ is the set of all proper Euclidean motions (translation and rotations) in the plane.

A definition of mean for classes of figures is then possible analogously to (8.22) with $d = h_C$. The practical determination of such mean value elements is even more complicated than that discussed above.

Covering function. A natural way of defining a mean for random compact sets is the use of the covering function as a mean. In this case the mean is not a set but a function taking values between 0 and 1. Perhaps one can consider $p_X(x)$ as the membership function of a fuzzy set $\mathsf{E}_f X$ (Dubois and Prade, 1980; Kruse and Meyer, 1987). In terms of the theory of fuzzy sets, $p_X(x)$ expresses the degree of membership of x in $\mathsf{E}_f X$. The mean area of X satisfies

$$\mathsf{E}A(X) = \int_{R^2} p_X(x) \, \mathrm{d}x. \tag{8.23}$$

If X is isotropic then $p_X(x)$ depends only on the distance $r = \|x\|$ of X from the origin o. The symbol p_X is used for the corresponding function $p_X(r)$:

$$p_X(r) = p_X(x) \quad (\|x\| = r).$$

The mean area can be written in the form

$$\mathsf{E}A(X) = 2\pi \int_0^\infty r p_X(r) \, \mathrm{d}r. \tag{8.24}$$

Figures 52 and 55 give examples of $p_X(r)$, and Fig. 74 presents a statistically determined covering function.

Vorob'ev's definition. Let X be a random compact set with $0 < \mathsf{E}A(X) < \infty$ and let there exist a positive p with the property that the compact set

$$S_p = \{x \in R^2 : p_X(x) \geq p\} \tag{8.25}$$

has area $\mathsf{E}A(X)$. (If p is not unique then take the infimum of all such p.) The *Vorob'ev mean* $\mathsf{E}_V X$ of X is then the set S_p (Vorob'ev, 1984). (While Vorob'ev considered sets with finitely many points, e.g. sets of pixels, which are of interest in image analysis, subsets of R^2 with positive area are considered here.) Vorob'ev's definition is not so artificial as it perhaps looks. The set $\mathsf{E}_V X$ satisfies the inequality

$$\mathsf{E}A(X \triangle \mathsf{E}_V(X)) \leq \mathsf{E}A(X \triangle B) \tag{8.26}$$

for all Borel sets B with $A(B) = \mathsf{E}A(X)$ (Stoyan, 1989b). Here $\triangle$ denotes the symmetric difference operator:

$$C \triangle D = C \backslash D \cup D \backslash C.$$

In this sense $\mathsf{E}_V X$ is that set of area $\mathsf{E}A(X)$ that is best fitted to the set X.

8.3.5   Variances of random compact sets

Until now there has been no generally accepted definition of the variance of random compact sets. (For the particular case of sets in R^1, Kruse (1987) suggested a variance definition that unfortunately has some disadvantages.)

It seems to be useful to proceed as in the case of random variables. For a random variable ξ the variance $\mathrm{var}\,\xi$ is defined by

$$\mathrm{var}\,\xi = \mathsf{E}(\xi - \mathsf{E}\xi)^2,$$

and it is well known that $\mathsf{E}\xi$ is the solution of the following optimization problem:

'Determine x such that $\mathsf{E}(\xi - x)^2$ is minimal'.

In the case of convex sets it is natural to use the support function and to interpret it as an element of the Banach space $C[0, 2\pi]$. Thus the following definitions make sense.

The *inf-variance* of a random compact convex set X is the quantity

$$\text{var}_I X = \inf_{K \in \mathcal{C}(\mathcal{K})} \|s_X - s_K\|2, \tag{8.27}$$

where s_K and s_X are the support functions of K and X and $\| \cdot \|$ is the supremum norm: $\mathcal{C}(\mathcal{K})$ is the set of all compact convex subsets of R^2. The quantity

$$\text{var}_I X = \mathsf{E}\|s_X - s_{\mathsf{E}X}\|^2 \tag{8.28}$$

is called the E-variance; $s_{\mathsf{E}_A X}$ is the support function of $\mathsf{E}_A X$.

It is easy to see that these variances may differ; the situation in the case of random sets is thus more complicated than for random variables.

Clearly,

$$\text{var}_I X \le \text{var}_\mathsf{E} X. \tag{8.29}$$

The E-variance is perhaps easier to calculate. Näther and Albrecht (1990) determined it for a random closed interval on the real axis having the form

$$X = [\xi - \delta, \ \xi + \delta],$$

where ξ and δ are independent random variables with finite variances. They give

$$\text{var}_\mathsf{E} X = \sigma_\xi^2 + \sigma_\delta^2 + 2\mathsf{E}|\xi|\mathsf{E}|\delta|. \tag{8.30}$$

In the sense of the theory of random variables with values in Banach spaces, var_I and var_E could be interpreted as 'strong' variances. 'Weak' variances are defined in terms of linear functionals. The following functions are of interest in the case of convex sets and when applying support functions — the variance function

$$\text{var}\, s_X(\varphi) = \mathsf{E}(s_X(\varphi) - \mathsf{E}s_X(\varphi))^2,$$

(i.e. the variance of the support function value at φ), for the first and the covariance function

$$K(\varphi, \psi) = \mathsf{E}(s_X(\varphi) - \mathsf{E}s_X(\varphi))(s_X(\psi) - \mathsf{E}s_X(\psi)).$$

for the second, If X is isotropic then $\text{var}\, s_X(\varphi)$ is constant, and $K(\varphi, \psi)$ depends only on the difference $\varphi - \psi$.

Of course, variances can be also defined starting from the covering function or the radius vector function (Stoyan 1989b).

The following variance corresponds naturally to the Fréchet mean:

$$\text{var}_F X = \inf_{a \in \mathcal{X}} \mathsf{E}d(X, a).$$

In the case of $\mathcal{X} = \mathcal{K}$ and $d = h$ this is identical to $\text{var}_I X$.

8.3.6 Medians of random compact sets

It is not difficult to define medians of random sets. Until now this notion appeared only (implicitly) in Vorob'ev (1984). The Vorob'ev median of a random compact set is defined by

$$\tilde{X} = \left\{ x \in R^2 : p_X(X) \geq \tfrac{1}{2} \right\}. \tag{8.31}$$

The name 'median' can be justified as follows.

(a) The median $m(x)$ of the value $1_X(x)$ of the indicator function of X at x is equal to one if $p_X(x) \geq \tfrac{1}{2}$ and zero otherwise. Obviously, $\tilde{X}$ is that set which has $m(x)$ as its indicator function.

(b) It is well known that the median of a random variable ξ minimizes the quantity $\mathsf{E}|\xi - a|$; $-\infty < a < \infty$. The set X has a similar property:

$$\mathsf{E}A(X \Delta \tilde{X}) \leq \mathsf{E}A(X \Delta B) \tag{8.32}$$

for every Borel set B (see Stoyan, 1989b). Here, as is (8.26), Δ denotes the symmetric difference operator.

Further median definitions are possible. For example, analogously to Aumann's mean definition a median could be defined by selection points taking the geometrical median instead of E. (The geometrical median can be defined as that point of R^2 which minimizes $\mathsf{E}\|\xi - a\|$ (Small, 1990).) Molchanov (1990) has also defined quantiles of random compact sets. A q-quantile of X is

$$X_q = \{ x \in R^2 : p(x) \geq 1 - q \} \qquad (0 < q < 1).$$

8.3.7 Some simple statistical methods

Estimation of the Aumann mean. Let $X_1, \ldots, X_n$ be a sample of compact convex sets. The X_i are assumed to be independent and identically distributed. If they are not convex then their convex hulls are taken. The sample mean

$$\tilde{X} = \frac{1}{n}(X_1 \oplus \cdots \oplus X_n) \tag{8.33}$$

is an unbiased and consistent estimator of $\mathsf{E}_A X$. If Vitale's mean is employed, then the same estimator may be used, but the X_i need not be convex.

Estimation of the covering function. Let $X_1, \ldots X_n$ be a sample as above. The function

$$\hat{p}_X(x) = \frac{1}{n} \sum_{i=1}^{n} {}^1X_i(x) \quad (x \in R^2). \tag{8.34}$$

Similarly, quantiles and probabilities $\Pr(X \cap K \neq \emptyset)$ can be estimated for compact K (Molchanov 1989, 1990).

Estimation of mean and median in the sense of Vorob'ev.

Estimators of $E_V X$ and $\tilde{X}$ can be obtained using estimators of the covering function. $\tilde{\tilde{X}}$ is an estimator of the median, where

$$\tilde{\tilde{X}} = \left\{ x \in R^2 : \hat{p}_X(x) \geq \tfrac{1}{2} \right\}. \tag{8.35}$$

Figure 74 shows an estimate obtained in this way.

Estimating $E_V X$ is more complicated. First the mean area of X has to be estimated, for example by $\overline{A(X)}$, the arithmetic mean of the areas of the X_i. Then that p has to be determined for which the set

$$\hat{S}_p = \{x \in R^2 : \hat{p}_X(x) \geq p\}$$

has area $\overline{A(X)}$. This set is an estimator of $E_V X$.

Estimation of mean and variance in the sense of Fréchet.

Estimators M and S^2 of $\mathcal{E}_F X$ and $\mathrm{var}_F X$ can be obtained as follows. (They are given here for a general metric space $\mathcal{X}$ with metric d.) The mean estimator M is an element of $\mathcal{X}$ with

$$\sum_{i=1}^n d(X_i, M)^2 = \inf_{a \in \mathcal{X}} \sum_{i=1}^n d(X_i, a)^2,$$

and S^2 is given by

$$S^2 = \frac{1}{n} \sum_{i=1}^n d(X_i, M)^2.$$

Both estimators are consistent in a certain sense (Ziezold, 1989).

8.4 FOUR FUNCTIONS FOR DESCRIPTION OF FIGURES

8.4.1 Introduction

As an alternative to the contour functions treated in Chapter 7, there are four further functions that describe planar figures in a quite different manner, of particular interest are the chord length distribution function $L(l)$ and the spherical erosion function $Q_s(r)$. All functions are independent of the positions of the figures in the (x, y)-plane, thus they coincide for congruent figures. It is not necessary to define 'centres' in the figures. That is why these functions are well suited to form statistics.

Both the spherical erosion function $Q_s(r)$ and the chord length distribution function $L(l)$ may be easily determined using image analysers. The determination of $L(l)$ has a long tradition of application to collections of figures forming structures in metal and ceramics. The series of formulae for chord length distribution functions may be of use in finding suitable models when empirical distribution functions are given. They also have an independent interest.

8.4.2   Chord length distribution functions

Random lines (Appendix H) generate chords of random length in convex sets. The corresponding distribution function is called the *chord length distribution function* and is denoted by $L(l)$. (Note that there are many other ways of defining random chords; see Solomon (1978) and Coleman (1989) and literature on the so-called Bertrand paradox.)

In the case of deterministic compact convex sets many formulae are to be found in the literature. Some will be given here. The mean chord length $\mathsf{E}\ell$ is given by

$$\mathsf{E}\ell = \frac{\pi A(X)}{U(X)}. \tag{8.36}$$

The third moment of the chord length is

$$\mathsf{E}\ell^3 = \frac{3A(X)^2}{U(X)} \tag{8.37}$$

(Santalo, 1976). For the second moment of the chord length there exists no general formula comparable to (8.36) or (8.37). Voss (1982) has given formulae for several figures X; for example,

$$\mathsf{E}\ell^2 \cdot U(X) = \tfrac{16}{3}\pi R^2$$

for a disc of radius R and

$$\mathsf{E}\ell^2 \cdot U(X) = \tfrac{4}{3}a\left(\sqrt{2} + \log\sqrt{3 + 2\sqrt{2}}\right)$$

for a square of side length a.

The chord length distribution function is known for many figures. The following formulae give the corresponding density functions $l(\ell)$:

(a) X = a disc of radius R,
(b) X = an ellipse with semiaxis lengths a and b $(a \geq b)$;
(c) X = a rectangle with side lengths a and b $(a \geq b)$;
(d) X = an equilateral triangle of side length a.

(a) Disc

$$l(\ell) = \frac{\ell}{2R\sqrt{4R^2 - \ell^2}} \quad (0 \leq \ell \leq 2R). \tag{8.38}$$

(b) Ellipse

$$l(\ell) = 16ab\ell\mathsf{E}\ell \int_g^{2a} \frac{1}{x^3\sqrt{x^2 - 4b^2}\sqrt{x^2 - \ell^2}\sqrt{4a^2 - x^2}}\, dx,$$

where $g = \max\{\ell, 2b\}$. This integral can be determined numerically, but this is not very easy. Figure 45 shows chord length density functions for two ellipses and a disc, and Fig. 46 those for a square and rectangles.

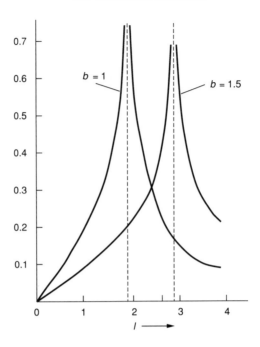

Figure 45 Chord length distributions for ellipses with $a = 2$, $b = 1$ and with $a = 2$, $b = 1.5$.

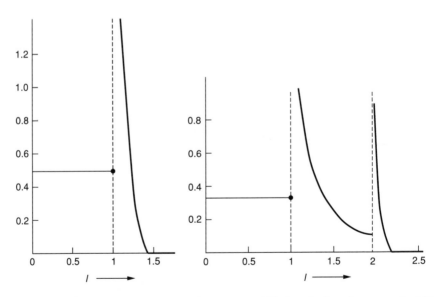

Figure 46 Chord length distributions for a square (side length 1) and a rectangle (side lengths 1 and 2).

(c) Rectangle (Gille, 1988)

$$
l(\ell) = \begin{cases}
\dfrac{1}{a+b} & (0 \le \ell \le b), \\[2ex]
\dfrac{ab^2}{\ell^2(a+b)\sqrt{\ell^2-b^2}} & (b < \ell \le a), \\[2ex]
\dfrac{ab}{(a+b)\ell^2}\left(\dfrac{a}{\sqrt{\ell^2-a^2}}+\dfrac{b}{\sqrt{\ell^2-b^2}}\right) - \dfrac{1}{a+b} & (a < \ell \le \sqrt{a^2+b^2}).
\end{cases}
$$

(d) Equilateral triangle (Sulanke, 1961)

$$
l(\ell) = \begin{cases}
\dfrac{\frac{1}{2}+\frac{\pi}{3}\sqrt{\frac{1}{3}}}{a} & (0 \le \ell \le \tfrac{1}{2}\sqrt{3}, \\[2ex]
-\dfrac{1}{\ell}\sqrt{1-\dfrac{3a^2}{4\ell^2}}+\dfrac{1}{a}\left[\dfrac{1}{2}-\dfrac{2\pi}{3\sqrt{3}}+\dfrac{2}{\sqrt{3}}\arcsin\left(\dfrac{\sqrt{3}a}{2\ell}\right)\right] & (\tfrac{1}{2}\sqrt{3}a < \ell \le a).
\end{cases}
$$

The form of the chord length density function is related to certain features of the corresponding figure. For example, poles of this function are related to parallel pieces of the contour and the form of $l(\ell)$ for l close to its maximum is essentially related to smaller details of the contour.

For examples of cases where $l(\ell)$ does not characterize the form of convex figures uniquely see Figure 1 in Mallows and Clarke (1970), but see also Gates (1982a,b, 1987), Waksman (1985) and Nagel (1993).

In the case of a random convex figure X the chord length distribution function $L(\ell)$ is the perimeter-weighted mean of the chord length distribution functions of the realizations. If, for example, X is a disc of random radius R then

$$
L(\ell) = \frac{1}{\mathsf{E}R}\int_0^\infty r L_r(\ell)\, \mathrm{d}F_R(r),
$$

where $F_R(r)$ is the distribution function of R and $L_r(1)$ is the chord length distribution function of a disc of radius r. (The perimeter weighting is here replaced by a radius weighting.)

More generally, let P_X be the distribution of the random compact convex set X. Then

$$
L(\ell) = \frac{1}{\mathsf{E}U(X)}\int U(\xi)L_\xi(\ell)P_X(\mathrm{d}\xi), \tag{8.39}
$$

where $L_\xi(\ell)$ is the chord length distribution function of the realization ξ of X and $U(\xi)$ is the corresponding perimeter.

Perimeter-weighted averaging is based on the following idea. Let there be a homogeneous and isotropic structure described by a germ–grain model (Appendix J). This means that the structure consists of particles whose centres form a homogeneous and isotropic point field. Particles shifted so that their centres are at the origin form random sets with the same distribution P_X. The structure is intersected by a line (e.g. the x-axis or a random line). The 'number distribution' of the chords obtained is the chord length distribution corresponding to the structure or the germ–grain model. (All chords are included in the number distribution with the same weight. A corresponding sample is obtained by taking n subsequent chords on the line of intersection. Another form of weighting is where long chords have a greater weight and have more influence on the mean chord length etc.) Since the probability of intersection of a convex set with a random line is proportional to its perimeter, the chord length distribution function corresponding to the number distribution coincides with the distribution function $L(\ell)$ given by (8.39).

There are close relations between the chord length distribution function and other functions describing random compact sets. Particularly important is the relation to the linear erosion function

$$Q_\ell(r) = \frac{1}{\mathsf{E}\ell} \int_0^r [1 - L(\ell)] \, d\ell, \tag{8.40}$$

where $\mathsf{E}\ell$ is the mean chord length (Stoyan *et al.*, 1987).

The relationship with the isotropized set covariance function $\bar{\gamma}_X(r)$ (§8.4.3) is also of note:

$$1 - L(r - 0) = -\frac{\pi}{\mathsf{E}U(X)} \frac{d}{dr} \bar{\gamma}_X(r) \quad (r \ge 0). \tag{8.41}$$

As Piefke (1978, 1979) has shown in the convex case there is a close connection between $L(\ell)$ and yet another distribution function, namely that of the distance between two independent uniformly distributed points in X. Let the corresponding density function be $P(r)$; sometimes this is called the distance distribution. It is given by

$$P(r) = \frac{2U(X)}{A(X)^2} r \int_r^{R_{\max}} (\ell - r) \, dL(\ell), \tag{8.42}$$

where $R_{\max}$ is the maximum distance between two points in X. The chord length moments and the moments corresponding to $P(r)$ satisfy

$$\int_0^R r^k P(r) \, dr = \frac{2U(X)}{(2+k)(3+k)A(X)^2} \int_0^R \ell^{k+3} \, dL(\ell) \quad (k = -1, 0, 1, \ldots). \tag{8.43}$$

Chords in non-convex sets

In non-convex sets intersection with a line can produce more than one line segment. If these segments are suitably connected then generalizations of the formulae for the moments of the chord lengths can be given (Miles, 1972, 1985).

The mean (total) chord length satisfies

$$\mathsf{E}\ell = \frac{A(X)}{M(X)}, \tag{8.44}$$

where $M(X)$ is the mean projection length of X on a random line. If X is connected, then

$$M(X) = U(\text{conv } X)/\pi,$$

where conv X is the convex hull of X.

Instead of the kth power of the total length ℓ of the segments, the so-called 'k-linc' is used (the word 'linc' comes from *line* section of *non-convex* domain). If the section figure consists of n segments and if ℓ_{ij} is the distance between P_i and P_j (Fig. 47) then the k-linc $[\ell^k]$ is given by

$$[\ell^k] = \sum_{i=1}^{2n} \sum_{j=i+1}^{2n} (-1)^{j-i+1} \ell_{ij}^k.$$

If $n = 1$ (i.e. if there is only one line segment), then $[\ell^k] = \ell^k$. In general, $[\ell^1] = \ell$ and $[\ell^2] = \ell^2$. If there are two line segments ($n = 2$) then $[\ell^3] = (\ell_1 + \ell_2)^3 + 6x\ell_1\ell_2$, where the ℓ_i are the lengths of the segments in X and x is the distance between P_2 and P_3.

Miles (1972) showed that

$$\mathsf{E}[\ell^3] = \frac{3}{\pi} \frac{A^2(X)}{M(X)}. \tag{8.45}$$

Statistical estimation of chord length distribution functions

Chord length distribution functions are mainly used in practical applications where the figures considered are particles in a structure as discussed on p. 119.

In this case the chords on test lines are considered. For a given ℓ the number $N(\ell)$ of chords longer than ℓ that are completely contained in the window of observation is determined. Then $L(\ell)$ is estimated as

$$\hat{L}(\ell) = 1 - \frac{N(\ell)t}{N(0)(t - \ell)}, \tag{8.46}$$

where t is the total length of the test lines. The term $t/(t - l)$ gives a boundary correction in favour of long chords (Ohser and Tscherny, 1988).

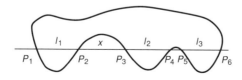

Figure 47   Representation of chords intersecting a non-convex set.

8.4.3   Isotropized set covariance function

The *set covariance* function or geometric covariogram of a deterministic compact convex set X is the function on R^2 defined by

$$\gamma_X(h) = A(X \cap (X + h))$$

or

$$\gamma_X(h) = \int_{R^2} 1_X(x) 1_X(x - h) \, dx \quad (h \in R^2). \tag{8.47}$$

Obviously, $\gamma_X(h)$ is the area of the intersection of X and X shifted by h (Fig. 48).

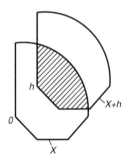

Figure 48   Definition of the set covariance function. The area of the shaded region is $\gamma_X(h)$.

Using polar coordinates $h = (r, \varphi)$, one can write

$$\gamma_X(h) = \gamma_X(r, \varphi).$$

By averaging over φ, the isotropized set covariance function $\bar{\gamma}_X(r)$ is obtained:

$$\bar{\gamma}_X(r) = \frac{1}{2\pi} \int_0^{2\pi} \gamma_X(r, \varphi) \, d\varphi. \tag{8.48}$$

This function is considered in detail in Matheron (1975). It is monotonically decreasing in the interval $(0, r_{max})$, where r_{max} is the smallest value of r for which $\bar{\gamma}_X(r)$ vanishes. For small r a good approximation is

$$\bar{\gamma}_X(r) \approx A(X) - \frac{U(X)}{\pi} r. \tag{8.49}$$

The relation to the chord length distribution function is given by (8.41). Thus $\bar{\gamma}_X(r)$ can in principle be determined using $L(\ell)$. Nevertheless, $\bar{\gamma}_X(r)$ is given here for two examples.

A disc of radius R

$$\bar{\gamma}_X(r) = 2R^2 \arccos\left(\frac{r}{2R}\right) - \tfrac{1}{2} r \sqrt{4R^2 - r^2} \quad (0 \le r \le 2R) \tag{8.50}$$

A very precise approximation is

$$\bar{\gamma}_X(r) \approx \pi R^2 - 2Rr + \frac{r^3}{6R}. \tag{8.51}$$

A rectangle of area A and side length ratio β (≥ 1)

$$\bar{\gamma}_X(r) = \frac{A}{\pi}
\begin{cases}
\pi - 2x - \dfrac{2x}{\beta} + \dfrac{x^2}{\beta} & (0 \leq x \leq 1), \\[2ex]
2 \arcsin\left(\dfrac{1}{x}\right) - \dfrac{1}{\beta} - 2(x - u) & (1 < x \leq \beta), \\[2ex]
2 \arcsin\left(\dfrac{\beta - uv}{x^2}\right) + 2u + \dfrac{2v}{\beta} - \beta - \dfrac{1 + x^2}{\beta} & (\beta < x < \sqrt{\beta^2 + 1}), \\[2ex]
0 & (x \geq \sqrt{\beta^2 + 1}),
\end{cases}$$

where

$$x = \frac{r}{\sqrt{A/\beta}}, \quad u = \sqrt{x^2 - 1}, \quad v = \sqrt{x^2 - \beta^2}.$$

As $L(\ell)$, also $\bar{\gamma}_X(r)$ does not uniquely characterize the set X. Nagel (1993) has shown that $\gamma_X(h)$ ($h \in R^2$) characterizes bounded convex polygons uniquely up to translations and central reflection. Because of

$$A(X \oplus \{o, h\}) = A(X \cup (X + h)) = 2A(X) - A(X \cap (X + h))$$

this is equivalent to the unique (up to translations and central reflections) characterization of X by the areas of X dilated by two-point sets. As Lešanovsky and Rataj (1990) and Rataj (1994) have shown, non-convex sets are not uniquely (up to translations and central reflections) determined by such areas. But, if dilations with three-point sets are considered, then the characterization is unique up to translations.

8.4.4 Erosion functions

The *spherical erosion function* $Q_s(r)$ is defined as

$$Q_s(r) = 1 - \frac{\mathsf{E}A(X \ominus b(o, r))}{\mathsf{E}A(X)} \quad (r \geq 0). \tag{8.52}$$

This definition is based on the following idea. According to the uniform distribution, a random point x is thrown into the random set X. Let x be the contour point of X nearest to x, and let $d(x)$ be the distance of x from $r(x)$. Of course, $d(x)$ is a random variable. Its distribution function is $Q_s(r)$.

The *linear erosion function* $Q_l(r)$ is given by

$$Q_l(r) = 1 - \frac{\mathsf{E}A(X \ominus s(o, r))}{\mathsf{E}A(X)} \quad (r \geq 0), \tag{8.53}$$

where $s(o, r)$ denotes a random line segment of length r whose endpoint lies in o. 'Random' here means that the orientation of the segment is uniform in $[0, 2\pi]$.

Analogously, the *pair erosion function* $Q_p(r)$ is defined as

$$Q_p(r) = 1 - \frac{EA(X \ominus \{o, r\})}{EA(X)} \quad (r \geq 0). \tag{8.54}$$

In this case, instead of the line segment $s(o, r)$, its endpoints appear.

Of course, $Q_l(r)$ and $Q_p(r)$ coincide for convex X.

The linear erosion function $Q_l(r)$ is closely connected with the chord length distribution function (see (8.40)), while the pair erosion function is up to a factor the same as the isotropized set covariance function. Erosion functions are in a certain sense complementary to the so-called 'contact distribution functions', (Stoyan, *et al.*, 1987).

The above erosion functions can also be explained in a way similar to the chord length distribution function $L(\ell)$ for random X. Let there be a structure consisting of grains that are (after shifting to the origin) random sets with the same distribution as X. A random point x is thrown into the structure. If it falls in one of the grains then the shortest distance $d(x)$ to the boundary of the grain is determined. The distribution function of $d(x)$ is called the spherical erosion function. It is an area-weighted average of the spherical erosion functions $Q_s(r)$ of the realizations of X:

$$Q_s(r) = \frac{1}{EA(X)} \int A(\xi) Q_{s,\xi}(r) P_X(d\xi) \quad (r \geq 0). \tag{8.55}$$

The function

$$f_X(r) = EA(X \ominus b(o, r)) \quad (r \geq 0),$$

which is used in the definition of the spherical erosion function (8.52), has been studied (Miles, 1974; Matheron, 1975; Weil, 1982b). The set $X \ominus b(o, r)$ is called the 'inner parallel set'. If the contour is sufficiently smooth then

$$f_X'(0) = -EU(X). \tag{8.56}$$

Under certain conditions, $f_X(r)$ is a quadratic polynomial in r. But the form

$$f_X(r) = EA(X) - rEU(X) + \pi r^2 \tag{8.57}$$

is not the only possibility. (It does not hold for a rectangle. Conditions for its validity have been given by Miles (1974).)

Like the erosion functions, functions that include 'opened' or 'closed' sets are also used, for example

$$G(r) = 1 - \frac{A((X \ominus b(o, r)) \oplus b(o, r))}{A(X)} \quad (r \geq 0).$$

(Note $\emptyset \oplus A = \emptyset$!). See Serra (1982, p. 333ff), Ripley (1988, Chap. 6) and Appendix D.

Statistical estimation of erosion function

For the statistical estimation of erosion functions image analysers that can carry out erosions should be used. For a sample $X_1, \ldots, X_n$ of sets the areas $A(X_i)$ of the eroded sets X_i are measured as functions of r. In the case of the spherical erosion function an estimator of $Q_s(r)$ is

$$\hat{Q}_s(r) = 1 - \frac{\sum_{i=1}^{n} A(X_i \ominus b(o, r))}{\sum_{i=1}^{n} A(X_i)} \qquad (r \geq 0). \qquad (8.58)$$

In the case of the linear or pair erosion function the areas of the eroded sets are determined for various directions (of line segments or point pairs) and then averaged. Frequently it appears that the X_i belong to a homogeneous and isotropic structure. Then the measurement can be carried out for the union U of the X_i. An estimator of $Q_s(r)$ is then

$$\hat{Q}_s(r) = 1 - \frac{A(U \ominus b(o, r))}{A(U)}. \qquad (8.59)$$

8.5 STOCHASTIC MODELS OF RANDOM COMPACT SETS

8.5.1 Poisson polygon

Let there be given a homogeneous and isotropic Poisson line field with parameter ρ (Appendix H). It divides the plane into convex polygons. These are shifted so that the centre of gravity is at the origin o. If all polygons are given equal weight ('number law') then the distribution of a random convex set X, called the *Poisson polygon*, is obtained. Many distributional characteristics of this random set are known.

The mean number of vertices is 4; the distribution of the number of vertices is given in Table 4. There

$$p_3 = 2 - \tfrac{1}{6}\pi^2, \quad p_4 = \pi^2 \log 2 - \tfrac{1}{3} - \tfrac{7}{36}\pi^2 - \tfrac{7}{2}(1 + 2^{-3} + 3^{-3} + \cdots).$$

The mean of the inner angle at any vertex is $\tfrac{1}{2}\pi$. the corresponding density function $f_a(\alpha)$ is

$$f_a(\alpha) = \tfrac{1}{2}\sin\alpha \quad (0 \leq \alpha \leq \pi).$$

Example. Figure 49 shows limestone fragments at several stages of abrasion characterized by the amount of time spent in a tumbling barrel. The original and other less rounded stones suggest that a description by convex polygons may make sense. Can one assume that they are Poisson polygons? The mean inner angle of the stones belonging to 0 and 0.5 miles is 110.3° with standard deviation 24.1°. The inner angles are thus greater than expected for Poisson polygons. (A mean

Table 4   Probabilities p_n for the number of corners of the Poisson polygon.

n	p_n
3	0.355
4	0.381
5	0.192
6	0.059
7	0.013
8	0.002

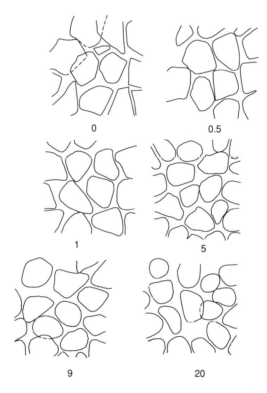

Figure 49   Limestone fragments abraded to different degrees (measured in miles) in a tumbling barrel; the longer the treatment the smoother the stones. (After Pettijohn (1975).)

inner angle of 90° is also obtained for polygons resulting from a homogeneous non-Poisson line process with the property that no more than two lines ever cross at the same point.) Two possible conclusions are:

(a)   in Fig. 49 small fragments are absent, e.g. triangular pieces with small inner angles;

(b)   the process in which the fragments are generated cannot be modelled by a line or plane process where fracture generation is uninfluenced by fractures already existing.

The means of the area and perimeter are

$$\mathsf{E}A(X) = \frac{4}{\pi \varrho^2}, \quad \mathsf{E}U(X) = \frac{4}{\varrho}.$$

The variances of area, perimeter and vertices number are

$$\mathrm{var}\, A(X) = \frac{8\pi^2 - 16}{\pi^2 \varrho^4}, \quad \mathrm{var}\, U(X) = \frac{2\pi^2 - 8}{\varrho^2}$$

and

$$\mathrm{var} N(X) = \tfrac{1}{2}\pi^2 - 4,$$

The correlation coefficients are

$$\varrho_{AU} = 0.864, \quad \varrho_{AN} = 0.487, \quad \varrho_{UN} = 0.564.$$

The density functions of area and perimeter have until now only been obtained by simulation (Crain and Miles, 1976). They are similar to density functions of exponential distributions.

The Poisson polygon is an isotropic random set and the Aumann mean is a disc of radius $2/\pi \varrho$. Thus for the Poisson polygon

mean of area = area of Aumann mean

(Mecke, 1987).

Matheron (1975, p. 183), has given the distribution of the width function $w(\varphi)$ defined on p. 67. (Because of the isotropy, the distribution is the same for all φ.) The chord length distribution function and the linear and spherical erosion functions of the Poisson polygon are easily obtained if properties of the Poisson line tessellation are used. Since it is known that the intersection of the lines of the process with the x-axis generates a linear Poisson point process of intensity ϱ,

$$L(\ell) = 1 - \mathrm{e}^{-\varrho \ell} \quad (\ell \geq 0).$$

Because of (8.40), the linear erosion function $Q_\ell(r)$ is also an exponential distribution function with parameter ϱ, i.e.

$$Q_\ell(r) = 1 - \mathrm{e}^{-\varrho r} \quad (r \geq 0).$$

The spherical erosion function $Q_s(r)$ can be obtained as follows. The quantity $1 - Q_s(r)$ is the probability that the disc $b(o, r)$ is not hit by one of the lines of the generating Poisson line process. Since in general the number of lines hitting a convex set K has a Poisson distribution with parameter $\varrho U(K)$,

$$Q_s(r) = 1 - e^{-2\pi\varrho r} \quad (r \geq 0).$$

The fact that both $L(\ell)$ and the erosion functions are exponential distribution functions can be used for testing the goodness-of-fit of Poisson polygons by a given sample of polygons.

The determination of functions such as the covering function or the contour covariance function is difficult if the centre of gravity of the polygon is the reference point. However, if one starts from the Poisson line tessellation, the origin is taken to be the reference point and one considers the polygon containing o then it is easier to obtain formulae. (The corresponding quantities can be interpreted as area-weighted variants of $p_X(x)$ or $x_R(\psi)$.) In particular,

$$p_X^0(r) = e^{-\varrho r} \quad (r \geq 0),$$

and the contour covariance function corresponding to the radius-vector function $\mathsf{E}\chi_R^0(\psi)$ is given by

$$\varrho^2 \mathsf{E}\chi_R^0(\varphi) = \begin{cases} 1 & (\varphi = 0), \\ -1 + \dfrac{4}{1+\cos\varphi}\left[1 - \dfrac{1-\cos\varphi}{1+\cos\varphi}\log\left(\dfrac{2}{1-\cos\varphi}\right)\right] & (0 < \varphi < \pi), \\ 0 & (\varphi = \pi) \end{cases}$$

(Cowan, 1987).

Finally, note that an effective method for simulating single Poisson polygons is described in George (1987).

8.5.2 Dirichlet polygons

Let there be given a Dirichlet tessellation with respect to a homogeneous and isotropic point field N of intensity λ (Appendix I). It divides the plane completely into convex cells where each contains just one point of N. Each cell is translated so that the original point process point lies at the origin. If all cells are given the same weight ('number law') then the distribution of an isotropic random compact convex set X is obtained, which is called the *Dirichlet polygon*.

The mean area of the Dirichlet polygon is

$$\mathsf{E}A(X) = \lambda^{-1}. \tag{8.60}$$

Its mean number of vertices is 6, with the exception of the degenerate case of a lattice N.

Further formulae are known almost only in the case that N is a Poisson process.

Table 5 Probabilities p_n for the number of vertices of the Poisson-Dirichlet polygon.

n	p_n
3	0.011
4	0.107
5	0.259
6	0.294
7	0.199
8	0.090

Poisson–Dirichlet polygon

The distribution of the number of vertices is given in Table 5. The mean perimeter is $4\lambda^{-1/2}$.

The variances of area, perimeter and number of vertices are

$$\operatorname{var} A(X) = 0.280\lambda^{-2}, \quad \operatorname{var} U(X) = 0.947\lambda^{-1}$$

and

$$\operatorname{var} N(X) = 1.782.$$

and the correlation coefficients of area, perimeter and number of vertices are

$$\varrho_{AU} = 0.953, \quad \varrho_{AN} = 0.568, \quad \varrho_{UN} = 0.502.$$

These numbers have been obtained by numerical integration or simulation (see also Stoyan *et al.*, 1987). Note that it is possible to simulate single Dirichlet polygons without constructing the whole tessellation (Hinde and Miles, 1980). The density functions of area and perimeter have until now only been simulated (Hinde and Miles, 1980).

For the areas a form was obtained similar to gamma distribution densities, while the distribution of perimeters can be approximated by a normal distribution.

The inner angles at the vertices have mean $\frac{2}{3}\pi$, variance $\frac{5}{9}\pi^2 - \frac{5}{6}$ and density function

$$f_a(\alpha) = \tfrac{4}{3} \sin\alpha(\sin\alpha - \alpha\cos\alpha) \quad (0 \leq \alpha \leq \pi).$$

The maximum of the density function is at 0.72π. The probability for the occurrence of angles less than $\frac{1}{2}\pi$ is $\frac{1}{6}$ (Icke and van de Weygaert, 1987). The joint distribution of all three angles at a vertex is given in Miles (1988), see also p. 157.

The mean chord length is

$$\tfrac{1}{4}\pi\lambda^{-1}.$$

Formulae and diagrams for the chord length distribution function $L(\ell)$ and the spherical erosion function $Q_s(r)$ are given in Muche and Stoyan (1992); Fig. 50 shows the density function for $L(\ell)$. The isotropized set covariance is discussed in Brumberger and Goodisman (1983).

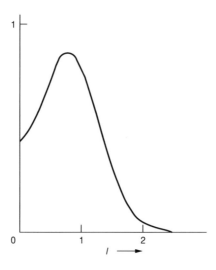

Figure 50   Chord length density function of the Poisson–Dirichlet polygon for $\lambda = 1$.

The covering function $p_X(r)$ is given by

$$p_X(r) = \exp(-\lambda \pi r^2) \quad (r \geq 0). \tag{8.61}$$

This formula can be obtained as follows. By definition, the origin o is the generating point of the Dirichlet polygon X. An arbitrary point $\mathbf{r}$ at distance r from o also belongs to X if the disc $b(\mathbf{r}, r)$ contains in addition to o no further point of the Poisson point process. This is an easy consequence of the Dirichlet tessellation construction principle. Because of the properties of the Poisson point field, this probability is equal to $\exp[-\lambda \times \text{area of } b(\mathbf{r}, r)]$.

The formula (8.61) yields the distribution function $F(r)$ of the radius-vector length:

$$F(r) = 1 - p_X(r) = 1 - \exp(-\lambda \pi r^2) \quad (r \geq 0). \tag{8.62}$$

Thus the radius-vector mean $\mathbf{E}_r X$ is the disc centred at o with radius $r_r = \frac{1}{2}\lambda^{-1/2}$.

It is also easy to calculate the means $\mathbf{E}_A X$ and $\mathbf{E}_V X$. Both are discs centred at o, with radii r_A and r_V:

$$r_A = \frac{2}{\pi}\lambda^{-1/2}, \quad r_V = \frac{1}{\pi}\lambda^{-1/2}.$$

Also the median $\tilde{X}$ is a circle centred at o; its radius is

$$\tilde{r} = \left(\frac{\log 2}{\pi}\right)^{1/2} \lambda^{-1/2}.$$

General case: N is not a Poisson point field

Hermann *et al.* (1990) have determined the distribution of the area of the Dirichlet polygon for various point field models *N*. As expected, the areas are more variable for cluster point fields than for hard-core point fields. Also for these models the density function $f(a)$ of $A(X)$ can be approximated relatively well by gamma distribution densities

$$f(a) = b^q a^{q-1} e^{-ba} / \Gamma(q) \quad (a \geq 0),$$

see also Lemaitre *et al.* (1993). If the statistically determined parameters *b* and *q* are represented for the various models in a diagram then Fig. 51 is obtained.

It is possible to calculate the covering function $p_X(r)$ for a general homogeneous and isotropic point field *N*. The basis of this is the idea in the proof of (8.61). Figure 52 shows the covering functions $p_X(r)$ for two non-Poisson point fields. Davy and Guild (1988) have found an approximation to the distribution function $F(r)$ of the random radius-vector length of the Dirichlet polygon for a further point field model. They considered the Poisson hard-core point field (§16.3). This field is a stochastic model for the centres of a random system of

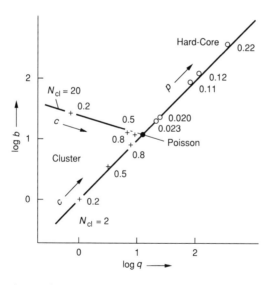

Figure 51 Dependence of the parameters *b* and *q* of the gamma distribution densities. These approximate the density functions of areas of Dirichlet polygons on the type of generating point field with fixed intensity. In the case of a cluster field N_{cl} is the mean number of points per cluster (Poisson-distributed number). The parameter *c* characterizes the distance of the daughter points from the cluster centre; it is inversely proportional to *c*. In the case of a hard-core field *p* characterizes the degree of order in the field, which increases with increasing *p*. See Hermann *et al.*, (1990).

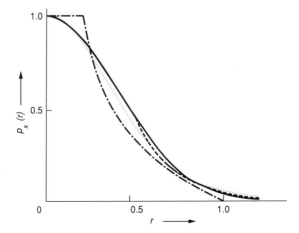

Figure 52 Covering functions for the Poisson–Dirichlet polygon and for Dirichlet polygons for two further point field models of intensity 1. The origin is always at the generating point; so that $p_X(0) = 1$. The dotted curve corresponds to a particular mixed Poisson point field, the dashed curve to a particular Gauss–Poisson field. (For an explanation of these point field types see Stoyan *et al.* (1987).) Both point fields are more variable than the Poisson field. The dot-dashed curve represents the approximation formula (8.63) of Davy and Guild (1988) with $p = 0.3$, $R = 0.309$ and $k = 0.531$.

non-overlapping discs of radius R. The approximation formulae are

$$F(r) = 1 - e^{-kv(r/R)} \qquad\qquad (r \geq R),$$
$$v(x) = \frac{2}{\pi} \left[(x^2 - 2) \arccos \frac{1}{x} + \sqrt{x^2 - 1} \right] \quad (x \geq 1). \qquad (8.63)$$

Here k is the solution of the equation

$$2 \int_1^\infty x e^{-kv(x)} \, dx = \frac{1 - p}{p},$$

where p is the area fraction of the homogeneous and isotropic random set that results from the union of the discs. If λ is the intensity of the point field then clearly

$$p = \lambda \pi R^2.$$

For $p = 0.1, 0.3$ and 0.5 the values of k are $0.1197, 0.5510$ and 1.4677 (Davy and Guild, 1988). The latter paper also contains further information on $F(r)$, and for the analogous three-dimensional case.

8.5.3 Rounded polygons

Sometimes objects are considered that look like polygons but have rounded vertices. Figure 49 shows examples. The rounding may be a result of natural or technical (e.g. abrasion or baking) processes. A rough approximation for a set X of this kind is

$$X = Y \oplus b(o, R),$$

where R is a random variable and Y is a random convex polygon. Y can be interpreted as the original particle and as the thickness of new outer layers. Some characteristics of X are given by

$$\mathsf{E}A(X) = \mathsf{E}A(Y) + \mathsf{E}R\mathsf{E}U(Y) + \mathsf{E}R^2, \qquad (8.64)$$

$$\mathsf{E}U(X) = \mathsf{E}U(Y) + 2\pi\mathsf{E}R, \qquad (8.65)$$

$$Q_s(r) = \int_0^\infty q_t(r)\,\mathrm{d}F(t), \qquad (8.66)$$

where $F(t)$ is the distribution function of R and $q_t(r)$ is the spherical erosion function of $Y \oplus b(o, t)$:

$$q_t(r) = \begin{cases} \dfrac{\mathsf{E}A(Y) + \mathsf{E}U(Y) + \pi r^2}{\mathsf{E}A(Y) + t\mathsf{E}U(Y) + \pi t^2} & (r \le t), \\[2ex] \dfrac{q_0(r)\mathsf{E}A(Y)}{\mathsf{E}A(Y) + t\mathsf{E}U(Y) + \pi t^2} & (r > t). \end{cases}$$

A further model is

$$X = [Y \ominus b(o, R_1)] \oplus b(o, R_2). \qquad (8.67)$$

Here again Y is a polygon, and R_1 and R_2 are random variables. To give formulae for this case is not easy. A particular case will be considered that is close to the situation

$$Y = \text{Poisson polygon} \quad \text{and} \quad R_1 = R_2 = r.$$

The formula (8.67) is not directly applicable to Poisson polygons, since $Y \ominus b(o, r) = \emptyset$ is possible with positive probability. But, of course, the empty set is not contained in samples of particles. A reasonable approach may be to consider only that polygons Y for which $Y \ominus b(o, r)$ is non-empty. As Matheron (1975) has shown, the collection of all polygons of a Poisson line tessellation eroded with $b(o, r)$ but not empty after the erosion has the same distribution as the polygons of the original tessellation. See also the discussion in Serra (1982, p. 516, on 'conditional invariance'). Thus, if (8.67) is interpreted in the 'conditional sense' then the rounded polygons considered have the same distribution as Poisson polygons dilated by addition of $b(o, r)$. That means, in the case of Poisson polygons, that (8.63) and (8.67) lead to practically the same figures.

It is easy to generate on an image analyser rounded polygons starting from normal polygons, since the required operations of Minkowski addition (for (8.63)) and opening (for (8.67)) are standard operations of image analysis.

As a generalization of the sets considered until now, let us also consider sets of the form

$$X = Y \oplus Z,$$

where Y and Z are random sets. Here Z can be interpreted as a random deformation of Y; Grenander (1976) speaks about 'Minkowski deformation'. If Y and Z are convex and Z additionally isotropic then fundamental characteristics of X can be easily given, using the generalized Steiner formula (8.11) and the additivity of support functions (§7.2.3). Statistical methods should exploit the particular geometrical shape of the figures. In the case $X = X \oplus b(o, r)$ the figures of a sample can be individually eroded by discs with increasing radii until polygons are obtained. So a sample of radii and polygons is obtained.

In the case $X = (Y \ominus b(o, R)) \oplus b(o, R)$ the original polygons can be reconstructed by elongating the edges.

If Y is deterministic and Z a convex random compact set with known Aumann expectation $E_A Z$, then an estimator of Y for a sample $X_1, X_2, \ldots$ is

$$\hat{Y} = \frac{1}{n}(X_1 \oplus \cdots \oplus X_n) \ominus E_A Z,$$

see Lin, Wei and Attele (1991).

8.5.4 Convex hulls of random figures

An intensively studied class of stochastic models of random compact sets comprises convex hulls of finitely many points, discs or other sets. Figure 53 shows two examples of such sets. Such figures must be realistic models for particles generated by abrasion processes; the points or discs stand for particularly resistant regions that are undamaged.

Surveys of the known mathematical results are given in Buchta (1985), Schneider (1988) and Weil and Wieacker (1993).

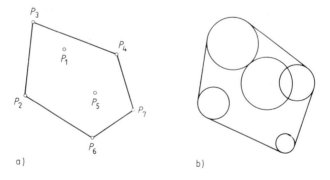

Figure 53 The convex hull of a set A is the smallest convex set conv A that contains A. The figure shows the convex hull of (a) a set of 7 points and (b) the union of 5 discs.

Convex hulls of points in convex sets

Let K be a compact convex set. In K, n independent points are scattered according to the uniform distribution. Their convex hull H_n is a random compact set. Beginning with work by Renyi and Sulanke (1963, 1964) the asymptotic properties of H_n have been most thoroughly investigated, and asymptotic formulae have been found for the means of vertex numbers, perimeter and area. For example, the area A of K and the mean area A_n of H_n are related by

$$A - A_n = \left(\tfrac{2}{3}\right)^{1/3} \Gamma\left(\tfrac{5}{3}\right) \left(\frac{A}{n}\right)^{2/3} \int_{\delta K} \kappa^{1/3} \, ds + O\left(\frac{1}{n}\right),$$

where Γ is the gamma function, δK is the contour of K and κ is the curvature of the contour (which is assumed to be sufficiently smooth[†]) and ds is the length element.

Exact formulae for A_n are known in some particular cases, for example for ellipses and convex polygons (Buchta, 1984b). Note also that other scattering principles for the points in K have been considered; for example random points on the contour of K (Affentranger, 1987) or rotationally symmetric normal distributions (Efron, 1965).

Convex hulls of points in discs

The case where K is a disc is well studied. It is sufficient to consider the unit disc $b(o, 1)$. Explicit formulae can be found for A_n ($= \mathsf{E}A(H_n)$), U_n ($= \mathsf{E}U(H_n)$) and e_n (= mean vertex number of H_n). We have, first,

$$\frac{A_n}{\pi} = 1 + \frac{2}{3}\frac{1}{(2\pi)^n} \int_0^{2\pi} (x - \sin x)^n \sin x \, dx;$$

see Buchta (1984c), who also gives formulae to help in the numerical calculation of the above integral. $\mathsf{E}A(H_n)$ can be calculated recursively from the values $\mathsf{E}A(H_j)$ ($j < n$) (Affentranger, 1988). For $n = 3$, 4 and 5 the means are $35/48\pi$, $35/24\pi$ and $175/72\pi - 23023/6912\pi^3$. Secondly,

$$U_n = 12\pi \binom{n}{2} \gamma_3^2 \sum_{k=0}^{n-2} (-1)^k \binom{n-2}{k} \gamma_1^{n-2-k} I(n-2-k, 5+k, k),$$

with

$$I(r, s, t) = \int_0^\pi x^r \sin^s x \cos^t x \, dx \quad (r, s, t = 0, 1, 2, \ldots),$$

$$\gamma_0 = \tfrac{1}{2}, \quad \gamma_{i+1} = [2\pi(i+1)\gamma_i]^{-1} \quad (i = 1, 2, \ldots).$$

[†]Asymptotic formulae are also known for convex polygons (Buchta, 1984a; Dwyer, 1988; Affentranger and Wieacker, 1989)

For $n = 2$, 3 and 4 the values are 1.81, 2.71 and 3.27. Finally,

$$e_n = 8 \binom{n}{2} \gamma_3 \sum_{k=0}^{n-2} (-1)^k \binom{n-2}{k} \gamma_1^{n-2-k} I(n-2-k, 4+k, k).$$

For $n = 4$ and 5 the values are 3.70 and 4.26.

These formulae are given in Buchta and Müller (1984), where the numerical calculation of the integrals is also explained.

Since ellipses can be obtained from discs by affine transformations, the formulae for area and vertices number also hold for the convex hulls of n points in arbitrary ellipses of area π.

Obviously, in the circular case H_n is an isotropic random set. Thus $E_A H_n$ is a disc $b(o, r_n)$. Its radius r_n can be obtained from (8.16) if the mean perimeter is known:

$$r_n = \frac{U_n}{2\pi}.$$

Little is known on further distributional characteristics of H_n.

The covering function $p_{H_n}(r)$ takes the value 0 on the boundary of the disc, and for the centre ($r = 0$) Wendel (1962) found

$$p_{H_n}(0) = 1 - \frac{n}{2^{n-1}}. \tag{8.68}$$

Lemma 2 in Buchta (1987) contains an integral formula that can be used for the calculation of $p_{H_n}(r)$ (also in the non-circular case):

$$p_{H_n}(r) = 1 - \tfrac{1}{2}n \int_0^{2\pi} \{\mu_r(\alpha)^{n-1} + [1 - \mu_r(\alpha)]^{n-1}\}\lambda_r(\alpha)\, d\alpha,$$

with

$$\mu_r(\alpha) = \begin{cases} \Lambda_r(\alpha) - \Lambda_r(\alpha + \pi) + 1 & (0 \leq \alpha \leq \pi), \\ \Lambda_r(\alpha) - \Lambda_r(\alpha - \pi) & (\pi < \alpha \leq 2\pi) \end{cases}$$

and

$$\lambda_r(\alpha) = \frac{d}{d\alpha}\Lambda_r(\alpha).$$

Here $\Lambda_r(\alpha)$ is the area of the hatched region in Fig. 54. For $0 \leq \alpha \leq \pi$ it satisfies

$$\Lambda_r(\alpha) = \frac{1}{2\pi}[\delta(\alpha) - r \sin \delta(\alpha)],$$

with

$$\delta(\alpha) = \alpha - \arcsin(r \sin \alpha),$$

and for $\alpha > \pi$

$$\Lambda_r(\alpha) = 1 - \Lambda_r(2\pi - \alpha).$$

Figure 55 shows $p_{H_n}(r)$ for $n = 3$, 5 and 10.

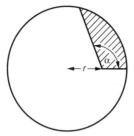

Figure 54   The function value $\Lambda_r(\alpha)$ is equal to the area of the hatched region in the unit disc.

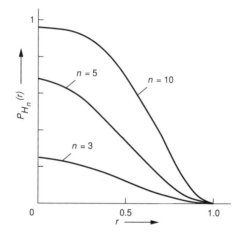

Figure 55   Covering functions for the convex hull H_n of n uniformly distributed independent random points in the unit disc. Since it is possible that the origin $(r = 0)$ does not lie in H_n, $p_{H_n}(0) < 1$.

Jewell and Romano (1985) describe a general method for calculating probabilities of the form $\Pr(K \subset H_n)$ for compact sets K.

Note that (8.63) shows that the median $\tilde{H}_3$ is an empty set, since $p_{H_3}(r) \leq p_{H_3}(0) = 0.25 < 0.5$ for all r.

Groeneboom (1988) and Cabo and Groeneboom (1994) have proved central limit theorems for A_n, U_n and e_n for K = disc and K = polygon.

Convex hulls of discs

Clearly in some cases the convex hull of n random discs is a realistic particle model. Unfortunately, only a little is known about this set. Its mean perimeter is equal to $2\pi r$ plus the mean perimeter of the convex hull of the centres if all discs are of the same radius r. Similarly the mean area can be calculated by the Steiner formula

$$A(X \oplus b(o, r)) = A(X) + U(X)r + \pi r^2.$$

Affentrager and Dwyer (1993) consider the problem of finding the convex hull of balls and study geometrical properties of this set.

8.5.5 Gaussian random sets

Since the normal distribution plays an important role in statistics, it would be interesting to investigate random sets related to this distribution. For example, consider the Gaussian random sets defined by Lyashenko (1983). A planar compact convex set X is called *Gaussian* if for any φ the value of the support function $s_X(\varphi)$ has a normal distribution. Lyashenko (1983) and Vitale (1984) have shown that Gaussian random sets have a very simple structure. Any such set X has the form

$$X = M + x, \tag{8.69}$$

where M is a deterministic compact convex set and x is a normally distributed random vector.

8.5.6 Inhomogeneous Boolean models

The random compact set X considered has the form

$$X = \bigcup_{x \in N} [S(x) + x].$$

Here N is an inhomogeneous Poisson point field with intensity function $\lambda(x)$ satisfying

$$\int_{R^2} \lambda(x) \, dx < \infty.$$

The $S(x)$ are independent, identically distributed random compact sets with finite mean area (e.g. discs). Figure 56 shows a sample of such a set.

The covering function $p_X(x)$ has the form

$$p_X(x) = 1 - \exp\left[-\int_{R^2} p(x, z)\lambda(z) \, dz\right] \quad (x \in R^2), \tag{8.70}$$

with $p(x, z) = \Pr(x \in S + z)$. Here S is a 'prototype' of the sets $S(x)$, i.e. a random set with the same distribution as the $S(x)$.

The formula (8.70) is proved like (3.1) in Stoyan *et al.* (1987). The proof is based on a theorem of Prekopa, which says that the independent position-dependent thinning of a Poisson field again yields a Poisson field. In the case $S = b(o, r)$

$$p(x, z) = \begin{cases} 1 & (\|x - z\| \leq r), \\ 0 & \text{otherwise.} \end{cases}$$

Analogously, probabilities of the form $\Pr(X \cap K \neq \emptyset)$ may also be determined.

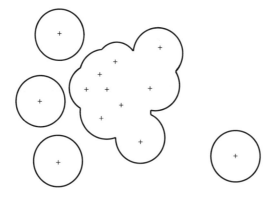

Figure 56 Sample of an inhomogeneous Boolean model with disc-shaped grains $S(x)$. The intensity function has its greatest value at the centre of the figure, which implies a high point density and nearly complete covering.

As an example, consider the tumour model of Cressie (1984, 1989), where $S = b(o, r)$ and $\lambda(x) = \lambda 1_A(x)$ for a given set A. Cressie and Hulting (1992) describe statistical methods for the estimation of model parameters, and show that the model is useful for the geometrical description of 'artificial' tumours.

8.5.7 Vorob'ev's forest-fire model

Vorob'ev (1984) studied Markovian processes $\{X_n\}$ whose states are subsets of the integer lattice. The processes start with $X_0 = \{0\}$. The random set X_{n+1} results from X_n by

$$X_{n+1} = \bigcup_{x \in X_n} (S_n + x).$$

Here the S_n are independent, for all x and n identically distributed random sets with $\Pr(0 \in S_n(x)) = 1$. The distribution of S_n is as follows: Let S be a discrete random compact set with the same distribution of the S_n, a 'prototype' of the S_n. The points of S lie in a lattice rectangle R around the origin o. The probability that the lattice point $y \in R$ belongs to S is $p(y)$, where $p(0) = 1$. The points of S are independent , i.e. for arbitrary $y_1, \ldots, y_k$ of the lattice

$$\Pr(y_1 \in S, \ldots, y_k \in S) = p(y_1) \cdots p(y_k).$$

Vorob'ev systematically studied such sets by simulation, and developed approximation methods for determining their characteristics. He used these processes as models for the spread of forest fires; X_n is the area aflame or in ashes at time n. There are connections with models of the theory of interacting particle systems (Richardson model) (Durrett, 1988a, b; Liggett, 1985). Durrett (1988b) has studied a still better forest-fire model in which the burnt-out areas do not spread the fire further.

Union-stable convex-stable random sets

Other models of random compact sets can be obtained as weak limits of scaled unions

$$a_n^{-1}(X_1 \cup \cdots \cup X_n)$$

or convex hulls of independent identically distributed "simple" random sets $X_1, \ldots, X_n$. These sets can be chosen to be points, balls, triangles etc. It any case their distributions should be regularly varying in a certain sense, see Molchanov (1993a, b). Say, for a random ball its center and radius must have a regularly varying density.

The corresponding weak limits for unions are said to be union-stable. In other words, a random set X is said to be union-stable if the union $X_1 \cup \cdots \cup X_n$ of any number of its independent copies has the same distribution as $a_n X$ for some $a_n > 0$, see Matheron (1975). A full characterization of such sets is given in Molchanov (1933b). In many interesting cases the capacity functionals and the Aumann expectations of such stable sets are easy to compute.

CHAPTER 9

Point Description of Figures

9.1 INTRODUCTION

It is assumed in this chapter that each of the figures of the sample considered can be described by l points, the so-called *landmarks*. These points $p_1, \ldots, p_l$ usually lie on the contour of the figure, but inner points (e.g. a centre) are also possible. Each of the landmarks has a certain biological meaning or is defined by geometrical properties. Possible choices are local maxima of curvature, spikes, points on an axis of symmetry; Mardia (1989) sketches a method for determining such 'mathematical landmarks'. Of course, the numbering of landmarks is the same for all figures of the sample, so that it makes sense to interrelate, for example, the kth points $(k = 1, \ldots, l)$. If the figures are skull profiles then p_1 might be the nose tip and p_2 the chin tip.

In statistical analyses either the l points are considered as a whole or geometrical quantities formed from them are studied. For example, distances d_{ij} between certain landmarks p_i and p_j or triangles formed by landmark triplets are frequently analysed. The following sections present methods for the exploratory analysis of landmark data. It will be shown how mean configurations can be calculated from samples of landmark configurations. Differences and similarities of form can be detected by Procrustean analysis. The case of triangles and triplets of points is studied in detail. Finally, a simple stochastic model of landmark configurations — the Bookstein model — is discussed.

The treatment here is based on Bookstein (1986, 1991) and Kendall (1984, 1989). Excellent surveys have been given by Small (1988) and Goodall (1991).

9.2 DESCRIPTION OF LANDMARK CONFIGURATIONS AND THEIR SIZE

In the following, landmark configurations are denoted by Roman capital letters P, Q etc. Thus P has the form

$$P = (p_1, \ldots, p_l),$$

with

$$p_k = (x_k, y_k) \ (x\text{- and } y\text{-coordinates}) \qquad k = 1, \ldots, l.$$

For the determination of the coordinates x_k and y_k there are various possibilities, depending on the nature of the figures analysed. For example, the figures can be arbitrarily placed in the plane, in which an (x, y)-coordinate system is given, and then landmark coordinates measured. In this case the absolute values of the coordinates are not so interesting — only the relative positions of landmarks are of importance. Configurations that are directly congruent (i.e. those that coincide after suitable translation and rotation) are considered as equal.

In other cases it makes sense to measure the landmark coordinates with respect to a coordinate system connected with the figures. For example, an axis of symmetry or the connecting line of two very important landmarks can be identified with the x-axis. (See the example in §10.2, which analyses the form of hands.) In this case the absolute values of the coordinates have a much greater meaning than above, and it may be inappropriate to rotate or translate the landmark configurations.

Sometimes it is useful to work with complex numbers. Then $p_k = (x_k, y_k)$ is replaced by

$$v_k = x_k + iy_k \tag{9.1}$$

and analogously q_k by w_k $(k = 1, \ldots, l)$.

The centre of gravity[†] of the landmark configuration P is denoted by

$$\bar{p} = (\bar{x}, \bar{y}),$$

$$\text{with } \bar{x} = \frac{1}{l}\sum_{k=1}^{l} x_k, \quad \bar{y} = \frac{1}{l}\sum_{k=1}^{l} y_k. \tag{9.2}$$

Sometimes the landmark configuration is translated so that its centre of gravity is at o. The resulting configuration P^0 has the form

$$P^0 = (p_1^0, \ldots, p_l^0), \quad \text{with } x_k^0 = x_k - \bar{x}, \ y_k^0 = y_k - \bar{y} \quad (k = 1, \ldots, l). \tag{9.3}$$

The size of the landmark configuration P is described by

$$s(P) = \left[\sum_{k=1}^{l}(x_k - \bar{x})^2 + (y_k - \bar{y})^2\right]^{1/2}. \tag{9.4}$$

This quantity is not changed by translations or rotations of the figure.

9.3 DISTANCES AND TRANSFORMATIONS OF LANDMARK CONFIGURATIONS

9.3.1 The general problem

Let P and Q be two planar configurations of points:

$$P = (p_1, \ldots, p_l) \quad \text{and} \quad Q = (q_1, \ldots, q_l).$$

[†]This centre of gravity does not necessarily coincide with that of the corresponding figure

The differences in the shape of the two configurations are to be measured, and it is to be shown how P may be transformed into a shape as close as possible to Q. Following Kendall (1984), a measure of deviations is considered that is symmetric in P and Q. (Sibson (1978) has considered other deviation measures, which are asymmetric in P and Q. He also used reflections as transformations. This leads to other formulae.) Let $\mathcal{T}$ be a set of admissible transformations for configurations (e.g. the set of all Euclidean motions that are combinations of translations and rotations). Every point of a configuration is transformed by the same transformation. Thus, by the transformation $T \in \mathcal{T}$, the configuration P is transformed into the configuration

$$T P = (T p_1, \ldots, T p_l).$$

If two configurations P and Q are given, one may ask for transformations T and U such that $T P$ and $U Q$ are 'close together'; that is, those that minimize

$$\sum_{k=1}^{l} \|T p_k - U q_k\|^2.$$

This question does not make sense in all cases — for example not if T and U are allowed to be dilations (see p. 144 and Appendix D). If both dilation factors are zero then P and Q are transformed into the same trivial configuration $(o, \ldots, o)$. Thus a more appropriate deviation measure is

$$\Delta_{T,U}(P, Q) = \frac{\displaystyle\sum_{k=1}^{l} \|T p_k - U q_k\|^2}{\displaystyle\sum_{k=1}^{l} \|T p_k - c\|^2 + \sum_{k=1}^{l} \|U q_k - c\|^2}. \tag{9.5}$$

Here c is a suitable reference point (e.g. the common centre of gravity of the transformed configurations). Transformations T and U in $\mathcal{T}$ that minimize $\Delta_{T,U}(P, Q)$ have to be determined. Depending on the choice of $\mathcal{T}$, different minima Δ and different minimizing T and U are obtained.

When using complex numbers, the transformations T and U are interpreted as transformations of complex numbers, and the deviation measure is then

$$\Delta_{T,U}(V, W) = \frac{\displaystyle\sum_{k=1}^{l} |T v_k - U w_k|^2}{\displaystyle\sum_{k=1}^{l} |T v_k - c|^2 + \sum_{k=1}^{l} |U w_k - c|^2}. \tag{9.6}$$

9.3.2 Formulae for some classes of transformations

The sets T of transforms considered in this book are algebraic groups. Such a group is either equal to one of the following four groups, or their elements are formed by combining transformations from them:

translations;
translations parallel to the x-axis;
rotations around the origin o;
dilations.

A *dilation* is a transform

$$(x, y) \rightarrow (\varrho x, \varrho y), \quad \varrho > 0.$$

An example of a set T of this type is the set of all transformations formed by combinations of x-translations and dilations. (Obviously, it does not make much sense to combine x-translations and rotations.) The choice of the group of transformations T depends on the problem and on the nature of the figures analysed. For example, translations parallel to the x-axis are of interest if the landmarks are measured with respect to a coordinate system such that the x-axis is an axis of symmetry. Note that 'affine' transforms

$$x \rightarrow ax + by, \quad y \rightarrow cx + dy$$

do not yield interesting results. For them the infimum of $\Delta_{T,U}(P, Q)$ is obtained for vanishing or infinitely large coefficients $a, \ldots, d$.

If T contains general translations then it is assumed that both configurations have their centre of gravity in the origin o. If this is not the case then P and Q are transformed using (9.3). (The '0' superscripts on p_k etc. are omitted in the following.) If the set T contains x-translations but no general translations then the figures have to be translated parallel to the x-axis so that the x-coordinates of the centres of gravity vanish; the x_k and x_k' are transformed as above, while the y_k and y_k' remains unchanged.

The reference point c is the common centre of gravity of the transformed configurations TP and UQ in the case of general translations. In the case of x-translations one sets $c = (\bar{x}, 0)$, where $\bar{x}$ is the x-coordinate of the common centre of gravity of TP and UQ.

It is possible to show that under the assumptions about centre of gravity and reference points, translations no longer need be considered. Further translations (in addition to those into the centre of gravity), do not improve the degree of fit. (The assumptions above have been made just to eliminate translations; see also the proof on p. 146.)

If the set T does not contain translations (e.g. if T is the group of all transforms formed by rotations and dilations) then the assumptions about the centres of gravity can be omitted. The formulae given below are then true if $c = 0$. Such sets T make sense if the landmarks are measured with respect to a coordinate system closely connected with the configurations (see the example in §10.2).

If $\mathcal{T}$ is a group of transforms formed by rotations around o then, independently of the other components of the transforms, the optimal rotation angle

$$\omega(P, Q) = \arctan\left[-\frac{\sum_{k=1}^{l}(x_k y_k' - y_k x_k')}{\sum_{k=1}^{l}(x_k y_k' + y_k x_k')}\right]. \tag{9.7}$$

The rotation parts of the transforms T and U can be chosen such that P is not rotated while Q is rotated by the angle $\omega(P, Q)$. Analogously, there exists a unitary optimal dilatation factor

$$\varrho(P, Q) = \left(\sum_{k=1}^{l}\|p_k\|^2\right)^{1/2}\left(\sum_{k=1}^{l}\|q_k\|^2\right)^{-1/2}, \tag{9.8}$$

or

$$\varrho(P, Q) = \left[\sum_{k=1}^{l}(x_k^2 + y_k^2)\right]^{1/2}\left[\sum_{k=1}^{l}(x_k'^2 + y_k'^2)\right]^{-1/2}.$$

The dilation part of the transform can be chosen so that P is not dilated while Q is dilated with factor $\varrho(P, Q)$. The dilation factors

$$\left(\sum_{k=1}^{l}\|p_k\|^2\right)^{-1/2} \quad \text{and} \quad \left(\sum_{k=1}^{l}\|q_k\|^2\right)^{-1/2}$$

can also be chosen. Using (9.7) and (9.8), the optimal values of $\Delta(P, Q)$ can be determined.

Dilations and translations

$$\Delta(V, W) = 1 - \frac{\sum_{k=1}^{l}\text{Re}(v_k \bar{w}_k)}{\left(\sum_{k=1}^{l}|v_k|^2 \sum_{k=1}^{l}|w_k|^2\right)^{1/2}} \tag{9.9}$$

($\bar{w}_k$ denotes the conjugate of w_k), or

$$\Delta(P, Q) = 1 - \frac{\sum_{k=1}^{l}(x_k x_k' + y_k y_k')}{\left[\sum_{k=1}^{l}(x_k^2 + y_k^2)\sum_{k=1}^{l}(x_k'^2 + y_k'^2)\right]^{1/2}}. \tag{9.10}$$

Because of the assumptions concerning the centre of gravity, this formula holds for both the general case and the case of x-transformations.

General translations and rotations

$$\Delta(V, W) = 1 - \frac{2\left|\sum_{k=1}^{l} v_k w_k\right|}{\sum_{k=1}^{l} |v_k|^2 + \sum_{k=1}^{l} |w_k|^2}, \tag{9.11}$$

or

$$\Delta(P, Q) = 1 - \frac{2\left\{\left[\sum_{k=1}^{l}(x_k x_k' + y_k y_k')\right]^2 + \left[\sum_{k=1}^{l}(x_k x_k' - y_k y_k')\right]^2\right\}}{\sum_{k=1}^{l}(x_k^2 + y_k^2 + x_k'^2 + y_k'^2)} \tag{9.12}$$

(Ziezold, 1989).

General translations, dilations and rotations

This case has been considered by Kendall (1984), and is the most complicated one. Here the derivation is sketched. As on p. 142, complex numbers are used.

The possible transforms are given by

$$v \to \alpha x + \lambda, \quad w \to \beta w + \mu$$

where α, β, λ and μ are complex numbers. The reference point c is the common centre of gravity of the transformed configurations; thus $c = \frac{1}{2}(\lambda + \mu)$. The deviation measure $\Delta_{T,U}$ is then

$$\frac{\sum_{k=1}^{l} |\alpha v_k - \beta w_k + (\lambda - \mu)|^2}{\sum_{k=1}^{l} \left|\alpha v_k + \frac{1}{2}(\lambda - \mu)\right|^2 + \sum_{k=1}^{l} \left|\beta w_k - \frac{1}{2}(\lambda - \mu)\right|^2}.$$

Taking squares yields

$$\frac{\sum_{k=1}^{l} |\alpha v_k - \beta w_k|^2 + |\lambda - \mu|^2}{\sum_{k=1}^{l} |\alpha v_k|^2 + \sum_{k=1}^{l} |\beta w_k|^2 + \frac{1}{2}|\lambda - \mu|^2}.$$

Since a function $f(x)$ of the form

$$f(x) = \frac{a + bx^2}{c + dx^2} \quad (a, b, c, d > 0)$$

takes its minimum at $x = 0$, the above quotient is minimized for fixed α and β at $\lambda = \mu$. It remains to determine the minimum of

$$\frac{\sum\limits_{k=1}^{l} |\alpha v_k - \beta w_k|^2}{\sum\limits_{k=1}^{l} |\alpha v_k|^2 + \sum\limits_{k=1}^{l} |\beta w_k|^2}$$

with respect to α and β. Dividing numerator and denominator by $|\alpha|^2$ (the case $\alpha = 0$ is uninteresting) yields

$$\frac{\sum\limits_{k=1}^{l} |v_k - \zeta w_k|^2}{\sum\limits_{k=1}^{l} |v_k|^2 + |\zeta|^2 \sum\limits_{k=1}^{l} |w_k|^2} = 1 - 2 \frac{\mathrm{Re}\left(\zeta \sum\limits_{k=1}^{l} v_k w_k \right)}{\sum\limits_{k=1}^{l} |v_k|^2 + |\zeta|^2 \sum\limits_{k=1}^{l} |w_k|^2},$$

where $\zeta = \beta/\alpha = \varrho e^{i\omega}$. Setting the first derivatives with respect to ω and ϱ of the last expression to zero yields the optimal dilation factor $\varrho(P, Q) = \varrho$, and the optimal rotation angle $\omega(P, Q) = \omega$:

$$\varrho = \left(\frac{\sum\limits_{k=1}^{l} |v_k|^2}{\sum\limits_{k=1}^{l} |w_k|^2} \right)^{1/2}, \quad \text{or} \quad \varrho(P, Q) = \left[\frac{\sum\limits_{k=1}^{l} (x_k^2 + y_k^2)}{\sum\limits_{k=1}^{l} (x_k'^2 + y_k'^2)} \right]^{1/2}$$

(cf. (9.8)) and

$$\omega = \arctan\left[-\frac{\mathrm{Im}\left(\sum\limits_{k=1}^{l} \bar{v}_k w_k \right)}{\mathrm{Re}\left(\sum\limits_{k=1}^{l} v_k w_k \right)} \right],$$

or

$$\omega(P, Q) = \arctan\left[-\frac{\sum\limits_{k=1}^{l}(x_k x_k' + y_k y_k')}{\sum\limits_{k=1}^{l}(x_k x_k' - y_k y_k')}\right]$$

(cf. (9.7)). Using the optimal ϱ and ω gives the minimal deviation

$$\Delta(V, W) = 1 - \frac{\left|\sum\limits_{k=1}^{l} v_k \bar{w}_k\right|}{\left(\sum\limits_{k=1}^{l}|v_k|^2 \sum\limits_{k=1}^{l}|w_k|^2\right)^{1/2}}. \qquad (9.13)$$

In terms of landmark coordinates, this is

$$\Lambda(P, Q) = 1 - \frac{\left\{\left[\sum\limits_{k=1}^{l}(x_k x_k' + y_k y_k')\right]^2 + \left[\sum\limits_{k=1}^{l}(y_k x_k' - x_k y_k')\right]^2\right\}^{1/2}}{\left[\sum\limits_{k=1}^{l}(x_k^2 + y_k^2)\sum\limits_{k=1}^{l}(x_k'^2 + y_k'^2)\right]^{1/2}}, \qquad (9.14)$$

with T and U chosen as follows:

$$T = \text{identity transformation;}$$
$$U = \text{dilation with factor } \varrho(P, Q)$$
$$\text{and rotation by an angle } \omega(P, Q).$$

Of course, the Δ-value for the case just considered is smaller than that in the case of translations and dilations only.

9.4 MEANS OF PLANAR CONFIGURATIONS OF POINTS

9.4.1 A metric for point configurations

The aim of this section is the determination of means of point configurations with respect to the definition by Fréchet (p. 112). For this, the following metric δ on equivalence classes of direct congruent configurations is used. Two configurations are equivalent if one can be transformed into the other by proper Euclidean motions (i.e. by translation and rotations).

The equivalence class to which the configuration P belongs is denoted by P_c; if P is described by complex numbers and denoted by V then the symbol V_c is used. It is assumed that the class representatives lie at o. Following Ziezold (1989), the metric δ is chosen as

$$\delta(V_c, W_c)^2 = \inf_{\varphi} \sum_{k=1}^{l} |v_k - e^{i\varphi} w_k|^2.$$

Methods of differential calculus yield the formula

$$\delta(V_c, W_c)^2 = \sum_{k=1}^{l} |v_k|^2 + \sum_{k=1}^{l} |w_k|^2 - 2 \left| \sum_{k=1}^{l} v_k \bar{w}_k \right|, \tag{9.15}$$

or

$$\delta(P_c, Q_c)^2 = \sum_{k=1}^{l} (x_k^2 + y_k^2 + x_k'^2 + y_k'^2)$$

$$-2 \left\{ \left[\sum_{k=1}^{l} (x_k x_k' + y_k y_k') \right]^2 + \left[\sum_{k=1}^{l} (y_k x_k' - x_k y_k') \right]^2 \right\}^{1/2}. \tag{9.16}$$

In these formulae v_k and w_k, and $x_k, \ldots, y_k'$, are the coordinates of the landmarks of arbitrary representatives V and W, and P and Q, of the equivalence classes V_c and W_c, and P_c and Q_c, respectively.

9.4.2   Calculation of mean configurations

Let $P_1, \ldots, P_n$ be n configurations. Let the corresponding equivalence classes be $P_{1c}, \ldots, P_{nc}$. Corresponding representatives are obtained by translating the centre of gravity of the P_i to the origin o (application of (9.2) and (9.3)). An equivalence class M_c with

$$\sum_{i=1}^{l} \delta(P_{ic}, M_c)^2 = \inf_{P_c} \sum_{i=1}^{l} \delta(P_{ic}, P_c)^2 \tag{9.17}$$

has to be determined. Here the infimum is taken over all equivalence classes P_c of configurations with l landmarks; δ is the metric introduced by (9.15) or (9.16). Each configuration M of M_c (or M_c itself) is called a mean configuration analogous to (8.22). The determination of M_c is an optimization problem, where $2l$ numbers (the x- and y-coordinates of a representative of M_c) have to be determined. If l is not too large then standard methods of numerical mathematics can be used; for example, the Gauss–Newton method, since δ is differentiable with respect to $x_k, \ldots, y_k'$. The following heuristic optimization method (suggested by Gower (1975) and Zierold (1989)) that uses properties of the mean is very simple. It proceeds as follows.

Start: Shift the centres of gravity of the P_i into the origin, replacing the P_i by the translated configurations. Choose a starting configuration M_o for the mean, for example

$$M_o := \frac{1}{n} \sum_{i=1}^{l} P_i \quad \text{or} \quad M_o := P_1.$$

$k := 0.$

1: Calculate the optimal rotation angles $\omega(M_k, P_i)$ $(i = 1, \ldots, n)$ using (9.8), and rotate the P_i by these angles. Replace the P_i by the rotated configurations.

2: Calculate the new configuration M_{k+1} as

$$M_{k+1} := \frac{1}{n} \sum_{i=1}^{l} P_i.$$

End: if M_{k+1} is close to M_k.
Else: $k := k + 1$, go to 1.

Practical experience shows that this algorithm frequently yields a minimum after only 3–5 steps (see also Goodall, 1991).

9.5 PROCRUSTEAN ANALYSIS

9.5.1 The problem

Procrustean analysis is a valuable method of exploratory data analysis for configurations of points. Its aim is to detect similarities in sets of point configurations and to determine the degree of similarity. Originally Procrustean analysis was used in factor analysis and in connection with multi-dimensional scaling (Sibson, 1978). There somewhat different approaches were used from those described here. A detailed exposition of methods of Procrustean analysis in shape statistics is given in Goodall (1991).

Consider n figures $X_1, \ldots, X_n$, with corresponding point configurations $P_1, \ldots P_n$. For any i and j $(i \neq j)$ one must determine how closely P_i can be transformed into P_j. P_i and P_j can be transformed as in §9.3, and the aim is to minimize the quantity $\Delta_{T,U}(P_i, P_j)$ as in (9.5).

The choice of the set T of transformations depends on the nature of the figures and the method of measurement of landmarks. If the position of the figures in the coordinate system is arbitrary then T should consist of arbitrary translations and rotations. If the scales differ for the figures then dilations should also be included. If, in contrast, the positions of the figures are uniform with respect to the coordinate system then maybe translation or rotation should be omitted. If the x-axis coincides with a symmetry axis of the figures then general translations do not make sense, and only x-translations should be considered. If the origin of the coordinate system is at a certain landmark then translations should be completely excluded. Also, if a further landmark always lies on the x-axis then rotations should not be considered.

Given a particular set $\mathcal{T}$ of transformations, the corresponding formulae from §9.3.3 can be used. If $\mathcal{T}$ contains translations then first the coordinates of landmarks have to be modified in such a way that the centre of gravity of all landmarks for every figure lies at o or has zero x-coordinate.

The formulae then yield the deviation measures $\Delta(P_i, P_j)$ and the transformation characteristics $\varrho(P_i, P_j)$ and $\omega(P_i, P_j)$.

9.5.2 Using the results of Procrustean analysis

Procrustean analysis can be used to describe deviations from an ideal configuration or for characterizing relationships between configurations. It is also possible to apply regression analysis to shape-changing processes (Goodall, 1990b, 1991).

Relating to reference configurations

Possible reference configurations are mean configurations or particular 'ideal' configurations. Such configurations are included in the Procrustean analysis together with the configurations to be analysed. Then series of Δ-values and ϱ-values are obtained:

$$\Delta_i = \Delta(\text{reference configuration}, i\text{th configuration}),$$

$$\varrho_i = \varrho(\text{reference configuration}, i\text{th configuration}) \quad (i = 1, \ldots, n).$$

These values can be analysed with the usual methods for samples of univariate data.

Relations between configurations

For n configurations Procrustean analysis yields two matrices $((\Delta_{ij}))$ and $((\varrho_{ij}))$, where

$$\Delta_{ij} = \Delta(i\text{th configuration}, j\text{th configuration}),$$

$$\varrho_{ij} = \varrho(i\text{th configuration}, j\text{th configuration}) \quad (i, j = 1, \ldots, n).$$

The Δ_{ij} behave like 'proximities' (Davison, 1988), where

$$\Delta_{ij} = \Delta_{ji} \qquad \Delta_{ii} = 0 \quad (i, j = 1, \ldots, n),$$

and a large value of Δ_{ij} implies a large difference in shape. By a suitable transformation, the same holds for the ϱ-values:

$$s_{ij} = |\log \varrho_{ij}| \quad (i, j = 1, \ldots, n).$$

(There $\varrho_{ij} = \varrho_{ij}^{-1}$, so $s_{ij} = s_{ji}$.)

The quantities ϱ_{ij} and s_{ij} characterize size differences. The analysis of configurations based on the matrices $((\Delta_{ij}))$ and $((s_{ij}))$ is performed in the same way. Thus in the following the neutral symbol t_{ij} is used.

Two methods of analysis of a matrix $((t_{ij}))$ of proximity values are useful for shape analysis: cluster analysis and multidimensional scaling.

Cluster analysis

The matrix of proximities $((t_{ij}))$ is interpreted as a distance matrix, for which the usual methods of cluster analysis are applied (see e.g. Dillon and Goldstein, 1984). As an example, the average linkage method is described here. This is a particular hierarchic agglomerative method. The corresponding algorithm has the following principal form:

1. Start with the finest partition, in which each configuration forms a single cluster, $N_c := n$.
2. Determine these clusters C_k and C_l of minimal distance t_{kl}, $t_{kl} := \min_{(ij)} t_{ij}$.
3. Combine C_k and C_l into a new (bigger) cluster, $N_c := N_c - 1$.
4. Determine the new distance matrix by deleting the ith line and row and replacing the elements of the kth line and row by

$$t_{ik} := \tfrac{1}{2}(t_{ik} + t_{il}).$$

5. If $N_c - 1$ then END; else go to 2.

An example of the application of this method to results of Procrustean analysis is given in §10.2.

Multidimensional scaling

The aim is to assign to each configuration a point in the plane (or in R^d with small d). There the distance d_{ij} of the points should be close to the given proximities t_{ij}. For the solution of this optimization problem many algorithms and programs exist (see e.g. Dillon and Goldstein, 1984; Davison, 1988). Various optimality criteria are used; for example

$$\sum_{(i,j)} (d_{ij} - t_{ij})^2 \to \text{Min},$$

or

$$\frac{\sum_{(i,j)} (d_{ij} - t_{ij})^2}{\sum_{(i,j)} d_{ij}^2} \to \text{Min}.$$

As an example, an application of multidimensional scaling is given in §10.2, where the first criterion is used.

9.6 SHAPE ANALYSIS FOR TRIANGLES AND POINT TRIPLETS

9.6.1 Introduction

An important and thoroughly investigated special case of landmark configurations is that of triangles or point triplets. The analysis of more general configurations can be reduced to the analysis of triangles and triplets (Bookstein and Sampson, 1990; Bookstein, 1991). The two cases are as follows.

Triangles. Here the numbering of vertices is uninteresting, and, besides translation, dilatations and rotations, reflections also preserve shape.

Triplets of landmarks. Now the numbering is important; two congruent triangles are considered as different if their vertices are differently numbered. In this case only translations, dilations and rotations preserve shape.

The shape theory of Kendall (1984, 1985, 1989) shows that it is useful to describe point triplets and triangles by points on the sphere $S^2(\frac{1}{2})$ of radius $\frac{1}{2}$ centred at the origin of R^3. A shape metric corresponds to this description, and very elegant shape distributions are obtained for important models of random triangles. This theory is not given here save for the resulting calculation techniques. It is explained how to obtain the corresponding points on the sphere $S^2(\frac{1}{2})$ for given triangles or point triplets. In this way samples of triangles or point triplets are transformed into point patterns.

The sphere has to be projected into the plane in order to plot such patterns. A possible method is the area-preserving Lambert or Schmidt transform. Thus the polar coordinates r and ψ are assigned to a triangle or point triplet.

9.6.2 Triangles

Let the triangle to be analysed have angles α, ω, and Ω, where ω is the smallest and Ω the largest.[†] Then define the angles φ and θ by

$$\varphi = \pi + \arctan\left[\frac{2\sin\omega\sin(\omega + \Omega)}{\sin(2\omega + \Omega - \pi)}\right],$$

$$\theta = 2\arctan\left[\frac{\sin\omega}{\sqrt{3}\sin(\varphi + \omega)}\right].$$

Figure 57 shows the geometrical meaning of φ and

$$m = \frac{\sin\omega}{\sin(\varphi + \omega)}.$$

[†]The degenerate cases with $\omega = 0$ are here not treated, as in Kendall's theory.

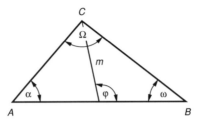

Figure 57 A triangle ABC, with angles ω and Ω, showing the variables φ and m.

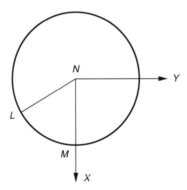

Figure 58 View on the sphere $S^2(\frac{1}{2})$. All points that correspond to triangles lie in the region LMN.

The angles φ and θ are transformed into the points X, Y and Z on the sphere $S^2(\frac{1}{2})$:

$$X = \tfrac{1}{2}\cos\theta, \quad Y = \tfrac{1}{2}\sin\theta\cos\varphi, \quad Z = \tfrac{1}{2}\sin\theta\sin\varphi.$$

If θ is interpreted as geographical latitude then the north pole of the sphere lies on the X-axis. All possible triangles lie in a sixth of the upper sphere, as shown in Fig. 58. The three vertices of the spherical triangle L, M and N shown in Fig. 58 have coordinates as given in Table 6. N corresponds to an equilateral triangle, while L and M correspond to isosceles triangles, where the angle between the sides is 0 for L and $180°$ for M.

Table 6 Coordinates of the three corners L, M and N.

	X	Y	Z	φ	θ
L	$\frac{1}{4}$	$-\frac{1}{4}\sqrt{3}$	0	π	$\frac{1}{3}\pi$
M	$\frac{1}{2}$	0	0	$\frac{1}{2}\pi$	0
N	0	0	$\frac{1}{2}$	$\frac{1}{2}\pi$	$\frac{1}{2}\pi$
	X	Y	Z	φ	θ

To plot point patterns, the spherical triangle LMN has to be mapped into the plane-preserving area. (Area preservation is necessary to correctly represent the densities.) By these means, the projection is chosen so that the north pole of the sphere is on the Z-axis. θ' is the corresponding geographic latitude ($= 0$ for the north pole), φ' is the corresponding geographical longitude ($= 0$ for the X-axis, $= 90°$ for the negative Y-axis). Thus

$$\theta' = \arccos(\sin\theta\sin\varphi), \tag{9.18}$$

$$\varphi' = \arcsin\left(-\frac{\sin\theta\cos\varphi}{\sin\theta'}\right). \tag{9.19}$$

The Lambert or Schmidt projection should be used, assigning the polar coordinates r and ψ to the angles θ' and ψ', and we have

$$r = \sin\tfrac{1}{2}\theta', \tag{9.20}$$

$$\psi = \varphi'. \tag{9.21}$$

In this way the triangle LMN is transformed into a sector of angle $60°$ and side length $\sqrt{\tfrac{1}{2}}$.

Since the arccos function makes some difficulties, an algorithm is given here that transforms ω and Ω into r and ψ. It uses only the arctan function (ATN). The angles are given in degrees.

```
9   S = 1/SQR (3): Y = SIN (ω)
10  X = SIN (2 * ω + Ω − 180): IF X = 0 THEN φ = 90: GOTO 20
11  φ = 180 + ATN (2 * Y * SIN (ω + Ω)/X)
20  X = SIN (φ + ω): IF X = 0 THEN θ = 2 * ATN(S): GOTO 30
21  θ = 2 * ATN (Y * S/X)
30  Y = SIN (φ):Z = SIN (θ):X = Y * Z:
    IF X = 0 THEN θ' = 90: GOTO 40
31  θ' = ATN (SQR (1 − X * X)/X)
40  X = SIN (θ'):IF X = 0 THEN φ' = 0: GOTO 50
41  X = Z * SQR (1 − Y * Y)/X:IF X = 1 THEN φ' = 0: GOTO 50
42  φ' = ATN (X/SQR (1 − X * X))
50  r = SIN(0.5 * θ'):ψ = φ'
```

The algorithm should yield $r = 0.1638$ and $\psi = 51.38$ for $\omega = 45$ and $\Omega = 70$. Figure 59 shows the positions of particular triangles in the sector. The upper vertex belongs to equilateral triangles, while the lower circular arc corresponds to triangles that have degenerated into line segments. The sides of the sector represent isosceles triangles. The right-hand side represents isosceles triangles for which the equal sides are shorter than the third, the left those in which the third is the shorter. The asymmetric form has been chosen in order to prepare for a more detailed description (see Fig. 63).

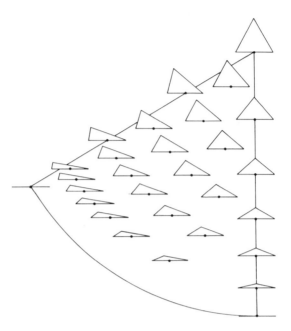

Figure 59   View of the region LMN of Fig. 58 after Schmidt projection. The corresponding triangles lie at the positions marked on the base lines of the triangles. The maximum angle is always at the upper corner. On the circular arc the triangles have degenerated into line segments.

Figure 60 shows line systems that may help to find the positions of triangles in the sector. The lines correspond to triangles with ω = const, and Ω = const.

Stochastic models for triangles

Kendall and his colleagues have studied various stochastic models of triangles (see e.g. Kendall, 1984, 1985, 1989). Of particular interest are triangles formed by three uniform points in a given set. The mathematical problem is to determine the probability density on the sphere or in the 60° sector.

Here only two simple cases are considered.

Normally distributed vertices.   Suppose that in the plane there is a symmetric two-dimensional normal distribution (centre at the origin o, equal variance of x- and y-coordinates). According to this distribution, each of the three independent points are generated and the corresponding triangles formed. Of course, their shape varies greatly. It is very interesting that the shape distribution of these triangles is the uniform distribution in the 60° sector. (The probability that the shape parameters r and ψ of such a triangle lie in an arbitrary domain of the sector is equal to the area of this domain divided by the area of the sector.) Dryden and Mardia (1990) have also given the shape distribution for the case of correlated normally distributed points. They also considered figures with more than three vertices.

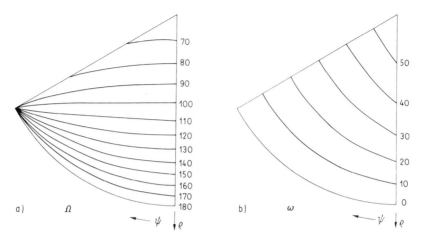

Figure 60 Isolines on which the triangles of fixed $\Omega(a)$ and $\omega(b)$ are positioned with respect to Kendall's parametrization.

Poisson–Delaunay triangles. To the Poisson–Dirichlet tessellation belong triangles that have their vertices at the points of the generating Poisson field and that do not contain further points of this field (Appendix I). Its shape has been investigated by Miles (1970) and Kendall (1983). For the joint distribution of the three angles at the vertices the density function has the following beautiful form:

$$f(\alpha, \beta, \gamma) = \tfrac{4}{9}(\sin^2 \alpha + \sin^2 \beta + \sin^2 \gamma)^2. \qquad (9.22)$$

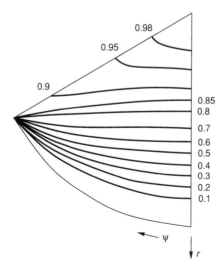

Figure 61 The shape density of the Delaunay triangles given by isolines.

This density has its maximum at $\alpha = \beta = \gamma = 60°$. Figure 61 shows the isolines of this density function in the $60°$ sector.

9.7 POINT TRIPLETS

Following Kendall, point triplets are transformed into points on the sphere $S^2(\frac{1}{2})$. The transformation is a generalization of that for triangles given above, and goes as follows.

Let the three points of the triplet be $A = (a_x, a_y)$, $B = (b_x, b_y)$ and $C = (c_x, c_y)$. The point triplet is moved and dilated such that the distance between A and B is 2 and the direction from A to B is the reference direction (i.e. the x-axis). The vector going from the centre of the side AB to C forms an angle φ with AB; its length is denoted by m (Fig. 62).

Then spherical coordinates φ and θ are introduced, where θ is obtained by

$$\theta = 2 \arctan \sqrt{\tfrac{1}{3}m}.$$

The corresponding point (X, Y, Z) on the sphere $S^2(\frac{1}{2})$ has coordinates

$$X - \tfrac{1}{2}\cos\theta, \quad Y = \tfrac{1}{2}\sin\theta\cos\varphi, \quad Z = \tfrac{1}{2}\sin\theta\sin\varphi.$$

Here the north pole of the sphere is given by the point $\theta = 0$. As in the case of triangles, the spherical coordinates are now transformed in such a way that the north pole comes to lie on the Z-axis. θ' is the corresponding geographical latitude ($= 0$ for the north pole) and φ' is the corresponding geographical longitude ($= 0$ for the X-axis, $= 90°$ for the negative Y-axis). The corresponding transformation formulae are refinements of (9.18) and (9.19):

$$\theta' \quad = \arccos(\sin\varphi\sin\theta), \tag{9.23}$$

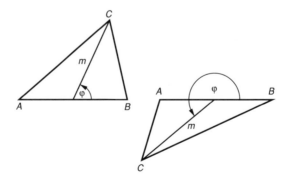

Figure 62 Two triangles and the corresponding variables φ and m. In contrast to Fig. 57, the maximum angle is not necessarily at the corner C.

$$\varphi' = \begin{cases} \beta & (\theta \le 90°), \\ 180° - \beta & (\theta > 90°), \end{cases} \tag{9.24}$$

$$\beta = \arcsin\left(-\frac{\sin\theta\cos\varphi}{\sin\theta'}\right).$$

If $\varphi' < 0$ then the value $360° + \varphi'$ should be used. The sphere is now mapped into the plane-preserving area by the Lambert or Schmidt transform, where the northern hemisphere ($\theta' < 90°$) and the southern hemisphere ($\theta' > 90°$) are each transformed into a disc of radius $\sqrt{\frac{1}{2}}$. The plane polar coordinates are denoted by r and ψ in both cases, and

$$r = \begin{cases} \sin\frac{1}{2}\theta' & (\theta \le 90°), \\ \sin\frac{1}{2}(180° - \theta') & (\theta > 90°) \end{cases} \tag{9.25}$$

and

$$\psi = \varphi'. \tag{9.26}$$

The corresponding algorithm is as follows (the angles are given in degrees):[†]

```
9    S = 1/SQR (3)
10   D1 = 0.5 * (aₓ + bₓ): D2 = 0.5 * (a_y + b_y):
     U = cₓ − D1: V = c_y − D2: X = U * U + V * V : Q = 0 : P = 0
12   U = bₓ − D1: V = b_y − D2: Y = U * U + V * V :
     IF Y = 0 THEN φ = 90 : θ = 90: GOTO 24
14   IF (c_y − a_y) * (bₓ − aₓ) < (b_y − a_y)(cₓ − aₓ) THEN Q = 180
16   Z = SQR (Y): M = SQR (X): U = cₓ − bₓ:
     V = c_y − b_y: R = U * U + V * V: X = (X + Y − R)/(2 * M * Z):
     m = M/Z: IF X = 0 THEN φ = 90 + Q: GOTO 22
18   φ = ATN (SQR (1 − X * X)/X):
     IF φ < 0 THEN φ = φ + 180
20   IF Q = 180 THEN φ = 360 − φ
22   θ = 2 * ATN (m * S)
24   Y = SIN (θ): X = Y * SIN (φ):
     IF X = 0 THEN θ' = 90: GOTO 28
26   θ' = ATN (SQR (1 − X * X)/X):
     IF θ' < 0 THEN θ' = θ' + 180: P = 1
28   X = SIN (θ'): IF X = 0 THEN ψ = 0: GOTO 33
29   X = −Y * COS (φ)/X: IF X * X = 1 THEN ψ = 90:
     IF X < 0 THEN ψ = 270: GOTO 33
30   IF X = 1 THEN GOTO 33
31   ψ = ATN (X/SQR (1 − X * X)):
```

[†] The degenerate cases, where A, B and C lie on one line, are not treated as in Kendall (1985, 1986).

IF $\theta > 90$ THEN $\psi = 180 - \psi$
32 IF $\psi < 0$ THEN $\psi = 360 + \psi$
33 $D = \theta'$: IF $\theta' > 90$ THEN $D = 180 - \theta'$
34 $r = \text{SIN}(0.5 * D)$

If $P = 0$ then the point lies on the northern hemisphere, while for $P = 1$ it is in the southern.

For $A = (0, 0)$, $B = (1, 1)$ and $C = (3, 0)$ the algorithm should yield $r = 0.419$, $\psi = 214.7°$ and $P = 1$ (southern hemisphere).

Figure 63 shows the position of 12 congruent triangles on $S^2(\frac{1}{2})$, where, with the northern hemisphere on the right and the southern hemisphere on the left, the side AB is always shown horizontally and A is its left endpoint. The 60° sector that is used for representing triangles as in Fig. 59 is shown in bold.

Another variant of describing triplets has been introduced by Bookstein. The triplet (A, B, C) is transformed and dilated as above. A coordinate system is introduced, the x-axis of which goes through the points A and B with the origin in the middle of the side AB. The transformed points then have the following coordinates:

$$A : (-1, 0)$$
$$B : (1, 0)$$
$$C : (\vartheta_x, \vartheta_y).$$

The quantities ϑ_x and ϑ_y describe the shape of the triangle ABC. They can be calculated as

$$\vartheta_x = \frac{2}{\delta^2}[(b_x - a_x)(c_x - a_x) + (b_y - a_y)(c_y - a_y)] - 1, \qquad (9.27)$$

$$\vartheta_y = \frac{2}{\delta^2}[(b_x - a_x)(c_y - a_y) - (b_y - a_y)(c_x - a_x)], \qquad (9.28)$$

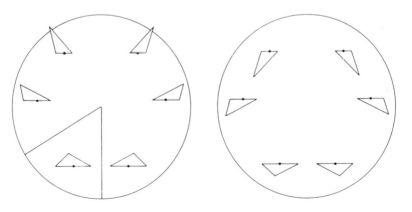

Figure 63 Positions of congruent triangles on the sphere $S^2(\frac{1}{2})$ in the Schmidt projection. The triangles differ with respect to the notation of the corners and their positions. The corners A and B always lie on the line ⟶, where A is the left endpoint.

with

$$\delta = \sqrt{(a_x - b_x)^2 + (a_y - b_y)^2}. \qquad (9.29)$$

Obviously ϑ_x and ϑ_y can take arbitrary real values: in Bookstein's shape description the whole of R^2 is used for describing the shape of triplets.

The parameters ϑ_x and ϑ_y can easily be interpreted geometrically. If the side AB is denoted as baseline then ϑ_y is equal to twice the ratio of height to length of baseline; ϑ_x describes the horizontal deviation of the vertex C from the right bisector of the baseline.

Since the shapes of triplets are given as points on the sphere by Kendall's shape description, it is clear that in Bookstein's shape description the 'shape points' often have greater distances. Therefore it is frequently better suited to visualizing shape differences between triplets or samples of triplets. In particular, this is true for obtuse triangles. For description of deviations from an equilateral basic shape, both variants give similar results. Bookstein (1991) has shown using his shape parameters how changes in growth processes can be described.

9.8 STATISTICS FOR THE BOOKSTEIN MODEL

9.8.1 The Bookstein model

Bookstein (1986) suggested the following model for planar random point configurations. Suppose there are deterministic fixed points $z_1, \ldots, z_l$ with x- and y-coordinates z_{kx} and z_{ky} $(k = 1, \ldots, l)$. These points are modified by independent random translations $d_1, \ldots, d_l$ such that the random configuration $P = (p_1, \ldots, p_l)$ is formed:

$$p_k = z_k + d_k, \qquad (9.30)$$

i.e.

$$x_k = z_{kx} + d_{kx}, \quad y_k = z_{ky} + d_{ky} \quad (k = 1, \ldots, l).$$

The random variables d_{kx} and d_{ky} are independent and normal with mean 0 and variance σ_k^2:

$$d_{kx} \sim N(0, \sigma_k^2), \quad d_{ky} \sim N(0, \sigma_k^2).$$

Thus the configuration P can be interpreted as a result of l independent diffusions described by Brownian motions in R^2 with starting points at $z_1, \ldots, z_l$.

This relatively simple model has been used in various statistical studies, for example by Bookstein (1986) and Mardia and Dryden (1989a,b). Frequently, however, it is too simple (§10.2). In any case it can be used as a 'null model': its validity is assumed and the nature of its deviations from the data are considered. The first step of generalization consists in using correlated d_k; anisotropic two-dimensional normal distributions for the d can also be used (Goodall, 1991; Goodall and Mardia, 1991). Such more general cases are also considered in Dryden and Mardia (1991, 1992), where approximations and examples are also discussed.

Goodall and Mardia (1993) present a systematic and extended treatment of shape densities using noncentral multivariate analysis.

Statistical analyses for the Bookstein model aim to get information about the σ_k^2 and the deterministic centres $z_1, \ldots, z_l$, starting with samples of identically independent distributed configurations $P_1, \ldots, P_n$, formed analogously to (9.30). This problem would be quite easy if the configurations were given by their coordinates in a fixed coordinate system. However, it is typical in shape statistics that the landmark positions are given only relatively to each other. The situation is — one may say — that the configurations are first generated according to the Bookstein model, but then arbitrarily shifted and rotated and finally given to the poor statistician. For her/him it is unreasonable to estimate $z_1, \ldots, z_l$; though it makes sense to ask for the distances d_{km} of the centres z_k and z_m, for the variances and for the form of triplets z_k, z_m and z_p of centres etc.

9.8.2 Estimation of distances and variances

In this section it is assumed that all configurations are given to the same scale. Thus it make sense to ask for the distances d_{km} and the variances. The problem of their estimation can easily be solved by statistical methods of the χ- and χ^2-distributions. Namely, let D be the random distance between the kth and the mth landmarks p_l and p_m.[†] Then

$$D = \left[\eta^2 + (\zeta + d_{12})^2\right]^{1/2},$$

where η and ζ are independent normal random variables with mean 0 and variance σ_{12}^2:

$$\sigma_{12}^2 = \sigma_1^2 + \sigma_2^2. \tag{9.31}$$

Consequently, D/σ_{12} has a so-called non-central χ-distribution with non-centrality parameter d_{km}^2/σ_{km}^2 and two degrees of freedom (Evans *et al.*, 1993). The density function of the distribution is

$$f(x) = x\sigma_{12}^{-2} \exp\left[-\tfrac{1}{2}\sigma_{12}^{-2}(x^2 + d_{12}^2)\right] I_0\left(\frac{xd_{12}}{\sigma_{12}^2}\right),$$

where $I_0(z)$ is the modified Bessel function of order zero. The second and fourth moments of D are

$$\mu_2 = d_{12}^2 + 2\sigma_{12}^2 \tag{9.32}$$

and

$$\mu_4 = d_{12}^4 + 8d_{12}^2\sigma_{12}^2 + 8\sigma_{12}^4. \tag{9.33}$$

These formulae lead to estimators of d_{12} and σ_{12}^2 by replacing the theoretical

---

[†]For simplicity of notation, $k = 1$, $m = 2$ and $p = 3$ are used without loss of generality.

moments with empirical ones:

$$\hat{d}_{12} = |2m_2^2 - m_4|^{1/4}, \tag{9.34}$$

and

$$\hat{\sigma}_{12}^2 = \tfrac{1}{2}|m_2 - \hat{d}_{12}^2|, \tag{9.35}$$

with

$$m_k = \frac{1}{n}\sum_{k=1}^{n} d_{12,i}^k.$$

Here $d_{12,i}$ denotes the distance between the landmarks p_{1i} and p_{2i} in the ith configuration.

As Anderson (1981) has shown, when $\sigma_{12} \ll d_{12}$, these estimators are practically of the same quality as the much more complicated maximum likelihood estimators. Simulations (Stoyan, 1990) have shown that Anderson's approximations for the variances of estimation of the maximum likelihood estimators can also be used for the estimators in (9.34) and (9.35):

$$\mathrm{var}(\hat{d}_{12}) \approx \frac{1}{n}\sigma_{12}\left(1 + \frac{\sigma_{12}^2}{d_{12}^2}\right), \tag{9.36}$$

$$\mathrm{var}(\hat{\sigma}_{12}^2) \approx \frac{2}{n}\sigma_{12}^4\left(1 + \frac{\sigma_{12}^2}{d_{12}^2}\right). \tag{9.37}$$

The bias (difference between means of estimators and true values) turns out to be very small.

The formula (9.35) at first gives only the sum of the variances σ_1^2 and σ_2^2. The single variances can be estimated by considering triplets of landmarks. That is, if in addition to p_1 and p_2 the landmark p_3 is also considered then three variance sums are available, which are connected with the single-landmark variances by

$$\sigma_{12}^2 = \sigma_1^2 + \sigma_2^2, \quad \sigma_{13}^2 = \sigma_1^2 + \sigma_3^2, \quad \sigma_{23}^2 = \sigma_2^2 + \sigma_3^2.$$

Of course,

$$\sigma_1^2 = \tfrac{1}{2}(\sigma_{12}^2 + \sigma_{13}^2 - \sigma_{23}^2)$$

etc. This leads to the estimators

$$\widehat{\sigma_1^2} = \tfrac{1}{2}(\widehat{\sigma_{12}^2} + \widehat{\sigma_{13}^2} - \widehat{\sigma_{23}^2}), \tag{9.38}$$

$$\widehat{\sigma_2^2} = \tfrac{1}{2}(\widehat{\sigma_{12}^2} + \widehat{\sigma_{23}^2} - \widehat{\sigma_{13}^2}), \tag{9.39}$$

$$\widehat{\sigma_3^2} = \tfrac{1}{2}(\widehat{\sigma_{13}^2} + \widehat{\sigma_{23}^2} - \widehat{\sigma_{12}^2}), \tag{9.40}$$

where $\widehat{\sigma_{12}^2}$, $\widehat{\sigma_{13}^2}$ and $\widehat{\sigma_{23}^2}$ are determined by (9.35). As simulations have shown, these estimators give acceptable results for samples of sufficient size ($n > 50$). However, the variances of estimation are much greater than for $\widehat{\sigma_{12}^2}$, in particular if σ_1^2, σ_2^2 and σ_3^2 have greater differences (Stoyan, 1990).

9.8.3 Shape statistics for the Bookstein model

Shape parameters are of value if not all configurations are given at the same scale and if the scales are unknown. Suppose for now that $l = 3$ and

$$\sigma_1 = \sigma_2 = \sigma_3.$$

For $l > 3$ see the literature; of course, any configuration of more than three landmarks can be divided into appropriate triangles.

The aim of the statistics is to estimate the shape parameters ϑ_x and ϑ_y for the ideal triplet z_1, z_2 and z_3. Furthermore, the quantity τ is estimated which is defined by

$$\tau^2 = \frac{\sigma^2}{\delta^2}, \tag{9.41}$$

where δ is the distance between z_1 and z_2. The starting point of the statistics is a sample of n identically independently distributed triplets $P_i = (p_{i1}, p_{i2}, p_{i3})(i = 1, \dots, n)$. The numbering of the points should be such that the side 1–2 is the (on average) longest. The quantities ϑ_{ix} and ϑ_{iy} are calculated for each i as

$$\vartheta_{ix} = \frac{2}{d_i^2}[(p_{i2x} - p_{i1x})(p_{i3x} - p_{i1x}) + (p_{i2y} - p_{i1y})(p_{i3y} - p_{i1y})] - 1, \tag{9.42}$$

$$\vartheta_{iy} = \frac{2}{d_i^2}[(p_{i2x} - p_{i1x})(p_{i3y} - p_{i1y}) - (p_{i2y} - p_{i1y})(p_{i3x} - p_{i1x})], \tag{9.43}$$

$$d_i = \|p_{i1} - p_{i2}\| \quad (i = 1, \dots, n). \tag{9.44}$$

In Mardia and Dryden (1989a) the distribution of the pair $(\vartheta_{ix}, \vartheta_{iy})$ in the Bookstein model was given. Furthermore, it was shown under which conditions on τ and on the shape of the triplet (z_1, z_2, z_3) the distribution is approximately normal (see also Bookstein, 1991). The general case of dependent normal d_{kx} and d_{ky} was considered in Goodall and Mardia (1991). The parameters of the approximating two-dimensional normal distribution are the mean-value vector

$$(\vartheta_x, \vartheta_y)$$

and the covariance matrix

$$\begin{pmatrix} \sigma^2 & 0 \\ 0 & \sigma^2 \end{pmatrix}$$

with

$$\sigma^2 = 2\frac{\sigma^2}{\delta^2}(3 + \vartheta_x^2 + \vartheta_y^2).$$

This approximation can be used if τ is 'small': for large values of $\vartheta_x^2 + \vartheta_y^2$ (≥ 3), τ should be smaller than 0.01. For small values of $\vartheta_x^2 + \vartheta_y^2$ larger τ-values are still admissible; for example, $\tau \leq 0.075$ for $\vartheta_x^2 + \vartheta_y^2 = 0$ (triangles degenerated to a line segment).

In statistical applications the approximation can still be used for larger τ-values up to 0.2 (Mardia and Dryden, 1989a).

In the case where the normal approximation can be used Mardia and Dryden (1989a) suggested the Bookstein estimators:

$$\bar{\vartheta}_x = \frac{1}{n}\sum_{i=1}^{n}\vartheta_{ix}, \tag{9.45}$$

$$\bar{\vartheta}_y = \frac{1}{n}\sum_{i=1}^{n}\vartheta_{iy}, \tag{9.46}$$

$$\widehat{\tau^2} = \frac{1}{4n(3 + \bar{\vartheta}_x^2 + \bar{\vartheta}_y^2)}\sum_{i=1}^{n}[(\vartheta_{ix} - \bar{\vartheta}_x)^2 + (\vartheta_{iy} - \bar{\vartheta}_y)^2]. \tag{9.47}$$

They are maximum likelihood estimators for the above normal distribution. The ϑ-estimators have infinite variance and a bias towards smaller values.

Mardia and Dryden (1989a) also gave the exact log-likelihood function and the corresponding likelihood equations. They suggested a numerical solution for determining maximum-likelihood estimators.

Using the estimators methods of §9.7.2, a further method of estimating ϑ_x and ϑ_y can be suggested (Stoyan, 1990). (However, it has been assumed that all configurations P_i are given at the same scale.) The method consists simply in using the estimators for the side lengths d_{12}, d_{13} and d_{23}, where (9.34) is used analogously. The shape parameters ϑ_x^m and ϑ_y^m of the triangle with these side lengths are estimators of ϑ_x and ϑ_y, which are of the same accuracy as $\bar{\vartheta}_x$ and $\bar{\vartheta}_y$ and the maximum likelihood estimates, as simulations have shown, see Stoyan and Frenz (1993). However, for larger values of τ, $\tau \geq 0.3$, the maximal-likelihood estimator should be used, see Mardia and Dryden (1994). The calculation uses

$$\vartheta_x^m = 1 - \frac{\hat{d}_{12}^2 + \hat{d}_{23}^2 - \hat{d}_{13}^2}{\hat{d}_{12}^2} \tag{9.48}$$

and

$$\vartheta_y^m = \left[\frac{4\hat{d}_{23}^2}{\hat{d}_{12}^2} - (1 - \vartheta_x^m)^2\right]^{1/2}. \tag{9.49}$$

Finally, this is a test that can be used to test the hypothesis that the form (shape) of the ideal figures given by the z are equal for two samples of landmark configurations, see Goodall (1991) and Ziezold (1990).

CHAPTER 10

Examples

10.1 THE FORM OF SAND GRAINS

The form of sand grains has been intensively studied for many years (see e.g. Mason and Folk, 1958; Friedman, 1961, 1967; Folk, 1964; Füchtbauer and Müller, 1970; Müller, 1970; Pettijohn, 1975; Engelhardt, 1979; Dowdswell, 1982; Willets *et al.*, 1982; Willets and Rice, 1983). An important aim of these studies has been a greater understanding of sedimentation conditions.

In particular, sand formations such as dunes have been considered by Bagnold (1941) and Barndorff-Nielsen *et al.* (1983). Barndorff-Nielsen (1986) and Barndorff-Nielsen and Christiansen (1988) studied the size of sand grains, and found that the so-called log-hyperbolic distribution (Barndorff-Nielsen, 1978) describes the mass and volume distribution of sand grains very well, both for aeolic and for fluvial and maritime sands. Nielsen (1985) found that superellipsoids (three-dimensional analogues of superellipses) are good models of the shape of sand grains. The process of formation of sand grains and pebbles by abrasion has also been studied, for example by Krumbein (1941) and Rayleigh (1942). By experiment, it could be shown that there is in these processes a tendency to approximate a sphere and that surface regions of greater curvature are abraded preferably. Firey (1974) studied the shape changes of three-dimensional convex sets in an abrasion process theoretically under the assumption that surface regions of greater curvature are preferentially abraded. He showed mathematically that these bodies converge towards balls in the Hausdorff metric. (see also Rogers (1976), who considered 'convergence' towards ellipsoids.)

Tables are used for the visual classification of sand grains (and analogously for pebbles and stones, for example those of Russell–Taylor–Pettijohn (Schneiderhön, 1954); see Fig. 64.[†] Using these tables, a rapid classification of sand grains is possible with respect to the degree of rounding or abrasion. Pettijohn (1975) has given the following relation for the roundness factor f_R in §8.2:

[†] Another series of particle shapes is given by the Hausner figures (Hausner, 1966; Underwood, 1980).

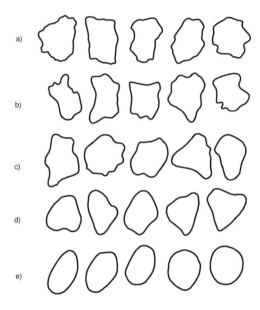

Figure 64 Degrees of roundness for sand grains and pebbles according to Russel–Taylor–Pettijohn: (a) angular; (b) subangular; (c) subrounded; (d) rounded; (e) well-rounded.

Degree of roundness	f_R
angular	0–0.15
subangular	0.15–0.25
subrounded	0.25–0.4
rounded	0.4–0.6
well rounded	0.6–1

Three sand grain samples

To demonstrate various statistical methods of form statistics of technical particles, three samples each of 24 sand grains will be studied. The starting point is a collection of photographs, i.e. images of orthogonal projections. It is assumed that the grains were photographed lying on their broadside. The sand grains have been taken from

> the Baltic Sea (a beach at Trassenheide);
> the River Selenchuk in the Caucasus;
> the Gobi Desert.

Figure 65 show the contours of the 72 sand grains. The expected form differences are visible. The river sand grains, for which the abrasion process is less advanced, are more angular than for the other two types. It is also not surprising that the sea

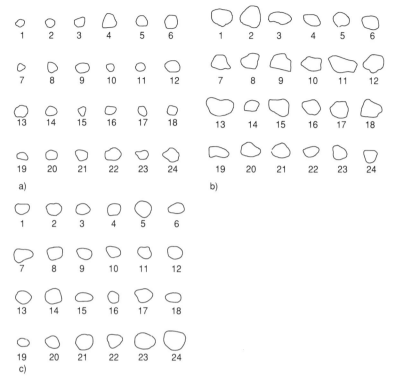

Figure 65 72 sand grains to scale 30 : 1; (a) Baltic Sea; (b) River Selenchuk; (c) Gobi Desert.

sand grains are relatively small. But it is obvious that the variation in the grains of all three types is so great that single sand grains may not always be classified. In contrast, it will be shown that samples can be classified very effectively. Many geometrical characteristics of the 72 sand grains have been measured and used to calculate shape factors such as

$$f_{AU}, \quad f_{\text{ell}} \quad \text{and} \quad f_C$$

(§8.2).

Taking the centres of gravity of the grains as reference points, the radius-vector functions are determined and the corresponding Fourier coefficients A_k are calculated. This yields further shape factors, namely

$$L_{20} = \sum_{k=1}^{20} \frac{A_k}{A_0}, \quad R_6^{20} = \sum_{k=6}^{20} \frac{A_k}{A_0},$$

$$\beta_1 = \frac{A_1}{A_0}, \quad \beta_2 = \frac{A_2}{A_0}, \quad \beta_3 = \frac{A_3}{A_0}.$$

Their means and standard deviations are shown in Table 7.

Table 7 Mean values and standard deviations of the eight shape ratios for the sand grains.

Ratio	Sea		River		Desert	
	$\bar{x}$	s	$\bar{x}$	s	$\bar{x}$	s
f_{AU}	0.862	0.031	0.808	0.063	0.860	0.050
f_C	0.988	0.017	0.965	0.028	0.987	0.015
f_{ell}	0.531	0.040	0.470	0.067	0.533	0.040
L_{20}	0.368	0.103	0.500	0.173	0.378	0.135
R_6^{20}	0.090	0.031	0.131	0.059	0.073	0.020
β_1	0.012	0.009	0.040	0.112	0.012	0.010
β_2	0.130	0.057	0.179	0.102	0.166	0.081
β_3	0.072	0.040	0.087	0.037	0.063	0.037

Table 8 Correlation coefficients for the eight shape ratios and A, U for the sand grains.

	f_{AU}	f_C	f_{ell}	L_{20}	R_6^{20}	β_1	β_2	β_3	A	U
f_{AU}	1	0.658	0.989	−0.905	−0.744	−0.397	−0.678	−0.417	−0.230	−0.315
f_C		1	0.636	−0.614	−0.671	−0.281	−0.284	−0.323	−0.358	−0.423
f_{ell}			1	−0.895	−0.718	−0.372	−0.663	−0.437	−0.203	−0.287
L_{20}				1	0.740	0.459	0.811	0.465	0.143	0.197
R_6^{20}					1	0.231	0.323	0.367	0.134	0.222
β_1						1	0.382	0.274	0.295	0.075
β_2							1	0.077	0.039	0.073
β_3								1	0.216	0.207
A									1	0.950
U										1

Obviously the shape factors are not useful for discriminating between sea and desert sand. But there are clear differences between river sand on one hand and sea and desert sand on the other. The differences in the values of f_{AU} and f_{ell} and the Fourier coefficients reflect the greater variability and length of the river sand grains.

Table 8 shows the correlation coefficients for these shape factors and for area and perimeter of the grain projections. There the 72 grains are considered as one sample in order to find general relations. (The coefficients of correlation for the various types in part differ remarkably from these 'mean' values.)

The correlations between f_{AU} and f_{ell} and between A and U are strong. As one would expect, the coefficient of correlation between f_{AU} and U is negative. The correlations between A and the shape factors are small. The relatively large correlation coefficients for L_{20} and R_6^{20}, and L_{20} and β_2, result from the fact that A_2 is rather larger in general, and clearly for large $A_6, \ldots, A_{20}$ and A_2 the sum of $A_1, \ldots, A_{20}$ is also large.

Principal-component analysis (see e.g. Dillon and Goldstein, 1984) yielded the results shown in Tables 9 and 10. The numbering of variables is the same as in Table 8.

Table 9 Eigenvalues and proportions of variance for the principal-component analysis.

Factor	Eigenvalue	% Proportion of variance
1	5.15	51.5
2	1.85	18.5
3	0.98	9.8
4	0.88	8.8
5	0.59	5.9
6	0.29	2.9
7	0.21	2.1
8	0.01	0.1
9	0.01	0.1

Table 10 Component loadings of the first three factors of the sand grain analysis.

Variable	Component loadings		
	1	2	3
1	0.95	−0.15	0.02
2	0.76	0.18	0.28
3	0.94	−0.17	0.05
4	−0.94	0.27	0.06
5	−0.78	0.11	−0.41
6	−0.50	0.02	0.57
7	−0.67	0.38	0.49
8	−0.51	−0.10	−0.34
9	−0.39	−0.89	0.22
10	−0.44	−0.86	0.05

The first component is dominated by the shape ratios f_{AU}, f_{ell} and L_{20} (variables 1, 3 and 4). Large absolute values of this component correspond to longish non-round grains. The second component is probably dominated by grain size. Here large values correspond to small grains. The third component is more difficult to interpret. Here the Fourier coefficients have a great influence. Large values seem to correspond to large deviations from circular shape.

Figure 66 shows the 10 variables in the coordinate system corresponding to the first two components. Figure 67 shows an independence graph (Whittaker, 1990) for the shape and size variables. It shows that they could be separated into three groups of more strongly correlated variables. It is interesting that the classical shape and size parameters (f_{AU}, f_{ell}, A, U, and β_1) form one group, while the Fourier coefficient shape parameters form the other, larger group. In the group of classical parameters there are two pairs of strongly correlated parameters: f_{AU} and f_{ell}, and A and U.

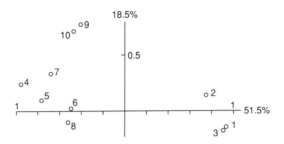

Figure 66 Result of the principal-component analysis: 10 variables in the coordinate system corresponding to the first two components. It is obvious that the first component is related to shape variables, and the second one to the size variables A and U.

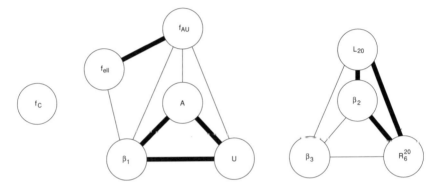

Figure 67 Independence graph of the shape and size parameters based on partial correlation coefficients greater than 0.4. The graph is not connected, but it consists of three subgraphs: the first one consists only of the vertex f_C, the second one has as vertices the classical size and shape parameters, while the third one is connected with the Fourier coefficient shape parameters. Thick edges mark partial correlation coefficients greater than 0.8.

The principal-component analysis suggests describing the form of sand grains by two variables: a size and a shape variable. For the following the variables A and f_{ell} are chosen; other possibilities could be pairs (A, f_{AU}) or (U, f_{AU}).

Figure 68 shows the 72 sand grains in the (f_{ell}, A)-plane. The discrimination between the three sand types is quite good, with the exception of extremely large or small grains. The lines shown are the discriminance lines corresponding to discriminant analysis (Dillon and Goldstein, 1984, Chap. 10 — assumption of normal distribution) if pairs of sand types are considered. They lead to the following simple classification of single sand grains into the three types:

Sea	if $A \leq 1.23 f_{ell} + 0.15$;
River	if $A \geq 5.56 f_{ell} - 1.63$;
Desert	otherwise.

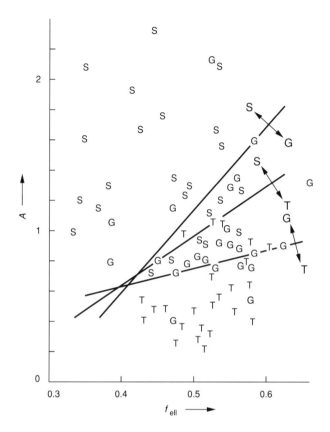

Figure 68 Plot of the 72 sand grains in the (f_{ell}, A)-coordinate system. The lines arise from discriminant analysis. T, Trassenheide Beach; S, Selenchuk River; G, Gobi Desert.

The two inequalities correspond to the discriminant lines sea–desert and river–desert.

Using this classification for the grains of the sample, the following misclassifications are observed:

> River sand grains: 8 classified as desert sand grains;
> Desert sand grains: 8 classified as sea sand grains,
>   3 classified as river sand grains;
> Sea sand grains: 2 classified as river sand grains.

As an alternative shape description, radial-rhombi are used. The formulae (7.32) and (7.33) in §7.5.2 were used to determine the lengths a and b of the semi-axis. Figure 69 shows the 72 sand grains in the (a, b)-plane. The means, standard deviations and correlation coefficients of the three sand types are given in Table 11.

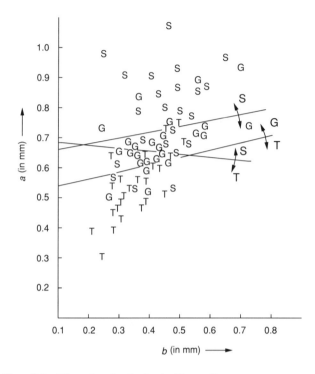

Figure 69 Plot of the 72 sand grains in the (a, b)-coordinate system corresponding to the rhombus description. As in Fig. 68, the lines come from discriminant analysis.

Table 11 Mean values $\bar{a}$, and $\bar{b}$, standard deviations s_a and s_b, and correlation coefficients r_{ab} for the radial-rhombus description of sand grains.

	$\bar{a}$	s_a	$\bar{b}$	s_b	r_{ab}
Sea	0.356	0.085	0.533	0.108	0.75
River	0.439	0.101	0.764	0.146	0.27
Desert	0.422	0.108	0.674	0.097	0.60

The one-dimensional distributions of a and b are similar to normal distribution for all three types. At least for the sea and desert grains the hypothesis of a two-dimensional normal distribution of (a, b) seems to be not bad. The nearly triangular shape of the cloud of points is particularly interesting. The sand grains approach a point at $a \approx 0.2$ and $b \approx 0.2$, i.e. a ball of radius 0.2 mm with increasing degree of abrasion.

The lines in Fig. 69 are discriminant lines for discriminant analysis just like those in Fig. 68. As with the description of the sand grains by A and f_{ell}, the following classification can be used:

Sea   if $a \leq 0.261b + 0.502$;
River   if $a \geq 0.212b + 0.628$;
Desert   otherwise.

For this classification method the following misclassifications are observed:

River sand grains: 9 classified as desert sand grains,
                                  2 classified as sea sand grains;
Desert sand grains: 7 classified as river sand grains,
                                  4 classified as sea sand grains;
Sea sand grains   5 classified as desert sand grains,
                                  1 classified as river sand grains.

Thus the success of the description by A and f_{ell} is greater than by radial-rhombi.

To further check the radial-rhombi model, the contour covariance functions have been determined and compared. These are estimates of $\mathsf{E}\chi(\varphi)$ for the radius-vector function, compared with theoretical functions for radial-rhombi calculated by (7.68). For the parameters $m_A, m_B, \ldots, \varrho_{AB}$ the empirical values $\bar{a}, \bar{b}, \ldots, r_{ab}$ are used. (Approximation of the empirical functions by contour covariance functions of radial-rhombi with parameters obtained by the least-squares method is no better.) Figure 70 shows the estimated mean contour covariance functions and their theoretical counterparts in normalized form. For values of φ below 45° the fit is acceptable, while for values greater than 90° the deviations are rather great. Besides essential deviations from the radial-rhombus model, they also result from asymmetries and local roughness.

Note that the mean empirical contour covariance functions shown in Fig. 70 differ distinctly for the three sand types. Thus these functions could be used for practical classifications. The differences are clearly connected to the shape of the grains: the sea grains are roundest (the function is close to 1) and least rough (nearly the same values for $\varphi = 0$ and $\varphi = \pi$). In contrast, the river grains are longish (the functions have a deep minimum near $\varphi = \frac{1}{2}\pi$) and very rough (the value for $\varphi = \pi$ is much smaller than $\varphi = 0$). The desert grains have a mean position.

In both variations of description (a and b, and A and f_{ell}) the success in the classifications of single grains is clearly not very great. But it seems to be useful to apply the above classification methods to the means of a and b, or A and f_{ell}, for small samples of sand grains of the same type. This was checked by an experiment for the sample sizes $n = 5$ and $n = 10$. A simple resampling method (see e.g. Efron and Tibshirani, 1993) has been used. By computer, using random numbers, new samples of sizes 5 and 10 were generated out the set of the 24 sand grains. (In this resampling process some grains may appear more than once. Such a sample of size 10 could for example consist of the grains of numbers 2, 4, 7, 7, 12, 15, 15, 16, 17 and 22.) For each new sample, using the value of a and b, or A and f_{ell}, in the memory of the computer, the means $\bar{a}$ and $\bar{b}$, or $\bar{A}$ and $\bar{f}_{\text{ell}}$, are calculated. According to the classification rules, the samples are then classified. For 1000 samples the results given in Table 12 are obtained. It is

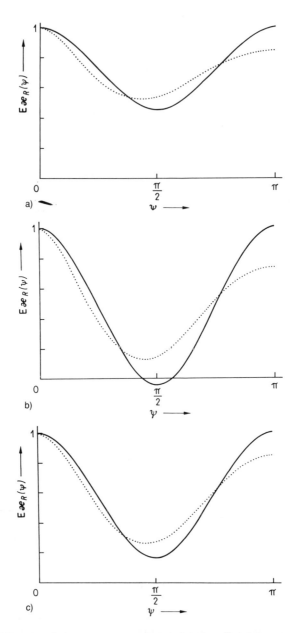

Figure 70 Estimates of the contour covariance function $\mathrm{E}\chi_R(\varphi)$ compared with the theoretical functions for radial-rhombi. The functions are normalized in such a way that they have the value one at $\varphi = 0$. (a) Baltic Sea; (b) River Selenchuk; (c) Gobi Desert; ——, theoretical; - - -, empirical.

Table 12   Results of the single grain classification after the radial-rhombus and the $f_{ell}-A$ classifications.

Radial-rhombus classification

	$n = 5$			$n = 10$		
	River	Desert	Sea	River	Desert	Sea
Sea	—	61	939	—	9	991
River	758	235	7	833	166	1
Desert	132	842	26	65	931	4

$f_{ell}-A$ classification

	$n = 5$			$n = 10$		
	River	Desert	Sea	River	Desert	Sea
Sea	2	—	998	—	—	1000
River	943	57	—	984	16	—
Desert	17	866	177	1	963	36

Table 13   Standard deviations, and minimum and maximum values of $Q_s(3)$ for the 24 sand grains.

	s	min	max
Sea	0.064	0	0.168
River	0.079	0.042	0.346
Desert	0.087	0	0.359

obvious that the classification method using A and f_{ell} is better than that based on radial-rhombi.

Finally, the spherical erosion functions $Q_s(r)$ of the sand grains are determined. Figure 72 shows the (area-) averaged curves for the three types. The curves decrease for small values of r at a nearly linear rate corresponding to the roundness of the grains. The different slopes at $r = 0$ correspond to different area : perimeter ratios, where the area weighting also plays a role. It is interesting that $Q_s(r)$ strongly discriminates between sea and desert sand grains (of course, the variations of $Q_s(r)$ for single grains are considerable). For example, the standard deviations and minimum and maximum values of $Q_s(3)$ are shown in Table 13. (here no area weighting was made).

Comparison of the curves of Figs. 71 and 72 shows that the contour covariance function $E\chi_R(\varphi)$ differentiates the sand type better than $Q_s(r)$. However, the effort in determining $E\chi_R(\varphi)$ is somewhat greater than for $Q_s(r)$.

Prod'Homme *et al.* (1991) studied in a similar way $BaTiO_3$ and Al_2O_3 powders, mainly by using shape factors.

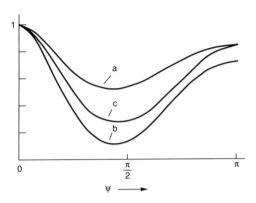

Figure 71 Plot of the three empirical contour covariance functions of Figs. 70(a–c). There are clear differences between these functions for the three types of sand.

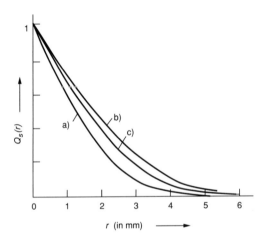

Figure 72 Spherical erosion functions $Q_s(r)$ for the three sand types. The curves are area-weighted means of each 24 curves.

10.2 THE FORM OF HANDS

The form of human organs is intensively studied by anthropologists, human geneticists, ergonomicists and designers. There is a vast literature on this topic. Usually certain lengths or perimeters are measured. Shape ratios have been used only rarely until now. For hands, the hand perimeter, the hand breadth and thickness, and finger lengths are often considered.

Fundamental facts on the form of hands are known by everybody. Male hands are usually bigger than female ones, and it is believed that female hands are more slender than male ones. In this section the form of hands will be studied.

20 student hands

The authors have studied the form of left hands of students aged between 20 and 24 : 10 males and 10 females. (These small sample sizes mean that the statistical analysis is largely a methodological example. For general statements on the form of students' hands they are much too small.)

The basis of the analysis is the hand contour. Since they have a natural longitudinal direction, it seems natural to use an adapted coordinate system as shown in Fig. 73.

Using this coordinate system, a set theoretic form analysis is possible. In particular, the covering function $p_X(x)$ can be estimated. Figure 74 shows this function given by some isolines separated for both sexes. It is obvious that there are remarkable deviations of form. The 0.5 isoline gives the Vorob'ev median. It clearly shows the differences of size, while shape differences are not so easily detected.

The shape differences are studied by the landmark approach. For each hand contour nine landmarks are chosen as in Fig. 73. The coordinates of these points are collated in Table 14. As a first step, quite elementary methods of form statistics are used. The hand size is characterized by the distance d_{38} of landmarks 3 and 8.

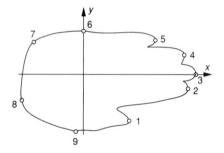

Figure 73   A left hand with nine landmarks and an adapted coordinate system.

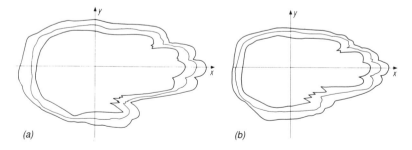

(a)   (b)

Figure 74   Plot of the empirical covering function $p_X(x)$ for (a) male and (b) female hands. The outer curves are the 0 isolines, the inner isolines the 1 lines and the mean isolines the 0.5 lines. The latter define the Vorob'ev medians.

Table 14 Coordinates of the nine landmarks for 20 hands. For an explanation of d_{38}, f and $s(P)$, see text.

	1	2	3	4	5	6	7	8	9	d_{38}	f	$s(P)$
Male												
1	52 −55	120 −18	131 0	117 17	84 34	0 46	−61 31	−71 −31	−13 −64	205	1.85	249.7
2	40 −58	110 −18	120 0	112 13	85 31	0 47	−62 38	−73 −38	−28 −62	196	1.75	247.1
3	51 −50	114 −19	125 0	114 16	83 34	0 46	−57 34	−69 −34	−18 −61	198	1.82	242.1
4	41 −45	103 −14	115 0	104 14	77 33	0 45	−66 28	−76 −28	−26 −57	193	1.83	234.8
5	43 −44	109 −18	122 0	111 16	80 30	0 43	−73 31	−82 −31	−29 −63	206	1.88	249.9
6	55 −58	116 −16	129 0	117 15	87 29	0 48	−79 21	−86 −21	−31 −60	216	1.92	264.0
7	35 −57	94 −20	107 0	99 14	69 28	0 40	−57 19	−75 −19	−27 −66	184	1.67	223.4
8	41 −63	104 −15	120 0	112 16	84 36	0 54	−66 39	−68 −39	−22 −63	192	1.61	247.3
9	40 −53	102 −19	112 0	99 16	75 20	0 47	−64 25	−70 −25	−25 −60	193	1.67	228.4
10	47 −53	108 −16	124 0	112 15	82 29	0 46	−67 31	−64 −31	−24 −56	190	1.81	240.0
Female												
11	41 −41	99 −14	110 0	101 13	75 25	0 39	−49 26	−57 −26	−18 −49	169	1.88	207.6
12	52 −39	109 −13	117 0	104 12	75 26	0 41	−50 26	−61 −26	−8 −48	180	2.01	214.7
13	40 −48	98 −16	108 0	87 14	70 29	0 42	−55 19	−65 −19	−21 −52	175	1.81	210.0
14	30 −51	86 −17	98 0	87 11	60 25	0 35	−54 26	−66 −26	−23 −53	166	1.83	200.8
15	36 −49	102 −17	110 0	99 9	74 24	0 38	−55 27	−64 −27	−28 −61	176	1.71	220.0
16	42 −45	98 −17	109 0	98 13	71 27	0 43	−40 31	−62 −31	−18 −52	175	1.80	208.3
17	36 −43	98 15	109 0	88 13	71 30	0 45	−59 33	−63 −33	−23 −49	175	1.81	214.3
18	39 −45	91 −14	95 0	88 11	61 23	0 35	−53 23	−61 −23	−20 −50	159	1.81	195.7
19	45 −34	94 −13	102 0	90 16	62 31	0 45	−40 25	−63 −25	−12 −46	167	1.82	195.6
20	35 −46	96 −15	105 0	97 12	71 30	0 43	−54 30	−65 −30	−27 −54	172	1.71	215.0

As the numbers in Table 14 show, even for the smallest male hand d_{38} is greater than for all female hands. The means and standard deviations are as follows:

$$\text{Male} \qquad \bar{d}_{38} = 197.3, \quad s_{38} = 9.3;$$
$$\text{Female} \quad \bar{d}_{38} = 171.4, \quad s_{38} = 6.2.$$

Table 14 also contains the size parameters $s(P)$ of the hands calculated by (9.4). They also show clearly the sex-specific size differences.

A useful shape ratio for hands is

$$f = \frac{d_{38}}{d_{69}},$$

where d_{69} is the distance between landmarks 6 and 9. Large values of f appear for slender hands. The f-values of the students' hands are also given in Table 14. Again they show sex-specific differences, but these shape differences are smaller than the size differences. The means and standard deviations of f are as follows:

$$\text{Male} \qquad \bar{f} = 1.78, \quad s_f = 0.10;$$
$$\text{Female} \quad \bar{f} = 1.82, \quad s_f = 0.08.$$

Thus female hands are slightly more slender than male ones.

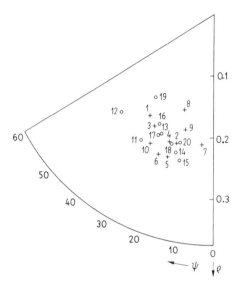

Figure 75   Plot of the 20 hands in (r, ψ)-coordinates. $+$, male, o, female.

A further statistical method is to use triangle analysis.[†] The shape of a hand is described by the triangle $\triangle_{369}$, whose vertices are landmarks 3, 6 and 9. Figure 75 shows the points in the (r, ψ)-diagram which correspond to the triangles (§9.6.2). Figure 76 shows the most extreme male and female hands. The latter is very slender, while the male hand is rather short. The triangle shape ratios ϑ_x and ϑ_y defined by (9.38) and (9.39) have been calculated for $A = p_9$, $B = p_3$ and $C = p_6$, yielding the following estimates $\bar{\vartheta}_x$ and $\bar{\vartheta}_y$:

	$\bar{\vartheta}_x$	$\bar{\vartheta}_y$
Male	-0.181,	1.138;
Female	-0.223,	1.143.

The differences are very small, but again the female hands appear to be more slender. (The female ϑ-values correspond to a more acute angle at the landmark corresponding to the tip of the middle finger.) Multidimensional scaling has been used to demonstrate geometrically the variability. The Δ_{ij} obtained by (9.13) have been used as proximities. The so-called ALSCAL algorithm of SAS gave Fig. 77. Here the points corresponds to hands, and the means of all male hands (m), all female hands (f) and of all hands (t) are also shown. The result is similar to the

---

[†] Here the numbering of the triangle corners is essentially as described in the case of a point triple on p. 158. That is, for all $\triangle_{369}$ the longest side is between landmarks 3 and 6, and the shortest between 6 and 9, so that the triangle lies in the distinguished sector of Fig. 63. This corresponds to the description of triangles as on p. 153.

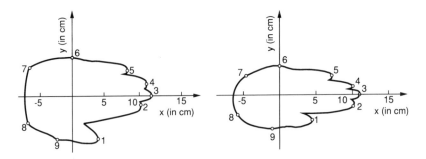

Figure 76   The two extreme hands: 8 (male) and 12 (female) (see Fig. 74).

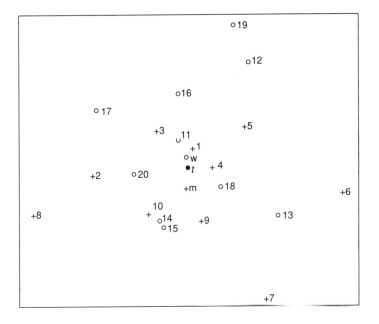

Figure 77   Result of multidimensional scaling applied to the Δ-proximities of the 20 hands.

(r, ψ)-diagram of Fig. 75 with respect to the fact that there is no clear discrimination between female and male hands; again the extremal hands of both sexes lie at great distances from the centre of the points cloud.

Further statistical work was based on Procrustean analysis. The hands are first compared with a 'mean' hand P. Its landmarks are the means of the landmarks of the 20 hands. Because of the choice of the hand-related coordinate system, the difference from the mean figure obtained by the algorithm in §9.4 is vary small; the differences are of the order of 0.1 mm, i.e. less than the precision of measurement. In Procrustean analysis three cases of transformations have been considered.

(a) dilations alone;
(b) x-translations and dilations;
(c) translations, dilations and rotations.

In the hand sample the numbers r and ψ for these three variants differ only a little.

The fact that in the given sample the male hands are bigger than the female ones is reflected by dilation ratios $\varrho_l = \varrho(P, P_l)$ that are greater than 1 for the male hands but less than 1 for the female. The ϱ_l are similar to the quantities $d_{38}/$(mean of all d_{38}). But for equal d_{38} the ϱ-values may be different. For the relatively long hand 4 and the relatively short hand 9 the same value $d_{38} = 1.93$ is obtained; the corresponding r-values are 1.043 and 1.020.

For the delta-values corresponding to case (c) (multiplied by 10^3) the following stem-and-leaf plot has been obtained:

$$
\begin{array}{r|l|l}
7 & 0 & 5567777 \\
(12) & 1 & 02256 \\
8 & 2 & 1124 \\
4 & 3 & \mathbf{5568}
\end{array}
$$

The bold numbers correspond to male hands. It is obvious that the male hands vary in shape more than the female ones. The most extreme hands with 3.6 and 3.8 are hands 8 and 6; while 8 is very broad and short, 6 is very long.

The Δ_{ij}-values were then analysed by cluster analysis. Figures 78 and 79 show the corresponding dendrograms. In case (b) there are many hands that form single clusters. For example, in the 7-cluster solution these are hands 6, 7, 8, 12, 17, and 19. Somewhat bigger clusters are obtained in case (c). The better structuring may perhaps be explained by the greater set of transformations and thus the possibility of a better fit. Figure 78 is an attempt to order the hands according to the shape ratio f. Figure 79 is an attempt to plot the dendrogram in such a way that female and male hands are separated. Again the female hands are more slender. But, since in the central part of the dendrogram female and male hands appear mixed, sex differences have only a small influence on the shape of hands, if described by our nine landmarks. There is a series of extremal hands forming single clusters. Hands 7, 8 and 9 are typical short strong male hands, hand 6 is a very long male hand, and hands 12, 17 and 19 are slender female hands.

Finally, the hand data have also been analysed using the Bookstein method. The formulae on p. 162 applied to the triangle Δ_{369} yielded for the male hands the values (distances, in mm)

$$\hat{d}_{36} = 129.0,$$
$$\hat{d}_{39} = 157.2,$$
$$\hat{d}_{69} = 110.2,$$

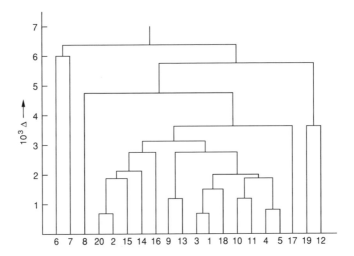

Figure 78 Dendrogram resulting from a cluster analysis of the hand data. The proximities used are the Δ-values corresponding to x-translations and dilations.

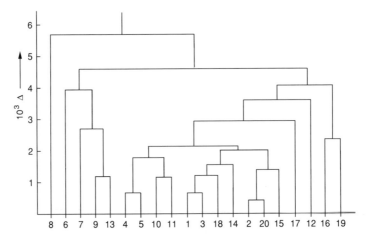

Figure 79 As Fig. 78, but the proximities are now the Δ-values corresponding to all Euclidean motions and dilations.

and (landmark variances, in mm)

$$\widehat{\sigma_3^2} = 37.4,$$
$$\widehat{\sigma_6^2} = 12.1,$$
$$\widehat{\sigma_9^2} = 2.9.$$

The large differences in the variances are remarkable. By simulation, it can be shown that they are not compatible with the assumption of equal variances (as assumed in the original Bookstein model) (Stoyan, 1990). Perhaps some readers will consider the great variance around the tip of the middle finger as plausible, since it reflects the fluctuations of hand lengths. Probably, the assumption of a radial normal distribution of the deviations from the landmark centre is not true. One of the more general models may fit the data better.

A quite different approach in that modelling hands is that of Chow *et al.* (1991) and Grenander and Keenan (1987), who considered the hands as closed curves consisting of curve pieces that are stochastically deformable. There Markovian dependences were assumed.

PART III
Point Field Statistics

CHAPTER 11

Fundamentals

Random point fields

Random point fields are mathematical models for irregular 'random' point patterns, like those shown in Figs. 80, 81, 82, 86 and 87. While mathematicians study point fields in spaces of arbitrary dimension, only planar point fields are considered in this book; this is the most important case in connection with geometrical problems. In the literature the term 'point processes' tends to be used rather than 'point field', even when time-independent phenomena are being considered.

There is a vast literature about point processes. Cox and Isham (1980) and Daley and Vere-Jones (1988) provide good introductions. Stoyan *et al.* (1987) also contains introductory chapters. The present text is meant to be an introduction for non-mathematicians, where the aim is to make accessible the fundamental notions and the important statistical methods.

The random point field studied is denoted by N. For a given Borel set B, $N(B)$ is the random number points of N contained in B. It is assumed that all point fields considered are 'simple', that is there are no multiple points. Thus the set of all points $\mathbf{x}_1, \mathbf{x}_2, \ldots$ is a random set.[†] It is written as

$$N = \{\mathbf{x}_n\} \text{ and } \mathbf{x} \in N.$$

'$\mathbf{x} \in N$' means that the point $\mathbf{x}$ is a point of the field. A further piece of notation must be explained. Suppose one is given a function $f(\mathbf{x})$. Consider the sum S of all function values $f(\mathbf{x}_n)$ as $\mathbf{x}_n$ runs through n. It may be written as

$$S = \sum_{\mathbf{x}_n \in N} f(\mathbf{x}_n) = \int_{R^2} f(x) N(\mathrm{d}x).$$

If the summation only ranges over the points in a Borel set B, it is written as

$$\sum_{\mathbf{x}_n \in N \cap B} f(\mathbf{x}_n) \quad \text{or} \quad \sum_{\mathbf{x}_n \in N} 1_B(\mathbf{x}_n) f(\mathbf{x}_n) \quad \text{or} \quad \int_B f(x) N(\mathrm{d}x).$$

[†]The points of the point field will be shown in bold, in contrast to variables or other points of R^2.

Here $1_B(\mathbf{x})$ is the indicator function of B.

$$1_B(x) = \begin{cases} 1 & (x \in B), \\ 0 & (x \notin B). \end{cases}$$

Probabilities connected with random point fields are written as

$$\Pr(N(B) = i) = \text{probability that } i \text{ points of} \\ N \text{ lie in the set } B.$$

As usual, means are denoted by the E-symbol:

$$\mathsf{E}N(B) = \text{mean number of points of } N \text{ in } B.$$

A random point field is a random variable in the sense of probability theory. Corresponding to this is a random mechanism, which, in principle, can generate infinitely many point patterns. Each is called a sample of the point field.

Marked random point fields

Marked random point fields[†] are a refinement of random point fields. Each point has a 'mark' distinguishing it to a greater extent. Many different quantities can be used as marks, for example

- qualitative marks to distinguish different groups; for example if forests are being investigated (0 = pine, 1 = spruce, 2 = birch);
- quantitative marks to characterize the object; for example 'stem diameter' for points thought of as trees or 'particle area' for points thought of as particle centres;
- descriptive marks for finer characterization of the objects; for example, let the marks be a series of form parameters where points represents particle centres.

Marked fields are denoted by N, as for unmarked fields. N, denoting the set of all marked points $[\mathbf{x}_1; m_1], [\mathbf{x}_2; m_2], \ldots$, is written as $N = \{[\mathbf{x}_n; m_n]\}$. The $\mathbf{x}_n$ are points of the plane and the m_n are the marks. The latter belong to the mark set M. This could have finitely many elements or it could be the set of all real numbers. Still more complicated cases are also possible, such as $M = \mathcal{K} = $ set of all compact subsets of R^2. (In the mathematical literature it is assumed that M is a so-called Polish space.) If B is a Borel set of R^2 and C is a subset of M then $N(B \times C)$ is the number of points in B with a mark in C.

Of course, any marked point field can be interpreted as a non-marked point field in a more general space, namely in the product space $S = R^2 \times M$. If A is a subset of S then $N(S)$ denotes the number of (marked) points in S. But, in the theory of point fields, the points and marks are treated differently. If a marked

---

[†] The adjective 'random' is often omitted.

point field is translated then only the points are changed, but not the marks. After translation by a vector $x \in R^2$, the marked point field $N = \{[\mathbf{x}_n; m_n]\}$ becomes $N = \{[\mathbf{x}_n + x; m_n]\}$. The situation is similar for rotations.

Finite and infinite point fields

All point patterns analysed by statisticians are finite. Nevertheless, it makes sense to use models that correspond to infinite point fields. The choice depends on the given situation. There are point patterns that are finite in principle, because they correspond to strictly locally limited phenomena. Some simple examples are

- the centres of bullet marks on a target;
- the positions of seedlings grown in one year from one plant;
- the centres of pores in the surface of a block of steel.

In other cases the given point pattern may be considered as a part of a much larger pattern in which the points are distributed according to the same law as in the window of observation. (At least it may be plausible to assume this.) Examples are

- the positions of trees in forests;
- the grain centres in planar sections of probes of materials such as metals or ceramics;
- the positions of seedlings that are generated in one year from a community of plants (e.g. a forest).

In these cases the point pattern is usually a small part of a much larger pattern. It is also possible that a forest analysed is only a small island in a landscape of meadows and fields. Even in such a case it may make sense to analyse tree positions by methods designed for infinite point fields, perhaps after excluding the trees at the boundary.

Usually, if one decides to analyse a point pattern assuming an infinite point field then further assumptions are made, namely those of homogeneity and frequently also of isotropy.

Homogeneity and isotropy

An assumption frequently made in the analysis of planar point patterns is that of *homogeneity* or *stationarity*[†]. It is fundamental for many statistical methods, and many characteristics of point fields make sense only for homogeneous point fields.

A point field $N = \{\mathbf{x}_n\}$ is called homogeneous if N and the translated field $N = \{\mathbf{x}_n + x\}$ have the same distribution for all $x \in R^2$. Less mathematically, homogeneity means the following. If the point field is observed from different regions of the plane then similar point configurations are observed. Differences only result from random fluctuations, which follow the same laws. Figures 81, 86, 87,

---

[†] In mathematical texts the term 'stationarity' is usually used.

96, 98, 118 and 119 show patterns of a form that mathematicians associate with the term 'homogeneous pattern'. But note that many physicists and material scientists make another use of the word 'homogeneous'. For them it means 'uniform'; thus they would not call the patterns in Figs. 87 and 118 homogeneous. But in our sense, if it can be assumed that the point clusters on these figures are scattered similarly on the whole plane, the word 'homogeneity' makes sense.

There are many possibilities for deviation from homogeneity.

- The point density may vary systematically. An example is the tree density in mountain forests, which decreases with increasing height.

- The frequency of occurrence of certain point configurations may be different in different regions of the plane. For example, it is possible that the points appear in clusters in one part of the plane but not in another, or that the cluster size is location-dependent.

The strict proof that a given point pattern behaves like a sample of a homogeneous point field is very difficult. If, as is usual in spatial statistics, only one pattern in a bounded window is given then such a proof is, in a strict sense, impossible. Namely, in a bounded window a homogeneous field may (with small probability) have samples that appear inhomogeneous.

Conversely, an inhomogeneous field may fool the statistician and show homogeneous-like behaviour in a bounded window. It is helpful to use special ecological or other scientific arguments to justify homogeneity. (In the case of a forest an argument could be that there are *unique* geological and climatic conditions.)

If more than one sample is given, and if this can be considered to be independent, then at least some aspects of homogeneity can be tested.

Example. In a fixed rectangular window W, n samples of a point field are given, which are assumed to be independent. It is suspected that in the upper part of the window the point density is greater than in the lower one. Let the distributional properties and observation conditions be such that theorems on asymptotic normality of point numbers (Ivanoff, 1982, Heinrich and Schmidt, 1985) are applicable. Then the test for comparing the means of two independent samples (Sachs, 1984) may be used. Here the values of the first sample are the point numbers in the upper part of the window, and those of the second sample the point numbers in the lower part, where both parts have the same area. (For example, each part could be a fifth of the whole area.)

In some situations it is possible to work with weaker homogeneity properties. A suitable notion is that of *statistical homogeneity* (Cowan, 1989). Here it is supposed that the mean point numbers are translation-invariant, i.e.

$$EN(B) = EN(B_x) \qquad (11.1)$$

for all compact B and all x.

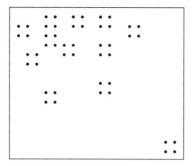

Figure 80 Inhomogeneous point pattern with a similar local point distribution around any point.

Matheron (1989) has suggested using 'local' homogeneity properties. Indeed, in small neighbourhoods of points there are often similar distributional situations in spite of a clearly inhomogeneous global distribution of points. A simple example is shown in Fig. 80. Obviously, the point pattern looks 'rather inhomogeneous'. But the nearest neighbour distances and the point numbers in small discs around points behave quite similarly. Therefore characteristics originally introduced for homogeneous fields can also be used in the inhomogeneous case if related to short interpoint distances.

The latter remark is more important than it first appears. Matheron (1989) has discussed comprehensively the possibility of contrasting quantities defined in the homogeneous case to such characteristics that

- also make sense in the inhomogeneous case, and
- are interpretable in the language of the data and are falsifiable in the Popperian sense.

As a simple example the *intensity* λ is discussed here. For a homogeneous point field this is the mean point number in any set B of unit area:

$$\lambda = \mathsf{E}N(B),$$

where the position and shape of B are irrelevant. In contrast, in the inhomogeneous case $\mathsf{E}N(B)$ may depend on the position and shape of B, and it is impossible to define an intensity.

Now it has to be considered that the notion of homogeneity is a mathematical idealization and the intensity a mathematical fiction. Nevertheless, if a real point pattern is given, one can speak of a 'mean point density', which is simply point number divided by area. If the pattern is 'sufficiently homogeneous' then this mean point density plays almost the same role that the intensity plays in the mathematical theory in the homogeneous case. But in the inhomogeneous case the notion 'mean point density' can also be given a meaning. Imagine a test disc of unit area. It is placed in the window W at random (this means that every location in W with the property that the disc centred at it is completely in W has the same chance to be chosen as centre). Then the number of points in the test disc is a random variable, whose mean is nearly the mean point density.

Isotropy is a property similar to homogeneity. Instead of translations, rotations around the origin o are considered. Let R_α be a rotation by the angle α ($0 \leq \alpha \leq 2\pi$) around o. If x is a point of R^2 with coordinates ξ and η then $R_\alpha x$ is the point with coordinates

$$\xi_\alpha = \xi \cos \alpha + \eta \sin \alpha,$$
$$\eta_\alpha = -\xi \sin \alpha + \eta \cos \alpha$$

(for positive α the rotation is anticlockwise). The point field N is called *isotropic* if $N = \{\mathbf{x}_n\}$ and $R_\alpha N = \{R_\alpha \mathbf{x}_n\}$ have the same distribution for any α.

If a point field is both homogeneous and isotropic then it is called *motion-invariant*. Then rotations around arbitrary points ($\neq o$) do not change the distribution either.

In the case of marked point fields the definitions of homogeneity and isotropy are analogous. Note that translations and rotations leave the marks invariant; the point $\mathbf{x}_n + x$ of the point field N shifted by x has the same mark as $\mathbf{x}_n$ of N etc.

Ergodicity

In point process statistics ergodicity is frequently assumed in addition to homogeneity (and isotropy). This property ensures that one sample (one point pattern) is sufficient to obtain statistically secure results, assuming that a sufficiently large window W is used. For example, if ergodicity holds then

$$\lim_{W \uparrow R^2} N(W)/A(W) = \lambda, \tag{11.2}$$

where λ is the intensity. Here $W \uparrow R^2$ is taken to mean that W contains a disc of radius r, such that r tends to infinity. The formal definition is not given here; see the literature on point processes.

A sufficient condition for ergodicity is the following *mixing* property. A homogeneous point field N is called mixing if for all $\mathcal{A}$ and $\mathcal{B}$

$$\Pr(\text{'}N \text{ has the property } \mathcal{A}, N_x \text{ has the property } \mathcal{B}\text{'})$$
$$\rightarrow \Pr(\text{'}N \text{ has the property } \mathcal{A}\text{'}) \Pr(\text{'}N_x \text{ has the property } \mathcal{B}\text{'})$$

as $\|x\| \rightarrow \infty$. For example, $\mathcal{A}$ may mean that the disc $b(o, r)$ does not contain a point and $\mathcal{B}$ may mean that the same disc contains two points. Then for mixing N

$$\Pr(N(b(o, r)) = 0, N(b(-x, r)) = 2)$$
$$\rightarrow \Pr(N(b(o, r)) = 0) \Pr(N(b(o, r)) = 2).$$

An example of an ergodic (and mixing) point field is the homogeneous Poisson field (Chapter 13). In contrast, a so-called mixed Poisson field (Stoyan *et al.*, 1987) is not ergodic. The lattice point fields where the points are at the nodes of a regular lattice, which may be randomly positioned in the plane, are also non-ergodic.

Matheron (1989) has used the notion 'local ergodicity' similarly to 'local homogeneity' as discussed on p. 193. If this property holds then many statistical methods originally developed for homogeneous and ergodic point fields can be used.

CHAPTER 12

Finite Point Fields

12.1 INTRODUCTION

This chapter deals with point fields that cannot be interpreted as parts of larger homogeneous, isotropic point fields. Rather, they are locally bounded and strictly inhomogeneous. For example consider the positions of point defects on the surface of a silicon wafer or geological objects of volcanic origin in a restricted region.

In statistical analyses two main cases have to be distinguished.

1. The sample consists of one point pattern. The aim of the analysis is a description of the point distribution and, if possible, a statistical model explaining the variation in the pattern.

2. The sample consists of m independent point patterns with the same probability distribution. The window of observation is always the same (e.g. the area of the wafer surface). In this case the probability distribution of the number n of points per pattern is of interest (e.g. the number of point defects per wafer). Then, as for case 1, the distribution of the points in the pattern is investigated, which may depend on n. An alternative aim is a stochastic model for the point field.

In the following, point fields of a fixed number of points are considered first. Then some theoretical distributions of random numbers are discussed briefly, for example, the Poisson and binomial distributions. Point fields with a random number of points are also considered. Two examples of the fixed-point-number case conclude the chapter; examples of the random point number case are considered in Chapter 13 in connection with inhomogeneous Poisson fields.

12.2 POINT FIELDS OF A FIXED NUMBER OF POINTS

12.2.1 Two stochastic models

(a) The binomial field

Let W be a window (e.g. a rectangle or another compact set with inner points), within which n points are independently and uniformly distributed. This means the following.

1. The n points $\mathbf{x}_1, \ldots, \mathbf{x}_n$ are stochastically independent, i.e. the probability that $\mathbf{x}_1$ lies in the Borel set $B_1 \subset W, \ldots$, that $\mathbf{x}_n$ lies in the Borel set $B_n \subset W$ satisfies the product formula

$$\Pr(\mathbf{x}_1 \in B_1, \ldots, \mathbf{x}_n \in B_n) = \Pr(\mathbf{x}_1 \in B_1) \cdots \Pr(\mathbf{x}_n \in B_n). \tag{12.1}$$

2. Each of the points $\mathbf{x}_1, \ldots, \mathbf{x}_n$ is uniformly distributed in W, i.e. for any $i = 1, \ldots, n$ and any Borel set $B \subset W$

$$\Pr(\mathbf{x}_i \in B) = \frac{A(B)}{A(W)}. \tag{12.2}$$

The probability that $\mathbf{x}_i$ lies in B is proportional to the area of B. The sequence of points $\mathbf{x}_1, \ldots, \mathbf{x}_n$ forms the binomial field.

Important formulae. The mean number of points per unit area is

$$\lambda = n/A(W). \tag{12.3}$$

The mean number of points in the Borel set B is

$$\mathsf{E}N(B) = \lambda A(B). \tag{12.4}$$

The one-dimensional number distributions are

$$\Pr(N(B) = k) = \binom{n}{k} p_B^k (1 - p_B)^{n-k} \quad (k = 0, 1, \ldots, n), \tag{12.5}$$

where $p_B = A(B)/A(W)$. Thus the number $N(B)$ of points in B has a binomial distribution — hence the name of this point field. The m-dimensional number distributions (for pairwise-disjoint B_i, $B_1 \cup \cdots \cup B_m = W$) are

$$\Pr(N(B_1) = k_1, \ldots, N(B_m) = k_m) = \frac{n!}{k_1! \ldots k_m!} \frac{A(B_1)^{k_1} \cdots A(B_m)^{k_m}}{A(W)^n}, \tag{12.6}$$

where $k_1 + \cdots + k_m = n$. Note that the point numbers $N(B_i)$ and $N(B_j)$ are not independent, even if B_i and B_j are disjoint.

Emptiness probability. This is given by

$$\Pr(N(B) = 0) = \frac{[A(W) - A(B)]^n}{A(W)^n}.$$

Simulation. First suppose the window W is the rectangle of side lengths a (parallel to the x-axis) and b (parallel to the y-axis) with left lower corner at the origin: $W = [0, a] \times [0, b]$. A point pattern that behaves like a sample of a binomial field of n points is obtained by generating a uniform point in W n times. A BASIC program which generates a list of the corresponding coordinates is as follows:

```
10   FOR I = 1 TO n
20   X = a * RND(0)
30   Y = b * RND(0)
40   PRINT X, Y
50   NEXT I
```

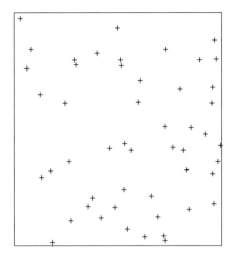

Figure 81 Simulated sample of a binomial field in a rectangle. The number of points, n, is 50.

Figure 81 shows a point pattern generated in this way. It is clear that a good random number generator is essential for 'quality' (the degree to which uniformity and independence properties are satisfied). Devroye (1986) and Ripley (1987) are recommended as references on random numbers. If W has another form then the rejection method is used. (Particular methods are used for circular W (Devroye, 1986).) A rectangle W^* is chosen that contains W. Then independent uniformly distributed points in W^* are generated. Each point that lies in W is accepted; points outside are rejected. This procedure is used until n points in W are obtained.

Statistics. If W and n are given then the statistician has only to test the validity of the model assumptions. This can be done using quadrat count methods described in §13.2.5. The L-test (p. 224) can also be used. (Ripley's and Koen's simulations have only been made for patterns of a fixed point number, that is, for a binomial field but not for a Poisson field.)

The two examples considered in §12.2.3 demonstrate the application of these tests.

(b) I–I-point field

Take a two-dimensional probability density function $f(x)$, $x = (x_1, x_2)$. The density of the two-dimensional normal distribution provides an example.

$$f(x_1, x_2) = \frac{1}{2\pi\sigma_1\sigma_2\sqrt{1 - \varrho^2}} \exp\left\{ -\frac{1}{2(1 - \varrho^2)} \right.$$

$$\times \left. \left[\frac{(x_1 - \mu_1)^2}{\sigma_1^2} - 2\varrho\frac{(x_1 - \mu_1)(x_2 - \mu_2)}{\sigma_1\sigma_2} + \frac{(x_2 - \mu_2)^2}{\sigma_2^2} \right] \right\}.$$

The point field is formed by n points that are stochastically independent ('I') and identically distributed ('I') with respect to $f(x_1, x_2)$. It is theoretically possible that the points lie in the whole of R^2.

Clearly, the binomial field is a particular case of the I–I-point field where

$$f(x_1, x_2) = \begin{cases} \dfrac{1}{A(W)} & (x \in W), \\ 0 & (x \notin W). \end{cases}$$

It is easy to see that the random number $N(B)$ of the points in the Borel set B has a binomial distribution. But its parameter p depends on the position of B.

The simulation of such point fields consists in simulating n independent random points with the density function $f(x)$. In the particular case of the normal distribution the following log-tri-algorithm is useful (Jansson, 1964); it yields the coordinates of a normally distributed point (x, y):

```
10 U1 = RND(0) : U2 = RND(0)
20 x = μ₁ + σ₁ * SQR(−2 * LOG (U1)) *
   (SQR(1 − ϱ²) *
   COS(2 * π * U2) + ϱ * SIN(2 * π * U2))
30 y = μ₂ + σ₂ * SQR(−2 * LOG(U1)) * SIN(2 * π * U2)
40 RETURN
```

Statistical analyses use methods of statistics for random vectors, which are well-known in the case of normal distribution; see the following example.

12.2.2 Two geological examples

(a) Basaltic formation in the area of the Swabian Alps

Figure 82 shows the centres of Tertiary basaltic formations in the Swabian Alps. Clearly, the points are randomly distributed and concentrated in the central region. Thus it is natural to try to describe them by an I–I-point field with a normal distribution density. It is known that the normal distribution density has ellipses as isolines; the equations of the ellipses are

$$\frac{(x_1 - \mu_1)^2}{\sigma_1^2} - 2\varrho \frac{(x_1 - \mu_1)(x_2 - \mu_2)}{\sigma_1 \sigma_2} + \frac{(x_2 - \mu_2)^2}{\sigma_2^2} = c^2,$$

where c is a constant. In the ellipse given by the constant c the probability mass $P(c)$ is given by

$$P(c) = 1 - \exp\left[-\frac{c^2}{2(1 - \varrho^2)}\right].$$

The coordinates belong to an a priori coordinate system. In the example the x_1-axis is the lower edge of the figure (W–E direction) and the x_2-axis is the left edge (S–N direction). The meaning of the parameters is simple; μ_1 and μ_2 are the means of the x_1- and x_2-coordinates respectively, σ_1^2 and σ_2^2 are the corresponding

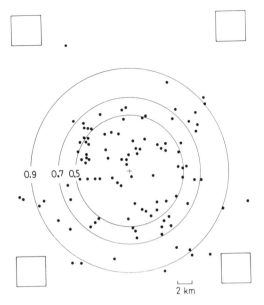

Figure 82 Distribution of Tertiary basalt bodies in the Swabian Alps. The centre of the point pattern is near the town Dettingen on the Erms. The circles are 50%, 70% and 90% lines of an approximating two-dimensional normal distribution. The squares are outer quadrats used in the dispersion index test.

variances, and ϱ is the correlation coefficient of the point coordinates. Estimators of these parameters are $\bar{x}_1$, $\bar{x}_2$, s_1^2, s_2^2 and r, i.e. the usual estimators of classical statistics.

The centre of the ellipses introduced above lies at the point (μ_1, μ_2). The major semi-axis lies on the line given by the equation

$$x_2 = \mu_2 + \text{sign}(\varrho)\frac{\sigma_2}{\sigma_1}(x_1 - \mu_1), \quad \text{sign}(\varrho) = \begin{cases} 1 & (\varrho > 1), \\ 0 & (\varrho = 0), \\ -1 & (\varrho < 0). \end{cases}$$

In the case $\varrho = 0$ circles are obtained. With $n = 105$ the example given here yields the values

$$\bar{x}_1 = 15.6, \; s_1 = 6.0,$$
$$\bar{x}_2 = 14.9, \; s_2 = 5.9,$$
$$r = 0.027$$

(the lengths are given in km as in Fig. 82). These values are used as estimates of the parameters $\mu_1, \ldots \varrho$. Figure 82 shows the density ellipses for the probabilities 0.5, 0.7 and 0.9. They are practically circles — as is perhaps expected by many readers. The point proportions for the areas between the ellipses and outside the 90% ellipse are close to the theoretically expected values.

Obviously, the points are not uniformly distributed. This can be shown by the dispersion index test (p. 221). That is, the point numbers in 72 squares of size

4 km × 4 km are counted. Figure 81 shows four external quadrats, and the reader can easily imagine the positions of the other ones. Mean and variance of point number per quadrat are

$$\bar{x} = 1.44 \quad \text{and} \quad s^2 = 4.81.$$

For $k = 72$ quadrats this gives 237.2 as the value of the dispersion index, which leads to rejection of the uniformity hypothesis. The large value of I corresponds to the clustered distribution of the points.

(b) Sinkholes in a region of the South Harz

Figure 83 shows the centres of 47 sinkholes in a region of the South Harz. Obviously, the points are randomly distributed with more disorder than for the basaltic formations. There are two types of points: 'old' and 'new' sinkholes.

The first step of the statistical analysis is to test the uniformity hypothesis. This is done by the dispersion index test (p. 221). The number k of subrectangles is 54, and the mean and variance of the point numbers are

$$\bar{x} = 0.85 \quad \text{and} \quad s^2 = 1.49.$$

This yields a dispersion index $I = 92.9$. The corresponding critical value is $\chi_{53;0.05} = 71.0$. Thus the uniformity hypothesis has to be rejected again — the clustering in the pattern is too strong.

Because of the small number of points, it is difficult to perform further statistical analysis. Figure 84 shows an isolines plot of the local point density, which was obtained by the methods of Chapter 13. Maybe this plot will tell some readers more than the original Fig. 83.

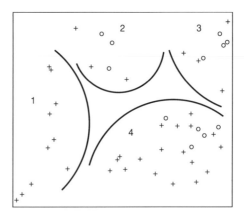

Figure 83 Distribution of sinkholes in a region of the South Harz, divided into four subregions. The length of the lower edge of the rectangle is 4.1 km. +, old sinkhole; ○, new sinkhole.

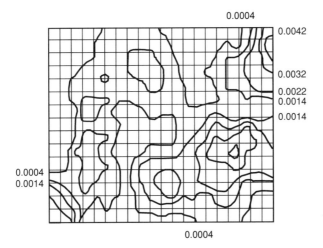

Figure 84 Isolines for the local point density of the sinkhole pattern. It was constructed using the method described on p. 235 with $h = 0.4$ km.

Finally, the distribution of the new sinkholes is discussed. One could make the hypothesis that it follows the same law as the old ones. A test of this hypothesis can be performed as follows. The region of observation is divided into four subregions, as shown in Fig. 83. For every subregion the numbers of new and old sinkholes are determined, O_i and N_i ($i = 1, \ldots, 4$). Then the test variable T is calculated:

$$T = on \sum_{i=1}^{4} \frac{1}{O_i + N_i} \left(\frac{O_i}{o} - \frac{N_i}{n} \right)^2,$$

where o and n denote the number of all old and new sinkholes respectively. (Quantities of this form have approximately a χ^2-distribution if o, n and the O_i and N_i are large. Despite the fact that these numbers are not large in this example, the χ^2-test is used.) For $\alpha = 0.05$ the uniformity hypothesis is rejected if T is greater than $\chi^2_{3;0.05} = 7.82$. Since $T = 3.60$, the hypothesis is not rejected. One may conclude that the new and old sinkholes are distributed according to the same law. (Of course, the way in which this result was obtained is somewhat dubious, and it should be treated with caution.)

12.3 POINT FIELDS WITH A RANDOM NUMBER OF POINTS

12.3.1 Introduction

Certain important aspects of the distribution of a point field with a random number of points can be analysed only if some independent samples can be analysed. This situation holds, for example, when analysing industrial products (the points are then centres of defects, pores etc.). It also holds in many biological studies, where the same organs of different individuals are considered. The points may be nuclei of interesting cells.

The basis of the statistical analysis is a collection of m point patterns $N_1, \ldots, N_m$ in the same window of observation W. The latter means that, for example,

- for m steel blocks the defects have to be determined in the corresponding regions of the blocks;
- for m specimens of organs the cell nuclei have to be considered in the same part of the organ and in regions of equal size and shape;
- from a very large homogeneous sample m directly congruent subsamples must be taken that are at such great distances that uniformity can be assumed (as in a large forest with uniform conditions of growth).

For each of the patterns the number of points is determined:

$$X_i = N_i(W) \quad (i = 1, \ldots, m).$$

The investigation of their distribution is usually the first step of statistical analysis. In §12.3.2 tools for this will be given. If the point number has a Poisson distribution then the methods of Poisson fields in Chapter 13 can be used.

The second step consists in the investigation of the distribution of the points in W as a function of the point number X_i. Usually one aims to find a uniform model for all numbers. A typical example is an I–I-point field with random n, but uniform density function $f(x)$. Of course, in general one has to assume that qualitative changes in the type of point distribution occur depending on the number of points. A typical example is a random packing of discs of fixed radius in a given area. For small n the discs may be distributed nearly randomly, but for large n they necessarily form an almost regular pattern.

12.3.2 Some distributions of random numbers

(a) The binomial distribution

The binomial distribution is closely connected with the so-called Bernoulli schema. It assumes the following. One makes n independent trials, and in each of them there are only two possible outcomes depending on chance: 'success' or 'failure'. The probability of success is p for all trials. The number of successes in n trials is denoted by X, and is given by

$$P(X = k) = \binom{n}{k} p^k (1 - p)^{n-k} \quad (k = 0, 1, \ldots, n). \tag{12.7}$$

One says that X has a binomial distribution in the parameters n and p. It is

$$\mathsf{E}X = np \tag{12.8}$$

and

$$\operatorname{var} X = np(1 - p). \tag{12.9}$$

If the notion of 'trial' is suitably interpreted then many situations can be described by this model. Two examples are as follows.

1. In a crystal lattice there are n sites that can be randomly occupied by foreign atoms. Let the probability of occupying a certain site be p and let it be the same for all sites. Then the total number X of occupied sites has a binomial distribution with parameters n and p.

2. In a region W, n independent points are uniformly distributed. Let X be the number of points that lie in a subregion B. It has a binomial distribution in the parameter n and

$$p = \frac{A(B)}{A(W)}$$

(p. 198).

Statistics for the binomial distribution. If n is known then only p has to be estimated. If m numbers $X_1, \ldots, X_m$ are given then p is estimated by

$$\hat{p} = \sum_{i=1}^{m} \frac{X_i}{mn}. \tag{12.10}$$

In this case formulae for confidence intervals for p are known (Sachs, 1984; Freund and Walpole, 1987). Furthermore, there are tests for the hypothesis that p has a given value p_0.

If n is unknown then it can be estimated as follows:

$$\hat{n} = \max \left\{ S^2 \frac{\phi^2}{\phi - 1}, X_{\max} \right\}, \tag{12.11}$$

where

$$\phi = \begin{cases} \dfrac{\bar{X}}{S^2} & \left(\bar{X} \geq \left(1 + \sqrt{\tfrac{1}{2}} \right) S^2 \right), \\[2ex] \max \left\{ \dfrac{X_{\max} - \bar{X}}{S^2}, 1 + \sqrt{2} \right\} & \text{otherwise.} \end{cases}$$

Here

$$\bar{X} = \frac{1}{m} \sum_{i=1}^{m} X_i,$$

$$S^2 = \frac{1}{m} \sum_{i=1}^{m} (X_i - \bar{X})^2,$$

$$X_{\max} = \text{greatest value from among } X_1, \ldots, X_m.$$

The estimator given by (12.11) is a stabilized form of the moment method estimator (Olkin *et al.*, 1981). The maximum-likelihood estimator also has to be stabilized. Sometimes the application of Bayesian methods is to be recommended, i.e. the use of a priori distributions of p (Carroll and Lombard, 1985; Günel and Chilko, 1989). Of particular practical interest is the case of a beta distribution with integer parameters a and b, where the density function of p is proportional to $p^a(1 - p)^b$. Then the maximum of $L(v)$ has to be determined, where

$$L(v) = \prod_{i=1}^{m} \binom{v}{X_i} \left[(mv + a + b + 1) \binom{mv + a + b}{a + \sum_{i=1}^{m} X_i} \right]^{-1} \quad (v \geq X_{\max}). \quad (12.12)$$

The v-value yielding the maximum is an estimator of n. Where n has been estimated by (12.11) or (12.12), p can be estimated by (12.10), where n is replaced by $\hat{n}$.

The goodness-of-fit for the binomial distribution can be tested by the χ^2-test. The same is also true for the other distributions in this section.

The Poisson distribution

The Poisson distribution plays an important role in the theory of point fields, because it appears in connection with the Poisson field. The number of points from a homogeneous Poisson field $X = N(B)$ in a given Borel set B, has a Poisson distribution.

A random variable X has a Poisson distribution if

$$\Pr(X = k) = \frac{\mu^k}{k!} e^{-\mu} \quad (k = 0, 1, \ldots; \mu > 0). \quad (12.13)$$

The parameter μ is equal to the mean and the variance of X:

$$\mathsf{E}X = \text{var } X = \mu. \quad (12.14)$$

The Poisson limit theorem may explain the occurrence of the Poisson distribution. Let there be a sequence of Bernoulli schemes with number of trials n and success probability $p_n (n = 1, 2, \ldots)$. Let the success probabilities tend towards zero with $np_n = \mu$ for all n. Then the number of successes X_n in the nth scheme satisfies

$$\lim_{n \to \infty} \Pr(X_n = k) = \frac{\mu^k}{k!} e^{-\mu} \quad (k = 0, 1, \ldots).$$

Therefore the term 'law of rare events' is sometimes used. For structures that contain relatively few irregularly distributed small objects, one may expect a Poisson distribution of the number of objects in a given test set.

Statistics for the Poisson distribution. Let there be a sample of m numbers $X_1, \ldots, X_m$. Then the parameter μ is estimated by

$$\hat{\mu} = \frac{1}{m} \sum_{i=1}^{m} X_i = \bar{X}. \tag{12.15}$$

An approximate confidence interval for the level $1 - \alpha$ is

$$\frac{1}{m} \left(\tfrac{1}{2} z_{\alpha/2} - \sqrt{m\bar{X}} \right)^2 \le \mu \le \frac{1}{m} \left(\tfrac{1}{2} z_{\alpha/2} + \sqrt{m\bar{X} + 1} \right)^2 \tag{12.16}$$

for large $m\bar{X}$. Here $z_{\alpha/2}$ is the $1 - \tfrac{1}{2}\alpha$ quantile of the normal distribution: $z_{\alpha/2} = 1.65, 1.96$ and 2.58 for $\alpha = 0.10, 0.05$ and 0.01 respectively.

For further statistical methods for the Poisson distribution see Crow and Gardner (1959), Sachs (1984) and Freund and Walpole (1987) (for example a test of the hypotheses that $\mu = \mu_0$ or $\mu_1 = \mu_2$ is given there).

The binomial and Poisson distribution are not sufficiently flexible for all applications. Thus the following two important classes of discrete distributions are presented that have many applications in spatial statistics (Pielou, 1977; Cliff and Ord, 1981).

Compound distributions

So-called compound distributions are obtained by making one (or more) parameters of a given distribution random. This is illustrated here by the Poisson distribution (there is a close connection to so-called Cox processes). It depends on the parameter μ, which is the mean. If now μ is a random variable then the random numbers are generated in a two-step process. First, by a random mechanism, the actual value of μ is determined (according to a density function $f(\mu)$). Then a Poisson-distributed random number is generated, where the parameter of the Poisson distribution is the newly determined μ. The probabilities of the corresponding distribution are given by

$$\Pr(X = k) = \int_0^\infty P_k(\mu) f(\mu) \, d\mu, \tag{12.17}$$

where $P_k(\mu) = (\mu^k / k!) e^{-\mu}$ $(k = 0, 1, \ldots)$. If the first and second moments of $f(\mu)$ are m_1 and m_2 respectively,

$$m_l = \int_0^\infty x^l f(x) \, dx \quad (l = 1, 2) \tag{12.18}$$

then

$$EX = m_1 \tag{12.19}$$

and

$$EX^2 = m_1 + m_2. \tag{12.20}$$

This implies

$$\operatorname{var} X = m_1 + m_2 - m_1^2. \tag{12.21}$$

An important particular case is where $f(\mu)$ is a gamma distribution density:

$$f(\mu) = \frac{\sigma^r}{\Gamma(r)} \mu^{r-1} e^{-\sigma\mu} \quad (\mu \geq 0, \quad \sigma > 0, \quad r > 0)$$

(the parameters of the gamma distribution are r and σ).

Here X has a so-called negative binomial distribution:

$$\Pr(X = k) = \frac{\Gamma(r+k)\sigma^r}{k!\Gamma(r)(1+\sigma)^{r+k}} \quad (k = 0, 1, \ldots), \tag{12.22}$$

$$\mathsf{E}X = \frac{r}{\sigma}, \tag{12.23}$$

$$\operatorname{var} X = \frac{r(1+\sigma)}{\sigma^2}. \tag{12.24}$$

These formulae can be used to estimate the parameters r and σ by the method of moments.

Another explanation of the negative binomial distribution is as follows. Consider, as in the case of the binomial distribution, a Bernoulli scheme. Let the success probability be $(1 + \sigma)^{-1}$. Denote by X_i the random number of success between the $(i-1)$th and ith failure; $i = 1, 2, \ldots$. Consider the random variable X,

$$X = X_1 + X_2 + \cdots + X_r.$$

It has a negative binomial distribution with the parameters r and σ.

(d) Generalized distributions

So-called generalized distributions can be used to describe numbers of objects that are formed by groups of random sizes. The corresponding random number X is assumed to be given in the form

$$X = \sum_{i=1}^{\nu} C_i,$$

where the C_i are non-negative integer identically and independently distributed random variables, which can be interpreted as group sizes. The number of the groups is then ν, where ν is a further integer random variable independent of the C_i.

The mean and variance of X satisfy the well-known formulae

$$\mathsf{E}X = \mathsf{E}\nu\,\mathsf{E}C_1 \tag{12.25}$$

and

$$\text{var } X = \text{E}\nu \text{ var } C_1 + (EC_1)^2 \text{ var } \nu. \tag{12.26}$$

The probabilities $p_k = \Pr(X = k)$ can be calculated by means of so-called generating functions:

$$H(z) = \sum_{k=0}^{\infty} p_k z^k,$$

$$G(z) = \sum_{k=0}^{\infty} \Pr(\nu = k) z^k,$$

$$g(z) = \sum_{k=0}^{\infty} \Pr(C_1 = k) z^k,$$

$$H(z) = G(g(z)) = \sum_{k=0}^{\infty} \Pr(\nu = k) \left[\sum_{j=0}^{\infty} \Pr(C_1 = j) z^j \right]^k.$$

If ν and the C_i are Poisson-distributed, with parameters μ_ν and μ_c respectively, the generalized distribution is called the Poisson–Poisson distribution. Here

$$\text{E}X = \mu_\nu \mu_c,$$
$$\text{var } X = \mu_\nu \mu_c (1 + \mu_c),$$
$$G(z) = e^{\mu_\nu(z-1)},$$
$$g(z) = e^{\mu_c(z-1)}.$$

Thus

$$H(z) = \exp(\mu_\nu e^{\mu_c(z-1)} - 1).$$

Let ν now have a Poisson distribution with parameter μ and let the C_i have a 'logarithmic' distribution:

$$\Pr(C_1 = k) = -\frac{\alpha^k}{k \log(1 - \alpha)} \quad (0 < \alpha < 1; \; k = 1, 2, \ldots).$$

Then X has a negative binomial distribution with parameters

$$\sigma = \frac{1 - \alpha}{\alpha} \quad \text{and} \quad r = \frac{\mu}{\log(1 + 1/\alpha)}.$$

Serfozo (1990) has discussed the problem of occurrence of negative binomial distributions in the theory of point processes.

CHAPTER 13

Poisson Point Fields

13.1 INTRODUCTION

Poisson point fields (or more briefly Poisson fields) are the simplest and most studied models of random fields. Their distribution satisfies very strong independent conditions. In particular, the point numbers in disjoint (mutually non-intersecting) sets are stochastically independent. It is clear that, under such assumptions, formulae for distributional characteristics are obtained relatively easily. Also, statistical problems can be solved with powerful and elegant methods.

In many applications it may be assumed that the distribution of points is *totally random* or at least that it is so with local variations in the point density. In such cases a Poisson field is a useful model. If such independence conditions do not hold then Poisson fields are still of value, either as rough *approximations* or as *null models*. That is, even a rough stochastic model is frequently useful instead of a purely deterministic approach, just to get a feeling for the influence of randomness. Then it is useful to have access to some theory and not merely to employ large computer simulations. In other cases the statistician can make, despite better knowledge or feeling, the assumption that the given field is a Poisson field. If this hypothesis is confirmed by statistical tests then he/she may conclude that it is unnecessary to look for complicated distribution laws in the pattern. Otherwise, the search for better models can begin.

Poisson fields are also important as components of more complicated models. Two important examples are Neyman–Scott fields and Boolean models. *Neyman–Scott cluster fields* are constructed as follows. In the plane so-called parent points are first scattered according to a Poisson field. Around each parent point, daughter points are scattered. The set of all daughter points forms the cluster field (see also §16.2). (A biological situation in which this model may perhaps be used is as follows: the parent points are positions of plants, and the daughter points stand for plants which result from seeds of the parent plants.)

The Poisson field of primary (or 'germ') plants is also the basis of the *Boolean model*. Around each primary point, a random set ('primary grain') is positioned. The set theoretic union of all these sets is a new random set called Boolean model. Two examples where it can be used as a stochastic model are:

Germ:	germ point of a trefoil plant in a grass-plot,
Primary grain:	area covered in a certain instant by those trefoils which stem from a given germ;
Germ:	germ point of a pore in bread dough,
Primary grain:	pore developed starting from a certain germ.

In the second example the independence assumption is probably true only at the beginning of the process of forming of the pores; later there will be interactions between them.

Clearly, the Neyman–Scott field is a special case of the Boolean model. Some basic facts about the Boolean model are given in Appendix F.

13.2 THE HOMOGENEOUS POISSON FIELD

13.2.1 Fundamental properties

The homogeneous Poisson field has two fundamental properties.

1. If k is any integer and if $B_1, \ldots, B_k$ are any disjoint (mutually non-intersecting) Borel sets then the random variables $N(B_1), \ldots, N(B_k)$ are stochastically independent.

2. The number $N(B)$ of points in any bounded Borel set B has a Poisson distribution with parameter $\lambda A(B)$. $A(B)$ denotes the Lebesgue measure of B, i.e. the area of B.

The parameter λ has quite a simple meaning: it denotes the intensity or the mean point density as given by

$$\mathsf{E} N(B) = \lambda A(B) \tag{13.1}$$

for any Borel set B. (Of course, it is always assumed that $0 < \lambda < \infty$.) It is clear that properties 1 and 2 imply that the point field is homogeneous and isotropic.

The following property is very important for understanding the kind of randomness of the homogeneous Poisson field. Let B be any bounded Borel set and suppose that the point field has exactly n points in B. Then these n points are uniformly and independently distributed in B. This fact is used in the simulation of homogeneous Poisson fields (p. 217).

The homogeneous Poisson field is already uniquely determined by property 1 above; there are no other homogeneous point fields with similar strong independence properties. More precisely, a point field that is simple (no two points have the same location in the plane), homogeneous and has the independence property 1 is necessarily a homogeneous Poisson field.

13.2.2 Some important formulae

One-dimensional number distributions:

$$\Pr(N(B)) = k) = \frac{[\lambda A(B)]^k}{k!} e^{-\lambda A(B)} \quad (k = 0, 1, \ldots; A(B) < \infty). \tag{13.2}$$

m-dimensional number distributions (pairwise-disjoint B_j of finite area):

$$\Pr(N(B_1) = k_1, \ldots, N(B_m) = k_m)$$
$$= \frac{[\lambda A(B_1)]^{k_1}}{k_1!} \cdots \frac{[\lambda A(B_m)]^{k_m}}{k_m!} \exp\left[-\lambda \sum_{i=1}^{m} A(B_i)\right] \quad (k_1, \ldots, k_m \geq 0). \tag{13.3}$$

Emptiness probabilities:

$$\Pr(N(B) = 0) = e^{-\lambda A(B)}. \tag{13.4}$$

Particular emptiness probabilities are given by the so-called contact distribution functions. These are defined as

$$H_B(r) = 1 - \Pr(N(rB) = 0) \quad (r \geq 0).$$

Here B is a compact '*test set*' with $A(B) > 0$ containing the origin o; rB denotes the dilation of B by a factor r: $rB = \{rx : x \in B\}$. Of particular interest is the spherical contact distribution $H_s(r)$. Here B is the unit disc, $B = b(o, 1)$, and $H_s(r)$ is given by

$$H_s(r) = 1 - \Pr(N(b(o, r)) = 1) = 1 - e^{-\lambda \pi r^2}. \tag{13.5}$$

The corresponding density function $h_s(r)$ is

$$h_s(r) = 2\lambda \pi r e^{-\lambda \pi r^2} \quad (r \geq 0).$$

Obviously, $H_s(r)$ is the distribution function of the distance from the origin to that point of the Poisson field closest to o.

The strong independence properties of the Poisson field also make it possible to calculate certain conditional probabilities. A typical example is the distribution $D(r)$ of the nearest-neighbour distance. Heuristically, this distribution can be described as that of the distance of a randomly chosen point to its nearest neighbour. There the term 'randomly chosen' is somewhat vague. Statistically, this means the following: if n points are given in a window of observation, then for each of them the nearest neighbour and the corresponding distance d_i is determined. The d_i form a sample, and the corresponding empirical distribution function

$$\hat{D}(r) = \frac{1}{n} \sum_{i=1}^{n} 1_{[o,r]}(d_i)$$

approximates $D(r)$. $D(r)$ is obtained by probabilistic methods as follows, studying the neighbourhood relationship of a single point. By the homogeneity of the point field, this can be done for a point close to the origin o. The conditional probability

$$D_\varepsilon(r) = 1 - \Pr(N(b(o, r)\backslash b(o, \varepsilon)) = 0 \mid N(b(o, \varepsilon)) = 1) \quad (r \geq 0),$$

is similar to what is wanted, namely the probability that the nearest neighbour of a point in a small disc $b(o, \varepsilon)$ lies at a distance not greater than r from o, where $r > \varepsilon$. As ε tends to 0, $D_\varepsilon(r)$ tends to the desired distribution function $D(r)$.

By the definition of conditional probability,

$$D_\varepsilon(r) = 1 - \frac{\Pr(N(b(o, r)\backslash b(o, \varepsilon)) = 0, N(b(o, \varepsilon)) = 1)}{\Pr(N(b(o, \varepsilon)) = 1)}.$$

The independence property 1 of p. 212 enables the simplification

$$D_\varepsilon(r) = 1 - \frac{\Pr(N(b(o, r)\backslash b(o, \varepsilon)) = 0)\Pr(N(b(o, \varepsilon)) = 1)}{\Pr(N(b(o, \varepsilon)) = 1)}$$
$$= 1 - \Pr(N(b(o, r)\backslash b(o, \varepsilon)) = 0)$$

which yields

$$D_\varepsilon(r) = 1 - \exp[-\lambda A(b(o, r)\backslash b(o, \varepsilon))]$$
$$= 1 - \exp[-\lambda \pi(r^2 - \varepsilon^2)] \quad (r \geq 0).$$

On letting $\varepsilon \to 0$,

$$D(r) = 1 - e^{-\lambda \pi r^2} \quad (r \geq 0). \tag{13.6}$$

The corresponding density function is

$$d(r) = 2\lambda \pi r e^{-\lambda \pi r^2} \quad (r \geq 0).$$

Comparison with (13.5) yields

$$H_s(r) = D(r) \quad (r \geq 0). \tag{13.7}$$

Because of the homogeneity of the Poisson field, this property can be expressed as follows: the distance to the nearest point of the point field seen from an arbitrary location of the plane has the same distribution as the distance to the nearest neighbour of a randomly chosen point of the field.

This statement is a particular case of a general property of the Poisson point field, which is known under the name 'Slivnyak's theorem' (Stoyan et al., 1987, p. 50). This theorem says that the point field seen from the position of a randomly chosen point, which for simplicity is translated to the origin o, has the same distribution as the original point field plus a point in o.

The mean and variance of $D(r)$ are

$$m_D = \frac{1}{2\sqrt{\lambda}}, \tag{13.8}$$

$$\sigma_D^2 = \frac{1}{\pi\lambda} - \frac{1}{4\lambda}. \tag{13.9}$$

The distribution functions $D_2(r)$, $D_3(r)$, ... of the distances to the 2nd, 3rd, ... nearest neighbours can also be given as

$$D_k(r) = 1 - \sum_{j=0}^{k-1} e^{-\lambda\pi r^2} \frac{1}{j!}(-\lambda\pi r^2)^j \quad (r \geq 0). \tag{13.10}$$

The corresponding density function is

$$d_k(r) = \frac{2(\lambda\pi r^2)^k}{r(k-1)!}e^{-\lambda\pi r^2} \quad (r \geq 0)$$

(Fig. 85).

The jth moment is

$$m_{k,j} = \frac{\Gamma\left(k + \frac{1}{2}j\right)}{(k-1)!(\lambda\pi)^{j/2}} \quad (j = 1, 2, \ldots).$$

Finally, the mode (maximum of density function) is

$$r_k = \sqrt{\frac{k - \frac{1}{2}}{\pi\lambda}} \quad (k = 1, 2, \ldots).$$

The function value $d_k(r_k)$ for small k is only weakly dependent on k: in the case $\lambda = 1$, it is 1.52 for $k = 1$ and 1.42 for $k = 10$.

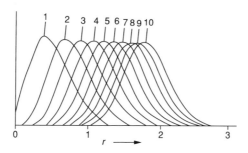

Figure 85 Density functions of the distances to the kth nearest neighbours ($k = 1, 2, \ldots, 10$) for a homogeneous Poisson field of intensity one.

13.2.3   Second-order characteristics

In the theory of point fields second-order characteristics play an important role, in particular for the description of distribution and in statistics. In the case of a Poisson field they have a particularly simple form.

The original aim is to determine quantities such as

$$EN(B_1)N(B_2) \quad \text{or} \quad \text{var } N(B).$$

The variance can be expressed as

$$\text{var } N(B) = EN(B)^2 - [EN(B)]^2,$$

and the formula for the variance of the Poisson distribution yields

$$\text{var } N(B) = \lambda A(B). \tag{13.11}$$

Simple calculations also yield $EN(B_1)N(B_2)$. If $B_1 \cap B_2 = \emptyset$ (i.e. B_1 and B_2 are disjoint) then

$$EN(B_1)N(B_2) = EN(B_1)EN(B_2) = \lambda^2 A(B_1)A(B_2) \tag{13.12}$$

because of the independence property 1 (p. 212). Otherwise, $B_1 = C_1 \cup D$ and $B_2 = C_2 \cup D$, where $D = B_1 \cap B_2$, $C_1 = B_1 \backslash D$ and $C_2 = B_2 \backslash D$. Then

$$N(B_1)N(B_2) = [N(C_1) + N(D)][N(C_2) + N(D)],$$

and, from the independence of the point numbers in disjoint sets,

$$
\begin{aligned}
&EN(B_1)N(B_2) \\
&= EN(C_1)EN(C_2) + EN(D)EN(C_2) + EN(C_1)EN(D) + EN(D)^2 \\
&= E[N(C_1) + N(D)]E[N(C_1) + N(D)] + EN(D)^2 - [EN(D)]^2 \\
&= EN(B_1)EN(B_2) + EN(B_1 \cap B_2)^2 - [EN(B_1 \cap B_2)]^2.
\end{aligned}
$$

The means in the last line can be calculated using (13.1); if additionally $EX^2 = \mu^2 + \mu$ is used for a Poisson-distributed random variable X with $\mu = \lambda A(B_1 \cap B_2)$ then finally

$$EN(B_1)N(B_2) = \lambda^2 A(B_1)A(B_2) + \lambda A(B_1 \cap B_2) \tag{13.13}$$

is obtained. This generalizes (13.12).

Now consider two disjoint infinitesimally small discs B_1 and B_2 of areas dF_1 and dF_2. Let the distance of the centres be r. Then

$$EN(B_1)N(B_2) = \lambda^2 \, dF_1 \, dF_2.$$

Since B_1 and B_2 are infinitesimally small,

$$EN(B_1)N(B_2) = P(N(B_1) = 1, N(B_2) = 1)$$

holds with negligible error. The formula (13.3) yields

$$\Pr(N(B_1) = 1, N(B_2) = 1) = \lambda^2 \, dF_1 \, dF_2 = \varrho^{(2)}(r) \, dF_1 \, dF_2.$$

In the language of the general theory of point fields, the factor $\varrho^{(2)}(r)$, is called the second-order product density (Chapter 14). The calculations above show that in the case of a Poisson field it has the trivial form

$$\varrho^{(2)}(r) \equiv \lambda^2.$$

The function $g(r) = \varrho^{(2)}(r)/\lambda$ is called the pair correlation function. For the Poisson field

$$g(r) \equiv 1.$$

The pair correlation function and the K-function are, in the general theory, the most important second-order characteristics. The K-function is defined such that $\lambda K(r)$ is the mean number of points in a disc of radius r centred at a 'randomly chosen' point of a point field, which is itself not counted. Because of the general property of Poisson fields mentioned on p. 213 (Slivnyak's theorem) this mean is equal to the mean number of points in the disc $b(o, r)$. The latter is equal to $\lambda \pi r^2$ by (13.1), which yields

$$\lambda K(r) = \lambda \pi r^2$$

and

$$K(r) = \pi r^2 \quad (r \geq 0). \tag{13.14}$$

In passing note that there is a general relation between $K(r)$ and $g(r)$:

$$g(r) = \frac{dK(r)}{dr} \Big/ 2\pi r,$$

which is of course satisfied in the case of the Poisson field.

In point field statistics the K-function is frequently not used, but rather a simpler function, the so-called L-function

$$L(r) = \sqrt{\frac{K(r)}{\pi}} \quad (r \geq 0)$$

is employed. One of the reasons for using this function is its simple form for the Poisson field:

$$L(r) = r \quad (r \geq 0). \tag{13.15}$$

13.2.4 Simulation of a homogeneous Poisson field

The simulation of Poisson fields is a very important problem. It is needed in Poisson field statistics (p. 227), and it is a starting point for the simulation of more complicated structures.

Here the case where the points lie in a rectangular window W of side lengths a and b is described. (If the window has another form then a rectangle that contains W may be used, and a Poisson field is simulated therein. The points in W then form a sample with the necessary properties.) The simulation requires two steps:

1. Generation of the number n of points in W, where n is a sample of a Poisson-distributed random variable with parameter $\mu = \lambda ab$;
2. Generation of a sample of a binomial field of n points in W.

Details of the first step are given in Devroye (1986) and Ripley (1987). If μ is not too large then a linear Poisson process is first simulated. If its intensity is unity then the number of points in the interval $(0, \mu]$ has a Poisson distribution with parameter μ. The simulation of a linear Poisson process of unit intensity is straightforward using the fact that the distances between subsequent points are independent and have an exponential distribution with parameter unity.

Thus exponentially distributed random numbers have to be generated first. This is possible using the inversion method: if z is a uniform random number on $[0, 1]$, then $e = -\log z$ is an exponentially distributed random number on the parameter unity. If (e_i) is a sequence of such random numbers, then the desired Poisson random number is the smallest n with

$$\sum_{i=1}^{n} e_i > \mu.$$

A BASIC program for generating Poisson distributed random numbers has the form

```
10    P = 1: N = 0: T = EXP(-μ)
20    Z = RND(0)
30    P = P * Z
40    IF P ≥ T THEN N = N + 1: GOTO 20
50    NZ = N: RETURN
```

The execution of step two has already been described in §12.2.1. Lewis and Shedler (1979) have suggested to perform both steps jointly: Generate a linear Poisson process of intensity λ in $[0, a]$ and take its points as x-coordinates. The corresponding y-coordinates are taken as uniform random numbers on $[0, b]$.

13.2.5   Statistics for the homogeneous Poisson field

In Section 13.1 it has been mentioned that the homogeneous Poisson field plays a fundamental role in the theory of point fields, as a basis for models and as a null model. Therefore there are various statistical methods for Poisson fields, particularly for testing the hypothesis that a given point pattern is part of a sample of a Poisson field. Some of them are described here (see also Cox and Lewis, 1966; Snyder 1975; Brillinger 1978; Ripley 1981, 1987; Diggle 1983; Karr 1986; Stoyan et al., 1987).

It is assumed here that the point pattern is given in a rectangular window W. Generalization to other window shapes is also possible. If more than one window can be analysed then the corresponding statistical quantities are obtained by area-weighted averaging.

(a) Estimation of the intensity

The distribution of a homogeneous Poisson field depends only on the parameter λ; if it is known then all interesting quantities and distributions can be calculated, as the formulae in §13.2.2 show. The maximum-likelihood estimator

$$\hat{\lambda} = \frac{N(W)}{A(W)} \tag{13.16}$$

is usually used to estimate λ. It is unbiased and consistent (i.e. its accuracy increases with increasing window area; see also (11.2)).

For some applications it is convenient to estimate λ using interpoint distances ('distance methods'). The interested reader is referred to Ripley (1981) and Diggle (1983).

It is possible to give a confidence interval for λ, which is based on the fact that the point number in the window, $N(W)$, has a Poisson distribution (and that this is asymptotically normal) of the parameter $\lambda A(W)$. Using (12.16), one obtains the interval

$$\left[\tfrac{1}{2}z_{\alpha/2} - \sqrt{N(W)}\right]^2 \leq \lambda A(W) \leq \left[\tfrac{1}{2}z_{\alpha/2} + \sqrt{N(W)+1}\right]^2 \tag{13.17}$$

at the confidence level $1 - \alpha$.

This confidence interval can be used in planning experiments to choose that window area $A(W)$ necessary for a given accuracy. Starting with a given breadth δ of the confidence interval, the required $A(W)$ is determined. The equation

$$\frac{\left[\tfrac{1}{2}z_{\alpha/2} + \sqrt{\lambda A(W)}\right]^2 - \left[\tfrac{1}{2}z_{\alpha/2} - \sqrt{\lambda A(W)}\right]^2}{A(W)} = \delta$$

yields the approximation

$$A(W) \approx \frac{4\lambda z_{\alpha/2}^2}{\delta^2}. \tag{13.18}$$

The unknown λ appears in the formula. An approximation must be used for λ, which is obtained by a priori knowledge or a pilot study.

Example. Figure 86 shows a system of nearly planar silver particles on a polished steel plate, which are treated as discs. Here the system of the disc centres is analysed.

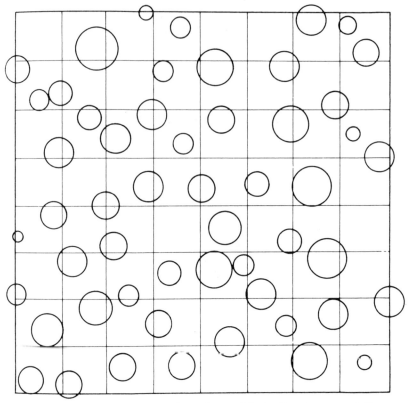

Figure 86   Silver particles on the surface of a steel body. See the text and Stoyan and Wiencek (1991). The centres of the particles approximated by discs form a point field of weak short-range order.

The area $A(W)$ of the quadratic window is 144 cm^2 magnified 2280-fold, and the number $N(W)$ of discs is 53. Thus an estimate of the intensity is given by

$$\hat{\lambda} = \frac{53}{144} = 0.368 \text{ cm}^{-2}$$

It was assumed that the point pattern belongs to a Poisson field, though some readers may doubt this, since the discs are relatively regularly distributed within the window. If (13.18) were used for $\alpha = 0.05$ and $\delta = 0.06$ for calculating the necessary window area then the result would be

$$A(W) = \frac{4 \times 0.368 \times 1.96^2}{0.06^2} = 1570.8 \text{ cm}^{-2}.$$

Thus a 40 × 40 window or eleven 12 × 12 windows are necessary. This result has been obtained assuming a Poisson distribution, which presupposes greater variability than that in Fig. 86. It may be assumed that with the calculated window area an error less than $\delta = 0.06$ may be obtained.

(b) Test of the distribution assumption

To test the assumption that a given point pattern can be interpreted as part of a sample of a homogeneous Poisson field, various methods are known (Ripley, 1981; Diggle, 1983; Stoyan et al., 1987). Here two groups of such methods are described, which are used particularly frequently: quadrat count methods and L-function tests.

Quadrat count methods. The window W is divided in k subregions of equal area (e.g. small quadrats). The number v_i of points in each subregion is determined. Assuming a Poisson field, the v_i are independent and identically distributed; the mean number per subregion is $\lambda A(W)/k$. Large deviations in the v_i from this value or a behaviour that is too uniform indicate deviations from the Poisson hypothesis.

Dispersion index test. One calculates

$$I = \frac{(k-1)s_v^2}{\bar{v}} \tag{13.19}$$

where k is the number of subregions, s_v^2 is the sample variance of point numbers in subregions, $\bar{v}$ is the mean number of points per subregion and is equal to $N(W)/k = n/k$. The Poisson hypothesis is rejected if

$$I > \chi_{k-1,\alpha}^2 \quad \text{or} \quad I < \chi_{k-1,1-\alpha}^2.$$

Here α is the probability of an error of type I and $\chi_{m,\beta}^2$ the $1 - \beta$ quantile of the χ^2-distribution with m degrees of freedom. (k should be greater than 6 and $\lambda A(W)/k$ greater than 1.)

In the first rejection case it may be assumed that the points appear in clusters. In contrast, the second one implies more regularity than is typical for a homogeneous Poisson field.

Remark. The dispersion index test can be interpreted as a χ^2-goodness-of-fit test. To understand this, the k subregions should be interpreted as k classes and the hypothesis should be tested that a uniform distribution is given (p. 212). Then the test variable of the χ^2-test is

$$\sum_{i=1}^{k} \frac{(v_i - n/k)^2}{n/k} = \frac{(k-1)s_v^2}{n/k} = (k-1)\frac{s_v^2}{\bar{v}}.$$

Example (silver particles, continued). The window is divided into $k = 64$ quadrats of size 1.5×1.5. The quadrat numbers are then

$$
\begin{array}{cccccccc}
1 & 1 & 0 & 0 & 1 & 1 & 1 & 1 \\
1 & 1 & 1 & 1 & 1 & 0 & 1 & 1 \\
0 & 1 & 1 & 2 & 1 & 1 & 1 & 1 \\
0 & 0 & 1 & 1 & 0 & 1 & 0 & 2 \\
0 & 1 & 1 & 1 & 0 & 1 & 1 & 0 \\
2 & 0 & 1 & 1 & 1 & 2 & 1 & 1 \\
0 & 1 & 1 & 1 & 1 & 0 & 0 & 3 \\
2 & 1 & 0 & 0 & 1 & 1 & 1 & 0
\end{array}
$$

This yields

$$
\bar{v} = \frac{53}{64} = 0.828, \quad s_v^2 = 0.399, \quad I = 63 \times \frac{0.399}{0.828} = 30.36.
$$

With $\chi^2_{63;0.05} = 82.5$ and $\chi^2_{63;0.95} = 45.7$, one obtains a clear rejection of the Poisson field hypothesis, and the value of I suggests a more regular distribution.

Greig–Smith test. The Greig–Smith test is a refined variant of the dispersion index test. Here not only the point numbers in the subregions are considered, but also the neighbourhood relations. It is assumed that the window is rectangular. By successive bisection of subrectangles, $k = 2^q$ subrectangles (SR) are obtained. The following quantities are then calculated:

$$
s_1^2 = \sum_{(i)} \left(\frac{\text{number in the } i\text{th}}{\text{SR}}\right)^2 - \frac{1}{2}\sum_{(m)} \left(\frac{\text{number in the } m\text{th}}{\text{pair of SR}}\right)^2,
$$

$$
s_2^2 = \sum_{(i)} \left(\frac{\text{number in the } i\text{th}}{\text{pair of SR}}\right)^2 - \frac{1}{2}\sum_{(m)} \left(\frac{\text{number in the } m\text{th}}{\text{quadruple of SR}}\right)^2.
$$

...

In the case $q = 6$ or $k = 64$ the SR can be numbered as on the chess square. Pairs of SR are then $(a1, a2), (a3, a4), \ldots, (b1, b2), (b3, b4)$ etc.; quadruples of SR are $(a1, a2, b1, b2), (a3, a4, b3, b4)$ etc.

The s_j^2 are used to calculate the quantities

$$
I_j = \frac{s_s^2}{\bar{v}}.
$$

Since the quantities $s_j^2/2^{q-j}$ are estimators for the variance of the point number in the SR as S^2, the I_j behave as I in the dispersion index test. But the number n of degrees of freedom for the corresponding χ^2-distributions is 2^{q-j}.

As in the dispersion index test, the Poisson field hypothesis can then be tested for $j = 1, 2, \ldots$. It may happen that the outcome is different for different j. This may lead to statements about the size of clusters or the scale of regularity. If, for example, I_3 is greater than the critical χ^2-value then there is clustering in the range of quadruples of SR. If then I_4 has a mean value, one may conclude that the quadruple-sized clusters are randomly distributed.

Example (silver particles, continued). The data for the SR-pairs, quadruples etc. are as follows:

Pairs									Quadruples			
2	2	1	1	2	1	2	2		4	2	3	4
0	1	2	3	1	2	1	3		1	5	3	4
2	1	2	2	1	3	2	1		3	4	4	3
2	2	1	1	2	1	1	3		4	2	3	4

Octuples				Groups of 16		Groups of 32	
5	7	6	8	12	14	25	28
7	6	7	7	13	14		

Table 15 contains the I_j and the critical χ^2-values. As expected, I_1 is smaller than $\chi^2_{32;0.95}$, which indicates a regularity tendency at small scale. For larger units, then, the I_j behave as in the case of a completely random distribution (if I_6 is ignored).

Example. For the sample of the cluster field shown in Fig. 87 the following data are obtained:

8	1	6	7	0	9	10	11
7	0	2	7	11	5	5	2
5	0	7	2	7	0	1	4
5	1	4	1	6	7	17	2
0	0	0	0	6	3	0	5
9	2	0	1	11	1	0	0
6	1	0	4	3	0	0	0
0	0	0	4	6	1	0	0

Calculations as in the preceding example yield Table 16. The I_j-values reflect the cluster structure of the point pattern very well.

Other tests analyse various distances; interpoint distances and distances to test points; see (13.7) (Ripley, 1981; Diggle, 1983; Shaw, 1990). In the latter paper the interpoint distances in subsquares are used, and boundary effects are considered.

Table 15 Results of the Greig-Smith tests for the silver particles.

j	Degrees of freedom	$\chi^2_{0.95}$	I_j	$\chi^2_{0.05}$
1	32	20.1	19.9	46.2
2	16	8.0	11.5	26.3
3	8	2.7	15.1	15.5
4	4	0.7	5.4	9.5
5	2	0.1	0.6	6.0
6	1	0.0	5.4	3.8

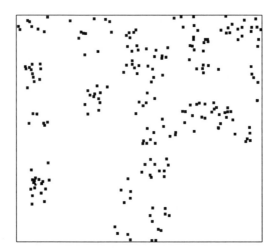

Figure 87 Simulated sample of a cluster field. The cluster centres (parent points) form a Poisson field of intensity 50. Each cluster consists of 5 points; they are uniformly distributed in a disc of radius 0.05. (The length of the lower edge of the rectangle is 1.) See §16.2 for details of cluster fields.

Table 16 Results of the Greig-Smith tests for the simulated cluster field.

j	Degrees of freedom	$\chi^2_{0.95}$	I_j	$\chi^2_{0.05}$
1	32	20.1	106.3	46.2
2	16	8.0	136.0	26.3
3	8	2.7	50.4	15.5
4	4	0.7	125.9	9.5
5	2	0.1	705.5	6.0
6	1	0.0	254.2	3.8

McKendrick (1991) suggested a test that uses the number of non-empty subregions of W. He compared its power by simulation with that of several other Poisson field tests.

The L-test The following test needs somewhat more calculations (namely the calculation of the empirical L-function (p. 279) but it handles the deviations from the distributional properties of a Poisson field in a more sensible way. It uses the fact that the L-function of a homogeneous Poisson field satisfies

$$L(r) = r \quad (r \geq 0).$$

Therefore the test variable

$$\tau = \max_{r \leq r_0} |\hat{L}(r) - r| \tag{13.20}$$

is used, where $\hat{L}(r)$ is an estimate of $L(r)$ and r_0 is an upper bound on the interpoint distance r.

If τ is very large then the hypothesis that the point pattern is a subsample of a Poisson point field must be rejected. (The alternative hypothesis is 'not a Poisson field', without further qualification.) For the case that $L(r)$ is determined by Ripley's estimator (see pp. 279, 282 and (15.7)), critical values for τ are determined by simulation. Ripley (1988) has suggested the critical value

$$\tau_{0.95} = 1.45 \frac{\sqrt{A(W)}}{N(W)} \qquad (13.21)$$

for $\alpha = 0.05$. This can be used if $N(W)U(W)r_0^3/A(W)^2$ is 'small'. The dependence of r_0 is said to be weak.

The results of Koen (1991) are more detailed (Tables 17 and 18). They are based on 1000 simulations and give critical values for a quadratic window of unit side length. In the case of side length a the values in the tables must be multiplied by a. The value of r_0 is connected with the auxiliary additional parameter ϑ by

$$r_0 = \frac{\vartheta a}{2\sqrt{N(W)}}.$$

Example (silver particles, continued). Figure 88 gives the empirical L-function and the 5% limits of Ripley and Koen. The formula (13.21) yields a critical value $\tau_{0.95} = 0.33$. For r smaller than 1.8 cm, $L(r)$ is clearly less than $r - 0.33$, so that the Poisson field hypothesis must also be rejected by the L-test.

Table 17 Approximate critical points $\tau_{0.95}$ for the L-test (after Koen, 1991).

n	ϑ					
	1.5	2.5	5.0	7.5	10.0	15.0
10	1.13E−1	1.24E−1				
20	5.59E−2	5.59E−1				
30	4.56E−2	4.56E−2	4.56E−2			
40	2.59E−2	2.84E−2	3.15E−2			
50	1.98E−2	2.13E−2	2.45E−2			
60	1.82E−2	1.87E−2	1.98E−2	2.45E−2		
70	1.40E−2	1.51E−2	1.72E−2	2.06E−2		
80	1.31E−2	1.38E−2	1.45E−2	1.72E−2		
90	1.10E−2	1.13E−2	1.29E−2	1.44E−2		
100	9.86E−3	1.08E−2	1.14E−2	1.31E−2	1.66E−2	
200	4.95E−3	5.27E−3	5.67E−3	6.21E−3	7.10E−3	
300	3.13E−3	3.39E−3	3.65E−3	3.85E−3	4.46E−3	7.40E−3

Read 3.13E−3 as 0.00313 etc.

Table 18 Approximate critical points $\tau_{0.99}$ for the L-test (after Koen, 1991).

n	ϑ					
	1.5	2.5	5.0	7.5	10.0	15.0
10	1.58E−1	1.58E−1				
20	6.05E−2	6.65E−1				
30	4.56E−2	4.56E−2	5.09E−2			
40	3.95E−2	3.95E−2	3.95E−2			
50	2.53E−2	2.63E−2	2.99E−2			
60	2.05E−2	2.12E−2	2.33E−2	3.05E−2		
70	1.85E−2	1.85E−2	2.03E−2	2.61E−2		
80	1.51E−2	1.59E−2	1.78E−2	2.14E−2		
90	1.38E−2	1.38E−2	1.48E−2	1.76E−2		
100	1.31E−2	1.37E−2	1.38E−2	1.61E−2	2.12E−2	
200	5.88E−3	6.18E−3	7.92E−3	7.71E−3	9.18E−3	
300	3.80E−3	3.96E−3	4.65E−3	4.73E−3	5.73E−3	9.70E−3

Read 1.58E−1 as 0.158 etc.

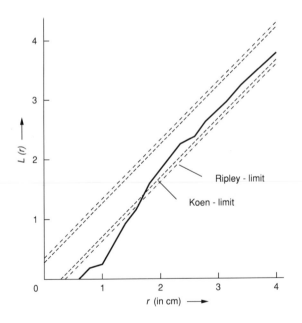

Figure 88 The empirical L-function for the centres of the silver particles of Fig. 87 is indicated by ———. The critical regions belonging to $\tau_{0.95}$ are indicated by − − − (Koen) and −·− (Ripley).

The calculation using Koen's tables is performed with $n = 60$ and

$$\frac{1}{2\sqrt{n}} = 0.0686.$$

If one chooses $\vartheta = 5.0$ then, since $a = 12$, a value 4.1 is obtained for r_0. The critical value here is $\tau_{0.95} = 0.0245a = 0.294$; it is slightly smaller than Ripley's value. Consequently, the invalidity of the Poisson field hypothesis is still clearer.

If the Poisson hypothesis is rejected, the form of the empirical L-function provides information about the type of point distribution in the pattern. Of particular importance are deviations of $L(r)$ from r for small r. If $L(r)$ is less than r then this suggests greater regularity in the point distribution than is typical for a Poisson field. Conversely, if $L(r)$ is greater than r then one must assume that there is a tendency for clustering of points.

If the variants of the L-test described above are not acceptable (whether because of doubts in the accuracy of (13.21) or the point number $N(W)$ not being in the tables, or because of greater deviations of the window shape from a square or a rectangle, and if a good computer is available, then a Monte Carlo test may be used. One chooses a suitable value of r_0, for example the half diagonal length for a rectangular window. Then independent binomial fields with exactly $N(W)$ points are simulated 999 times in a window that is congruent to the given window W. For each sample the L-function is estimated, and then the quantities

$$\tau^{(i)} = \max_{r \leq r_0} |\hat{L}_i(r) - r| \quad (i = 1, \ldots, 999)$$

are determined. The $\tau^{(i)}$ and the corresponding value τ of the empirical L-function are ordered according to magnitude:

$$\tau^{*(1)} \leq \tau^{*(2)} \leq \cdots \leq \tau^{*(1000)}.$$

If the index of τ in this series is greater than 950 or 990 then the Poisson process hypothesis is rejected for $\alpha = 0.05$ or 0.01 respectively.

Test for homogeneity

In general, only certain aspects of homogeneity can be tested. This is particularly problematic if only one sample is given.

Differences in the point density can be tested as follows. Let W_1 and W_2 be two subregions of the window W. They have to be chosen a priori, i.e. before determining the points (examples include the boundary region or interior of W, lower and higher regions of a forest, head and foot regions in the example of p. 232). It would be methodically incorrect to choose the subregions a posteriori, when the point pattern is already known. This might be done in such a way that the point density is large in one of them and low in the other.

Under the assumption of a homogeneous Poisson field, the quantity

$$F = \frac{A(W_1)(2n_2 + 1)}{A(W_2)(2n_1 + 1)}$$

has an approximate F-distribution (Sachs, 1984). The degrees of freedom are $2n_1 + 1$ and $2n_2 + 1$, where n_i is the number of points in W_i: $n_i = N(W_i)$ ($i = 1, 2$). There the indices are such that $F > 1$.

The homogeneity assumption is rejected if

$$F > F_{2n_1+1, 2n_2+1; \alpha/2},$$

where α is the probability of an error of type I.

Example. The point pattern considered here is that of Fig. 90. W_1 and W_2 are the upper and lower halfs of the observed rectangle, so that

$$n_1 = 10, \quad A(W_1) = 1485, \quad n_2 = 27, \quad A(W_2) = 1485.$$

Thus $F = 2.619$. For $\alpha = 0.05$ the critical F-value is

$$F_{2n_1+1, 2n_2+1; \alpha/2} = F_{21, 55; 0.025} = 2.20,$$

i.e. the difference in point density has to be considered significant.

13.3   INHOMOGENEOUS POISSON FIELDS

13.3.1   Fundamental properties

Inhomogeneous Poisson fields are stochastic models of point patters with deterministic differences in the point density. Instead of the intensity λ an intensity function $\lambda(x)$ or an intensity measure Λ is used. These fields have the same independence property 1 (p. 212), as the homogeneous Poisson field. Property 2 is modified as follows.

2'.   The number $N(B)$ of points in a bounded Borel set B has a Poisson distribution of the parameter $\Lambda(B)$. Here Λ is a measure, the so-called *intensity measure*. Λ is diffuse, i.e. there is no x with $\Lambda(\{x\}) > 0$ because otherwise there would be multiple points. In the homogeneous case

$$A(B) = \lambda A(B). \tag{13.22}$$

There is often a density function (with respect to the Lebesgue measure) $\lambda(x)$ of Λ so that

$$A(B) = \int_B \lambda(x) \, dx. \tag{13.23}$$

The function $\lambda(x)$ is called the *intensity function*. The mean number of points in B is $\Lambda(B)$. The probability that in an infinitesimally small disc of centre x and area dF there is a point belonging to the field is $\lambda(x) \, dF$.

If $\Lambda(R^2) = \nu < \infty$ then the point field has only finitely many points; their number has a Poisson distribution with parameter ν.

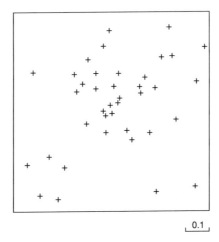

Figure 89 Simulated inhomogeneous Poisson field. The point density is proportional to $\exp(-\sqrt{x_1^2 + x_2^2})$, where the origin is at the centre.

Example. Let the intensity function $\lambda(x)$ have the form

$$\lambda(x) = a \exp\left(-b\sqrt{x_1^2 + x_2^2}\right), \quad x = (x_1, x_2).$$

This means that the point density is maximal at the origin o and decreases exponentially with increasing distances from the origin. Figure 89 shows a simulated point pattern with parameters $a = 5$ and $b = 1$. The mean number of points in the disc $b(o, r)$ centred at o with radius r is

$$\mathsf{E} N(b(o, r)) = \pi a \int_0^r e^{-bu} u \, du = \pi \frac{a}{b^2}[1 - (1 + br)e^{-br}],$$

using polar coordinates ($u^2 = x_1^2 + x_2^2$). This intensity function satisfies

$$\Lambda(R^2) = \nu = \frac{\pi a}{b^2}.$$

13.3.2 Important formulae

Analogous formulae hold for the number distribution as in the homogeneous case; though $\lambda A(B)$ should be replaced by $\Lambda(B)$. The same is true for emptiness probabilities and contact distribution functions. It is also possible in principle to define distance distribution functions corresponding to $D(r)$. They depend, however, on the position of the reference point. (In the above example the nearest-neighbour distances for points close to o are smaller than for points very distant from o.) They can be obtained by limiting procedures as in §13.1.2.

The formulae for second-order characteristics are also similar to those in the homogeneous case. In particular,

$$\operatorname{var} N(B) = \Lambda(B) \tag{13.24}$$

and

$$\mathsf{E} N(B_1) N(B_2) = \Lambda(B_1) \Lambda(B_2) + \Lambda(B_1 \cap B_2). \tag{13.25}$$

Product densities can be expressed in terms of the intensity function $\lambda(x)$. The probability that there is a point in each of two infinitesimally small discs centred at x_1 and x_2 with areas dF_1 and dF_2 is

$$\varrho^{(2)}(x_1, x_2) = \lambda(x_1) \lambda(x_2) \, dF_1 \, dF_2.$$

Poisson fields have an important thinning property. Let $p(x)$ be a function with $0 \le p(x) \le 1$ and consider a Poisson field of intensity function $\lambda(x)$. It is decided randomly and independently of the others whether or not each individual point is eliminated, where the probability for a point at x of being retained is $p(x)$. The result of this thinning procedure (called a 'location-dependent independent' thinning) is a new point field. It is, as perhaps expected, a Poisson field with intensity function

$$\lambda_p(x) = \lambda(x) p(x). \tag{13.26}$$

The corresponding intensity measure satisfies

$$\Lambda_p(B) = \int_B p(x) \lambda(x) \, dx = \int_B p(x) \Lambda(dx).$$

13.3.3 Simulating an inhomogeneous Poisson field

An elegant way of simulating an inhomogeneous Poisson field with intensity function $\lambda(x)$ is to thin a homogeneous Poisson field. One assumes that $\lambda(x)$ is bounded, i.e. there is a number λ^* such that

$$\lambda(x) \le \lambda^*$$

for all x. The function

$$p(x) = \frac{\lambda(x)}{\lambda^*}$$

is used as location dependent thinning function.

In the first step of the simulation a homogeneous Poisson field of intensity λ^* is generated in the window (§13.2.4). Then it is independently decided whether or not the points thus obtained are eliminated. To do this, $p(x)$ is calculated for every point x, and a random number z is generated. If

$$z > p(x),$$

the point x is rejected; otherwise it is a member of the sample of the inhomogeneous Poisson field.

13.3.4   Statistics for an inhomogeneous Poisson field

The most important task of statistics for inhomogeneous Poisson fields is the determination of their intensity functions. This is the sole problem to be considered here. The statistical methods are described for the case of a single point pattern. If there are k independent patterns (belonging to the same intensity function and the same window) then the superposition of the k patterns can be formed, and the corresponding intensity function can then be determined by the methods below. It is then divided by k to yield the required intensity function.

Parametric methods for estimating the intensity function

A parametric form for $\lambda(x)$ frequently make sense, for example

$$\lambda(x) = \lambda(x_1, x_2) = \mu e^{-\alpha x_1}, \quad x_1, x_2 \geq 0, \quad x = (x_1, x_2). \quad (13.27)$$

Here the point density decreases exponentially in the direction of the x_1-axis (see the example on p. 232). This function depends on the two-dimensional parameter $\theta = (\mu, \alpha)$.

The determination of such an unknown parameter θ of an intensity function $\lambda(x, \theta)$ of a Poisson field can be done using the maximum likelihood method. The corresponding likelihood function is

$$L(x_1, \ldots, x_n; \theta) = \lambda(x_1; \theta) \cdots \lambda(x_n; \theta) \exp\left[-\int_W \lambda(x; \theta) \, dx\right] \quad (13.28)$$

(Snyder, 1975; Karr, 1986). The parameter θ is chosen such that the likelihood function or its logarithm are maximized. This leads to the following mathematical problem:

maximize the quantity

$$\sum_{i=1}^{n} \log \lambda(x_i; \theta) - \int_W \lambda(x; \theta) \, dx$$

by a suitable choice of θ.

Frequently the solution can be obtained by differentiating with respect to θ and setting the derivatives equal to zero. The following example demonstrates the solution for the particular case (13.27).

First a heuristic explanation of (13.28) will be given. The quantity $L(x_1, \ldots, x_n; \theta) \, dx_1 \cdots dx_n$ may be interpreted as the probability that a point of the field lies in each of the infinitesimally small area elements of area $dx_1 \cdots dx_n$. No further points should lie outside these areas. These events are independent on account of the independence of the Poisson field. Their probabilities are equal to $\lambda(x_i; \theta)\,dx$ respectively. The probability that no further points lie outside these area elements is

$$\exp\left[-\int \lambda(x; \theta)\,dx\right],$$

where the integration is taken over

$$W\setminus\bigcup_{i=1}^{n} \text{(area elements)}.$$

Clearly the area elements are negligible and may be forgotten. Thus the integration is over the whole of W.

In the modern statistical literature so-called likelihood quotients are used instead of likelihood functions with respect to the distribution of a homogeneous Poisson field (Karr, 1986, p. 118ff). But this leads to the same estimates as the method above.

Example (Siegel *et al.*, 1990). Figure 90 shows the distribution of $n = 37$ defect pores on one side of a cut steel block of length $L = 90$ cm and breadth $b = 33$ cm. The pores are randomly distributed in the observation area, with an increased density near the so-called head. It seems natural to choose an inhomogeneous Poisson field with an intensity function of the form (13.27). The parameters μ and α are to be estimated. Instead of μ, the quantity μ^* is estimated, where

$$\mu^* = \mu b$$

and the originally two-dimensional problem is considered as one dimensional.

The log-likelihood function is

$$l(\mu^*, \alpha) = \int_0^L \mu^* e^{-ax}\,dx + \sum_{i=1}^{n} \log(\mu^* e^{-ax_i})$$

$$= -\frac{\mu^*}{\alpha}(1 - e^{-\alpha L}) + n \log \mu^* - \alpha \sum_{i=1}^{n} x_i'.$$

The x_i' are the x_1-coordinates of the pore centres. Setting the derivatives with respect to α and μ^* equal to zero gives

$$0 = -\frac{1 - e^{-\alpha L}}{\alpha} + \frac{n}{\mu^*}$$

and

$$0 = -\sum_{i=1}^{n} x_i' - \mu^* \frac{\alpha L e^{\alpha L} - (1 - e^{-\alpha L})}{\alpha^2}. \tag{13.29}$$

This yields the following formula for μ^*:

$$\mu^* = \frac{\alpha n}{1 - e^{-\alpha L}}.$$

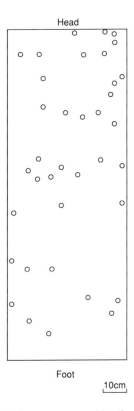

Figure 90 Defect pores on a side of a steel block.

Combining this and (13.29) yields the following equation for α:

$$0 = -\sum_{i=1}^{n} x_i' + \frac{1 - (1 + \alpha L)e^{-\alpha L}}{\alpha(1 - e^{-\alpha L})} n.$$

This can be solved numerically, for example, by *regula falsi*. The following estimates are obtained for the example: $\hat{\alpha} = 0.031$ cm^{-1} and $\hat{\mu}^* = 1.16$ cm^{-1}. This gives the value for μ: the estimate $\hat{\mu} = 0.035$ cm^{-1}. Exactly the same solution would be obtained when considering the problem as a two-dimensional one: the integral in the log-likelihood is then a double integral over the rectangular block.

To check the hypothesis of an inhomogeneous Poisson field, one conducts the following test (Ogata, 1988). It is based on a 'time' transform for a linear non-stationary (inhomogeneous) Poisson process $\{t_i\}$ of intensity function $\lambda(t)$:

$$t_i \to \tau_i = \Lambda(t_i),$$

with $\Lambda(t) = \int_0^t \lambda(u)\, du$. If the t_i are points of a linear Poisson process then the τ_i have positions in the interval $[0, \Lambda(\infty)]$ and behave like independent uniform points ordered in ascending order.

Table 19   Class frequencies after the 'time' transformation.

Classes	Frequencies
0–4.9	4
5–9.9	4
10–14.9	4
15–19.9	5
20–24.9	5
25–29.9	6
30–34.9	9
35–37.4	—

In the particular case of the pores in the steel block the x_1-coordinates $\mathbf{x}_i'$ are transformed according to

$$\tau_i = \frac{\hat{\mu}^*}{\hat{\alpha}}(1 - e^{-\hat{\alpha}\mathbf{x}_i'}).$$

Under the Poisson process hypothesis, a uniform distribution of the τ_i in $[0, \hat{\mu}^*/\hat{\alpha}]$ = $[0, 37.4]$ can be expected. The true class frequencies are given in Table 19. If one ignores the values in the two final classes, then the result is in good agreement with the uniformity hypothesis.

If the intensity function $\lambda(x)$ satisfies the condition

$$\int_{R^2} \lambda(x)\,dx = v < \infty$$

then $\lambda(x)$ can be written as

$$\lambda(x) = vf(x).$$

v is the mean total number of points per point pattern, and $f(x)$ is a two-dimensional density function.

Statistically, v is estimated from k point patterns by

$$\hat{v} = \frac{1}{k}\sum_{j=1}^{k} N_j(R^2) \tag{13.30}$$

where $N_j(R^2)$ is the total number of the jth point pattern. The density function $f(x)$ is then determined by methods for estimating two-dimensional density functions. For example, if $f(x)$ is a normal distribution density as in §2.2.2 then the formulae given there can be used to estimate the parameters of the distribution.

Example (defect pores, continued). This example can be also considered with the classical method of statistics, if the 'censoring' resulting from the finiteness of the block length L is ignored. The known methods for the exponential distribution give

$$\hat{\alpha} = \frac{1}{\bar{x}},$$

where $\bar{x}$ is the mean of the x_1-coordinates of the defect pores. For the example $\bar{x} = 31.89$ yields $\hat{\alpha} = 0.031$ cm^{-1}, which was already obtained on p. 233. A natural estimate of $\hat{\mu}^*$ is

$$\hat{\mu}^* = \hat{\alpha}n.$$

For large α and L this differs little from the formula given on p. 232 for μ^*. For the example the value $\hat{\mu}^* = 1.16$ cm^{-1} is again obtained.

(b) Non-parametric methods for estimating the intensity function

Kernel estimators   The choice of the form of $\lambda(x, \theta)$ as above is sometimes unnatural, particularly if there is no physical or biological law that determines the point density. Therefore it is desirable, in particular for exploratory data analysis, to obtain estimates for $\lambda(x)$ that are free of such decision. A powerful method for solving this problem is the use of kernel estimators (see also Appendix L). A quite simple form is the estimator given by Diggle (1985):

$$\hat{\lambda}_h(x) = \frac{N(b(x, h))}{\pi h^2}. \tag{13.31}$$

Here the intensity function at the location x is estimated by the mean point density in a disc centred at x with radius h. It is clear that the parameter h has great influence on the form of the function $\hat{\lambda}_h(x)$. If h is large then the function $\hat{\lambda}_h(x)$ is smooth and local point density fluctuations are compensated. Conversely, $\hat{\lambda}_h(x)$ may be very rough and variable for small h. Thus the same data lead to different results for different choices of h, and the question arises as to which h is the 'right' or 'optimal' value.

Example (defect pores, continued).   Figure 91 shows isolines of $\hat{\lambda}_h(x)$ for the defect pores of Fig. 90 with values $h = 10$ and $h = 20$. Near the boundary of the window W the formula

$$\hat{\lambda}_h(x) = \frac{N(W \cap b(x, r))}{A(W \cap b(x, r))}$$

has been used. (This is a form of edge-correction. Sometimes it leads to local maxima, if there are points quite close to the boundary.)
    A more general approach uses a kernel function $k_h(z)$. Here

$$\hat{\lambda}_{k_h}(x) = \sum_{i=1}^{n} k_h(x - \mathbf{x}_i) \tag{13.32}$$

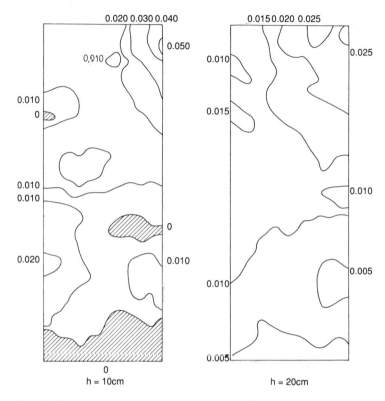

Figure 91 Isoline plots of the estimated intensity function for the pores of Fig. 90. Obviously, the choice of the smoothing parameter h has great influence on the result. The plot for $h = 20$ appears to be acceptable.

is an estimator for $\lambda(x)$. The particular choice

$$k_h(z) = \frac{1_{b(o,h)}(z)}{\pi h^2}$$

gives the estimator of (13.31). Another possibility is the planar Epanečnikov kernel

$$k_h(z) = \frac{8}{3\pi h} e_h(\|z\|),$$

where

$$e_h(t) = \begin{cases} \dfrac{3}{4h}\left(1 - \dfrac{t^2}{h^2}\right) & (|t| < h), \\ 0 & \text{otherwise.} \end{cases}$$

Also a 'quadratic' version of the Epanečnikov kernel has been used:

$$k_h((Z', Z'')) = e_h(Z')e_h(Z'') \qquad (Z = (Z', Z'')).$$

The simple notation $\lambda_h(x)$ will be used from now on, even in the case of a general kernel function.

Choice of parameter h. Frequently a reasonable way of choosing h is to determine the functions $\hat{\lambda}_h(x)$ for different h and to choose that which gives the most pleasing or reasonable result.

An easy numerical method is the likelihood cross-validation method. The explanation given in Silverman (1986, pp. 52, 53), can be transferred to the situation considered here as follows. Suppose y is a further member of the point field in addition to the points given in W. If $\lambda(x)$ is the true (unknown) intensity function then the value of the log-likelihood function belonging to y is

$$\log \lambda(y) - \int_W \lambda(x)\,dx.$$

Here it is natural to replace $\lambda(x)$ by the estimator $\hat{\lambda}_h(x)$ and to consider h as a parameter to be estimated.

But there is no additional point y. Thus, one proceeds so that first the point x_i is removed from the set of n original points. For the remaining points an estimate $\hat{\lambda}_{i,h}(x)$ of the intensity function is determined for the kernel with parameter h. This yields for x_i the log-likelihood value

$$\log \hat{\lambda}_{i,h}(x_i) - \int_W \hat{\lambda}_{i,h}(x)\,dx.$$

Since all points are of the same kind, it is averaged and the quantity

$$PL(h) = \frac{1}{n} \sum_{i=1}^n \left(\log \hat{\lambda}_{i,h}(x_i) - \int_W \hat{\lambda}_{i,h}(x)\,dx \right)$$

is considered. This quantity is maximized by optimal choice of h. Since the integrals are obviously equal to $n-1$ for all i, it suffices to maximize the sum

$$\sum_{i=1}^n \log \hat{\lambda}_{i,h}(\mathbf{x}_i) = \sum_{i=1}^n \log \sum_{\substack{j=1 \\ j \neq i}}^n k_h(\mathbf{x}_i - \mathbf{x}_j).$$

This can be easily done on a computer by determining the double sums for a series of h-values and choosing that h which yields the biggest double sum. Unfortunately, this method frequently gives rather small values of h. Thus the optimization is modified by introducing so-called penalty functions, which ensure sufficiently smooth results (see e.g. Silverman, 1986; Ogata and Katsura, (1988).

Berman and Diggle (1989) have chosen the smoothing parameter h in (13.31) such that a mean quadratic error is minimized. For the quadratic Epanečnikov kernel (p. 236), Diggle (1981) has recommended the value

$$h = 0.68 n^{-0.2}$$

for a quadratic window W. (As always, n is the number of points observed in W.) The same author has also suggested edge corrections (see Diggle, 1985).

It is clear that there is a close connection with estimating two-dimensional density functions. Thus the methods used there (Silverman, 1986) can be applied here as well.

Example (defect pores, continued). As above, it is assumed that $\lambda(x) = \lambda(x_1)$, i.e. that the intensity function depends only on the x_1-coordinate, namely the distance from the head side. Thus to estimate $\lambda(x)$, only the x_1-coordinates of the defect pores are used. Using the Epanečnikov kernel, $\lambda(x_1)$ is estimated as

$$\hat{\lambda}_h(x_1) = \sum_{i=1}^{n} e_h(x_1 - \mathbf{x}'_i),$$

where $\mathbf{x}'_i$ is the x_1-coordinate of the ith point.

The function $PL(h)$ has a clear maximum at $h = 0.8$; thus the likelihood cross-validation estimate of h is 0.8. Unfortunately this value is too small, and $\hat{\lambda}_{0.8}(x_1)$ does not look reasonable (Fig. 92). Since the function $PL(h)$ is monotonically decreasing in h for $h > 0.8$, the smallest value for h should be taken at which additional smoothness and positivity properties of $\lambda(x_1)$ hold.

Figure 93 shows several estimated intensity functions for different values of h. For $h = 6$ a function is obtained that is practically positive, but this function is

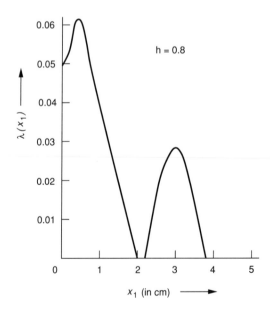

Figure 92   Estimated intensity function $\lambda(x_1)$ for the pores of Fig. 91 with $h = 0.8$. Here the Epanečnikov kernel was used. The meaning of h differs from that in Fig. 91.

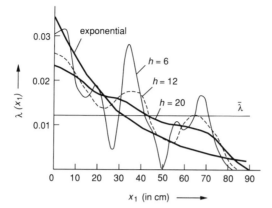

Figure 93 Several estimates of the intensity function $\lambda(x_1)$ for the pores of Fig. 91. Again the Epanečnikov kernel was used. The parametric estimate of p. 233 is labelled 'exponential'. $\bar{\lambda}$ is that value which belongs to the assumption of a homogeneous Poisson field.

too still rough. A smooth curve is obtained only for a bandwidth $h = 20$; it is interesting to see that it clearly differs from an exponential function.

CHAPTER 14

Fundamentals of the Theory of Point Fields

14.1 INTRODUCTION

The aim of this chapter is to present the reader with the important facts of the theory of point fields in an elementary manner. The aim is to help in understanding those characteristics that are of importance in point process statistics. Furthermore, a formalism is introduced in which the statistical estimators of Chapter 15 are given.

The first- and second-moment measures and the related characteristics play a central role. In some sense they correspond to mean and variance of random variables and to the mean value and covariance function of stochastic processes. Furthermore, distance distributions are considered.

More detailed expositions of the theory are given by Cox and Isham (1980), Stoyan *et al.* (1987), Daley and Vere-Jones (1988), Reiss (1993), Karr (1986) and Kerstan *et al.* (1976), in increasing order of difficulty.

14.2 THE INTENSITY MEASURE

The *intensity measure* plays a similar role in the theory of point fields as the mean does for random variables. It is defined by

$$\Lambda(B) = \mathsf{E}N(B) \tag{14.1}$$

for any Borel set B; thus $\Lambda(B)$ is the mean point number in the set B. Often there exist a density function $\lambda(x)$ for Λ with

$$\Lambda(B) = \int_B \lambda(x)\,\mathrm{d}x. \tag{14.2}$$

The function $\lambda(x)$ is called the *intensity function*. If the point field N is homogeneous then the intensity λ completely describes the intensity measure:

$$\Lambda(B) = \lambda A(B). \tag{14.3}$$

This is a simple consequence of $EN(B) = EN(B_x)$ (because of homogeneity) or $\Lambda(B) = \Lambda(B_x)$ and the fact that the Lebesgue measure A is the only translation-invariant measure on R^2 up to constant factors that assigns a finite value to the unit square.

In taking B to be of unit area, it becomes clear that λ may be interpreted as a point density or a mean point number per unit area. A local interpretation of λ is also possible. Let C be a small disc of infinitesimal area dF. Then the probability that C contains exactly one point of the point field is

$$\lambda \, dF + o(dF).$$

The probability that C contains more than one point is $o(dF)$; see the literature for a rigorous mathematical formulation of these facts. Using the intensity measure, important means may be written elegantly. In particular, the *Campbell theorem* holds.

Let there be a measurable function $f(x)$, which assigns to each point x of R^2 a non-negative number. One wishes to determine the mean

$$E \sum_{x \in N} f(\mathbf{x}) = E \int f(x) N(dx).$$

For each point $\mathbf{x}$ of N the value $f(\mathbf{x})$ is found, and then the sum of the function values is considered. The result is a random variable. Then

$$E \int f(x) N(dx) = \int f(x) \Lambda(dx). \tag{14.4}$$

Using the intensity function $\lambda(x)$, the right-hand side can be rewritten as

$$\int f(x) \lambda(x) \, dx.$$

If the point field is homogeneous then (14.4) simplifies further:

$$E \int f(x) N(dx) = \lambda \int f(x) \, dx. \tag{14.5}$$

In the case of marked point fields the mean value theory is somewhat richer. First the field may be considered without marks and the intensity measure and intensity can be defined. Alternatively, the full *intensity measure* Λ_M of the marked point field can be used. It is defined by

$$\Lambda_M(S) = EN(S),$$

where S is a subset of $R^2 \times M$. In particular,

$$\Lambda_M(B \times C) = EN(B \times C)$$

is the mean number of points in B with a mark in C.

If the marked point field is homogeneous, Λ_M simplifies as in the case without marks. Then

$$\Lambda_M(B \times C) = \lambda A(B)\mathbf{M}(C). \qquad (14.6)$$

Here λ is the intensity, and $\mathbf{M}$ is the so-called *mark distribution*. $\mathbf{M}(C)$ may be interpreted as the probability that the mark of an 'arbitrarily' chosen point lies in the mark set C. ('Arbitrarily' means that a selection scheme is used in which every point has the same chance to be selected.) Another interpretation is that $\mathbf{M}(C)$ is the probability that the mark of the 'typical' point lies in C.

In many applications the marks are real-valued and one requires a density function $\mathbf{m}(m)$ for the mark distribution. Then it makes sense to speak about a mean mark, which is calculated as

$$\bar{m} = \int_{-\infty}^{\infty} m\mathbf{m}(m)\,\mathrm{d}m.$$

A *Campbell theorem* also holds for marked point fields. In the homogeneous case

$$\mathsf{E} \sum_{[\mathbf{x};m] \in N} f(\mathbf{x}, m) = \lambda \int \int f(x, m)\mathbf{M}(\mathrm{d}m)\,\mathrm{d}x. \qquad (14.7)$$

Here $f(\mathbf{x}, m)$ is a function that depends on both a point and its mark. If the marks are real numbers and there is a density function $\mathbf{m}(m)$ then the right-hand side of (14.7) can be written as

$$\lambda \int \int f(x, m)\mathbf{m}(m)\,\mathrm{d}m\,\mathrm{d}x.$$

If the marks of a marked point field N are positive numbers then it has an associate random measure, the so-called *mark sum measure* S_N, defined by

$$S_N(B) = \sum_{[\mathbf{x};m] \in N} 1_B(\mathbf{x})m.$$

Here the marks of all points in B are added. For any B, $S_N(B)$ is a random variable. If N is homogeneous then so is S_N, and

$$\mathsf{E}S_N(B) = \lambda \bar{m} A(B). \qquad (14.8)$$

14.3   EMPTINESS PROBABILITIES

If K is a compact subset of R^2, then the corresponding emptiness probability is given by

$$v(K) = \Pr(N(K) = 0).$$

Emptiness probabilities play an important role in the theory of point fields: The system of all $v(K)$ for compact K determines the distribution of N uniquely. Thus the function $v(K)$ has a meaning similar to that of the distribution function for random variables.

In practice emptiness probabilities are used for homogeneous point fields of positive intensity λ, and also when only particular 'test sets' K are used. The case where K is the disc centred at o with radius r is particularly important:

$$K = b(o, r).$$

It is related to the so-called *spherical contact distribution function*

$$H_s(r) = 1 - \Pr(N(b(o, r)) = 0) \quad (r \geq 0) \tag{14.9}$$

The function $H_s(r)$ is monotonically increasing, and

$$H_s(0) = 1 - \Pr(o \notin N) = 0,$$
$$H_s(\infty) = 1 - \Pr(N(R^2) = 0) = 1.$$

The function $H_s(r)$ can be interpreted as the distribution function of the distance from the origin o to the nearest point of the point field. Another interpretation is that of the distribution function of the distance of a random test point independent of the point field to its nearest neighbour.

Formulae for $H_s(r)$ are known for Poisson, Cox and Neyman–Scott fields (pp. 213 and 312).

14.4 SECOND-ORDER CHARACTERISTICS

14.4.1 Definitions and formulae

Second-order characteristics describe variation and correlation in point fields. There is some similarity to correlation functions of stochastic processes.

In this section it is assumed that the point field is homogeneous; in addition isotropy is often (and explicitly) supposed.

The exposition begins with Ripley's K-function. Each point of the point field $N = \{x_n\}$ is taken as the centre of a disc of radius r. The number of points in the disc $b(x, r)$ excluding the point x is denoted by $n_r(x)$ and can be used as a mark, and thus the field may be regarded as a homogeneous marked point field $\{[x_n; n_r(x_n)]\}$. Let the mean mark be $\mathbf{n}_r$. This gives *Ripley's K-function* by

$$\lambda K(r) = \mathbf{n}_r \quad (r \geq 0). \tag{14.10}$$

Heuristically speaking, one says that $\lambda K(r)$ equals the mean number of points that have a distance smaller than r from the 'typical' point (see also §14.9). Usually, the function $K(r)$ is of order r^2 for large r. In particular, for a homogeneous Poisson field

$$K(r) = \pi r^2 \quad (r \geq 0). \tag{14.11}$$

Graphs of K-functions may be difficult to interprete because of their parabolic form. It is convenient to use the so-called L-function, which is given by

$$L(r) = \sqrt{\frac{K(r)}{\pi}} \quad (r \geq 0). \tag{14.12}$$

In the case of the Poisson field it takes the simple form

$$L(r) = r \quad (r \geq 0). \tag{14.13}$$

Since Poisson fields are frequently used as null models, this is a natural definition of the L-function. In Chapters 15 and 16 examples of its use in statistical analysis are given.

Instead of discs, other sets can also be used. Let B be an arbitrary Borel set. To the point $\mathbf{x}$ the number $n_B(\mathbf{x})$ of points in the set $B_{\mathbf{x}} = B + \mathbf{x}$ is assigned as a mark, where $\mathbf{x}$ is once again not counted (if it is in B):

$$n_B(\mathbf{x}) = N(B_{\mathbf{x}} \backslash \{\mathbf{x}\}).$$

Thus a marked point field $\{[\mathbf{x}_n, n_B(\mathbf{x}_n)]\}$ is obtained. The mean mark of this marked point field is written in the form

$$\bar{n}_B = \lambda \mathfrak{R}(B). \tag{14.14}$$

Here $\mathfrak{R}$ is a Borel measure, the so-called *reduced second-moment measure*. Clearly

$$K(r) = \mathfrak{R}(b(o, r)) \quad (r \geq 0).$$

Some readers might wonder why the characteristics introduced are called second-order characteristics, although they are defined as means. The explanation is as follows. The original second-order quantities are the so-called *moment measures*. The *second-moment measure* is given by

$$\mu^{(2)}(B_1 \times B_2) = \mathsf{E} N(B_1) N(B_2). \tag{14.15}$$

An equivalent is

$$\mu^{(2)}(B_1 \times B_2) = \mathsf{E} \sum_{\mathbf{x}_1 \in N} \sum_{\mathbf{x}_2 \in N} 1_{B_1}(\mathbf{x}_1) 1_{B_2}(\mathbf{x}_2). \tag{14.16}$$

B_1 and B_2 are Borel sets. (Standard constructions of measure theory lead to a measure on the product space $\mathcal{B}^2 \otimes \mathcal{B}^2$.) In terms of the means of $\mu^{(2)}$ and Λ, the variance of point number $N(B)$ in the Borel set B can be expressed as

$$\operatorname{var} N(B) = \mu^{(2)}(B \times B) - [\Lambda(B)]^2$$

and in the homogeneous case

$$\operatorname{var} N(B) = \mu^{(2)}(B \times B) - [\lambda A(B)]^2. \tag{14.17}$$

A similar formula is true for the covariance of the numbers $N(B_1)$ and $N(B_2)$:

$$\text{cov}(N(B_1), N(B_2)) = \mathsf{E}N(B_1)N(B_2) - \mathsf{E}N(B_1)\mathsf{E}N(B_2)$$
$$= \mu^{(2)}(B_1 \times B_2) - \lambda A(B_1)\lambda A(B_2).$$

In the theory of point fields the *factorial moment measure* $\alpha^{(2)}$ is frequently used instead of $\mu^{(2)}$. This is given by

$$\alpha^{(2)}(B_1 \times B_2) = \mathsf{E}\sum_{\mathbf{x}_1 \in N}\sum_{\substack{\mathbf{x}_2 \in N \\ (\mathbf{x}_2 \neq \mathbf{x}_1)}} 1_{B_1}(\mathbf{x}_1)1_{B_2}(\mathbf{x}_2).$$

This is the mean of all point pairs with one member in B_1 and the other in B_2 where both members are different. Obviously,

$$\mu^{(2)}(B_1 \times B_2) = \alpha^{(2)}(B_1 \times B_2) + \mathsf{E}\sum_{\mathbf{x} \in N} 1_{B_1}(\mathbf{x})1_{B_2}(\mathbf{x})$$
$$= \alpha^{(2)}(B_1 \times B_2) + \Lambda(B_1 \cap B_2)$$
$$= \alpha^{(2)}(B_1 \times B_2) + \lambda A(B_1 \cap B_2).$$

The name 'factorial moment measure' is used since

$$\alpha^{(2)}(B \times B) = \mathsf{E}N(B)[N(B) - 1].$$

For an integer random number X the quantity $\mathsf{E}X(X - 1)$ is called the second factorial moment.

The last formula for $\mu^{(2)}$ gives

$$\text{var}\, N(B) = \alpha^{(2)}(B \times B) - [\lambda A(B)]^2 + \lambda A(B). \tag{14.18}$$

For any measurable non-negative function $f(x_1, x_2)$

$$\mathsf{E}\sum_{\mathbf{x}_1 \in N}\sum_{\mathbf{x}_2 \in N} f(\mathbf{x}_1, \mathbf{x}_2) = \int f(x_1, x_2)\mu^{(2)}(d(x_1, x_2)) \tag{14.19}$$

and

$$\mathsf{E}\sum_{\mathbf{x}_1 \in N}\sum_{\substack{\mathbf{x}_2 \in N \\ (\mathbf{x}_2 \neq \mathbf{x}_1)}} f(\mathbf{x}_1, \mathbf{x}_2) = \int f(x_1, x_2)\alpha^{(2)}(d(x_1, x_2)). \tag{14.20}$$

The summation is over all pairs of points of N with distinct members.

Frequently, α^2 has a density function $\varrho^{(2)}(x_1, x_2)$ so that

$$\alpha^{(2)}(B_1 \times B_2) = \int_{B_1}\int_{B_2} \varrho^{(2)}(x_1, x_2)\, dx_2\, dx_1.$$

Using this function, the right-hand of (14.20) can be written in the form

$$\int\int f(x_1, x_2)\varrho^{(2)}(x_1, x_2)\, dx_2\, dx_1.$$

The density function $\varrho^{(2)}(x_1, x_2)$ is called the *second-order product density*.

In particular, consider

$$f(x_1, x_2) = 1_{B_1}(x_1)1_{B_2}(x_2 - x_1) = 1_{B_1}(x_1)1_{B_2+x_1}(x_2).$$

By (14.20),

$$\int 1_{B_1}(x_1)1_{B_2}(x_2 - x_1)\alpha^{(2)}(\mathrm{d}(x_1, x_2))$$

$$= \mathsf{E}\sum_{x_1 \in N}\sum_{\substack{x_2 \in N \\ (x_2 \neq x_1)}} 1_{B_1}(\mathbf{x}_1)1_{B_2}(\mathbf{x}_2 - \mathbf{x}_1)$$

$$= \mathsf{E}\sum_{x_1 \in N}\sum_{\substack{x_2 \in N \\ (x_2 \neq x_1)}} 1_{B_1}(\mathbf{x}_1)1_{B_2+\mathbf{x}_1}(\mathbf{x}_2)$$

$$= \mathsf{E}\sum_{x_1 \in N} 1_{B_1}(\mathbf{x}_1)N(B_2 + x_1\backslash\{\mathbf{x}_1\}).$$

The last mean equals the mean of the mark sum measure for B, if the mark of the point $\mathbf{x}$ is the number $N(B_2 + x\backslash\{\mathbf{x}\})$. The formulae (4.8) and (4.14) lead to the value $\lambda^2 A(B_1)\mathcal{R}(B_2)$, thus

$$\mathsf{E}\sum_{x_1 \in N}\sum_{\substack{x_2 \in N \\ (x_2 \neq x_1)}} f(\mathbf{x}_1, \mathbf{x}_2) = \lambda^2 A(B_1)\mathcal{R}(B_2).$$

This formula is a particular case of the following which is true for all non-negative measurable functions $f(x_1, x_2)$:

$$\mathsf{E}\sum_{x_1 \in N}\sum_{\substack{x_2 \in N \\ (x_2 \neq x_1)}} f(\mathbf{x}_1, \mathbf{x}_2) = \lambda^2 \int\int f(x, x + h)\mathcal{R}(\mathrm{d}h)\,\mathrm{d}x. \tag{14.21}$$

(In the particular case considered above $f(x, x + h) = 1_{B_1}(x)1_{B_2}(h)$.)

Now that it is clear that for given λ the measure $\mathcal{R}$ determines uniquely the second-order factorial moment measure and, because of the relation between $\alpha^{(2)}$ and $\mu^{(2)}$, the second-order moment measure as well. Also, the name 'reduced second-order moment measure' is now justified on account of (14.21) with the factorization of $\alpha^{(2)}$.

If the point field N is isotropic then λ and $K(r)$ already determine uniquely the second-order factorial moment measure. Therefore Ripley's K-function, the L-function and $\mathcal{R}$ are indeed second-order characteristics.

The variance of the point number $N(B)$ in the Borel set B satisfies

$$\mathrm{var}\,N(B) = \lambda^2 \int \gamma_B(h)\mathcal{R}(\mathrm{d}h) + \lambda A(B) - [\lambda A(B)]^2. \tag{14.22}$$

Here $\gamma_B(h)$ denotes the area of the intersection of B and B_h:

$$\gamma(h) = A(B \cap B_h).$$

The formula (14.22) follows easily from (14.18) and (14.21), with

$$f(x_1, x_2) = 1_B(x_1) 1_B(x_2).$$

The integral on the right-hand side of (14.21) then takes the form

$$\lambda^2 \int \int 1_B(x) 1_B(x + h) \, dx \, \Re(dh).$$

The inner integral is equal to $\gamma_B(h)$, which is called the *set covariance function* of B, see p. 122.

In the isotropic case, and analogously to (14.22),

$$\operatorname{var} N(B) = \lambda^2 \int_0^\infty \bar{\gamma}_B(r) \, dK(r) + \lambda A(B) - [\lambda A(B)]^2. \tag{14.23}$$

Here $\bar{\gamma}_B(r)$ is the isotropized set covariance function, which was considered in detail in §8.4.3. There formulae are also given for particular sets B.

The quantities $\Re$, $K(r)$ and $L(r)$ together with λ determine all second-order moments involving point numbers. In statistics they are used extensively, in particular for goodness-of-fit tests. Their interpretation is not always simple, and therefore they are not suitable for exploratory data analysis. Other functions are used that have the character of densities; particularly important is the so-called pair correlation function.

The product density $\varrho^{(2)}(x_1, x_2)$ has already been mentioned above. (Its existence is not always evident. For example, in the case of lattice point processes there is no product density — only in the sense of the theory of generalized functions.) It is considered now in more detail.

Its meaning can be understood intuitively as follows (see also p. 242). Let C_1 and C_2 be two infinitesimally small discs centred at x_1 and x_2 and of areas dF_1 and dF_2 (because of the smallness of the discs it can be assumed that they are disjoint). Then the probability that in each of the discs there is a point of the point field N is approximately

$$\varrho^{(2)}(x_1, x_2) \, dF_1 \, dF_2.$$

If the point field N is homogeneous then $\varrho^{(2)}(x_1, x_2)$ depends only on the difference $h = x_2 - x_1$. The distance between the points x_1 and x_2 and the direction of the line through x_1 and x_2 are the only important considerations. Therefore the value of the product density corresponding to h is written in the simple form $\varrho^{(2)}(h)$. It is sometimes useful to use polar coordinates

$$h = (r, \varphi) \quad (0 \le \varphi < \pi),$$

and to employ the symbol $\varrho^{(2)}(r, \varphi)$. In the homogeneous case the symbols $\varrho^{(2)}(x_1, x_2)$, $\varrho^{(2)}(h)$ and $\varrho^{(2)}(r, \varphi)$ mean the same.

In the isotropic case the direction of the line through x_1 and x_2 is uninteresting. The product density depends only on the distance r of the points x_1 and x_2 or on the length of h. For simplicity, the symbol is not changed: $\varrho^{(2)}(r)$ denotes the product density, which depends only on r.

It is useful to normalize the function $\varrho^{(2)}(r)$ and to define a new function

$$g(r) = \frac{\varrho^{(2)}(r)}{\lambda^2} \quad (r \geq 0).$$ (14.24)

$g(r)$ is called the *pair correlation function*.

In the homogeneous case, when using the function $\varrho^{(2)}(h)$, the moment formulae above simplify, starting from (14.20),

$$\mathsf{E} \sum_{x_1 \in N} \sum_{\substack{x_2 \in N \\ (x_2 \neq x_1)}} f(\mathbf{x}_1, \mathbf{x}_2) = \int f(x_1, x_2) \alpha^{(2)}(\mathrm{d}(x_1, x_2))$$

$$= \int \int f(x_1, x_2) \varrho^{(2)}(x_1, x_2) \, \mathrm{d}x_1 \, \mathrm{d}x_2$$

$$= \int \int f(x, x + h) \varrho^{(2)}(h) \, \mathrm{d}h \, \mathrm{d}x.$$

Thus

$$\mathsf{E} \sum_{x_1 \in N} \sum_{\substack{x_1 \in N \\ (x_2 \neq x_1)}} f(\mathbf{x}_1, \mathbf{x}_2) = \int \int f(x, x + h) \varrho^{(2)}(h) \, \mathrm{d}h \, \mathrm{d}x.$$ (14.25)

Comparison with (14.21) shows that

$$\lambda^2 \mathfrak{K}(\mathrm{d}h) = \varrho^{(2)}(h) \, \mathrm{d}h.$$

In the isotropic case one has analogously

$$\lambda^2 \, \mathrm{d}K(r) = 2\pi r \varrho^{(2)}(r) \, \mathrm{d}r.$$

Consequently

$$g(r) = \frac{1}{2\pi r} \frac{\mathrm{d}}{\mathrm{d}r} K(r) \quad (r \geq 0).$$ (14.26)

Finally, (14.23) can be written in the form

$$\mathrm{var} \, N(B) = 2\pi \lambda^2 \int_0^\infty \bar{\gamma}_B(r) r g(r) \, \mathrm{d}r + \lambda A(B) - [\lambda A(B)]^2.$$ (14.27)

It is obvious that the variance of point number in B is determined by two aspects of the problem: by the variability of the point field N, expressed by $g(r)$, and by the geometry of the set B, expressed by $\bar{\gamma}_B(r)$.

For 'large' B ('spherically infinite' in the sense of Girling, 1982) the approximation

$$\frac{\mathrm{var}\, N(B)}{A(B)} \approx \lambda + 2\pi\lambda^2 \int_0^\infty (g(r) - 1)r\, \mathrm{d}r \tag{14.28}$$

can be used.

14.4.2   Interpretation of pair correlation functions[†]

An important tool for exploratory statistical analysis of point fields is the pair correlation function. This section aims to help the reader in interpreting empirical pair correlation functions. First, the relationship between the form of the pair correlation function and the point distribution is discussed. Later, pair correlation functions are described for important point field models and typical point patterns.

First,

$$g(r) \geq 0,$$

and

$$\lim_{r \to \infty} g(r) = 1$$

for mixing point fields. The value unity is thus an asymptote of the pair correlation function.

Large values of $g(r)$ show that point pairs of distance r appear frequently; small values occur if point pairs with this distance are rare. $g(r)$ is exactly zero if the interpoint distance r never occurs. Pair correlation functions may have poles:

$$\lim_{r \to r^*} g(r) = \infty.$$

In particular, poles for $r^* = 0$ are not rare; for example, they appear for cluster fields.

Figure 94 shows a pair correlation function with a form typical in many applications. Its form will now be explained.

First, it is obvious that the corresponding point field must have the hard core distance r_0. Interpoint distances smaller than r_0 are impossible. If the points are centres of mutually non-intersecting discs then their diameters have to be smaller than r_0.

The maximum of $g(r)$ at $r = r_1$ characterizes the range of the most frequent short interpoint distances. From the point of view of a randomly chosen point of

---

[†] Here only the pair correlation functions of planar point fields are discussed. Three-dimensional point patterns and their pair potentials are discussed in the physics literature. Their interpretation is similar to that of the planar case.

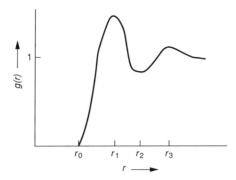

Figure 94 Typical form of the pair correlation function of a point field with weak short-range order. The meanings of r_0, r_1, r_2 and r_3 are explained in the text.

Figure 95 Neighbourhood relations at a point •. The points denoted by + are the direct neighbours of •; those denoted by ○ could be called 'second order neighbours'.

the field, the maximum refers to its direct neighbours (Fig. 95) (one of them is the nearest neighbour). The existence of a clear maximum of the pair correlation function shows that the direct neighbours have a relatively uniform distance to the reference point. The more uniform the distance is, the sharper is the maximum; its sharpness is thus a measure of the order in the point pattern.

The second maximum at $r = r_3$ is connected with those points that lie behind the direct neighbours (they are the direct neighbours of these points). Again the sharper the second maximum is, the more ordered is the point distribution. Its existence is already an indicator of short-range order. Typically $r_3 < 2r_1$. Finally, the minimum at $r = r_2$ results from a gap in the point distribution; behind the direct neighbours there are relatively few points before the next points appear. The deepness of the maximum is again connected with the degree of short-range order in the point pattern.

A simple parameter for characterizing the degree of short-range order is

$$M = \frac{g(r_1) - g(r_2)}{r_2 - r_1}. \tag{14.29}$$

The greater M is, the more short-range order one can expect. For a Poisson field M is precisely 0; this corresponds to the case of no short-range order. However, note

20 µm

Figure 96 Planar section through a Cd–Zn eutectic with rod-shaped Zn. The centres of gravity of the sections of the Zn rods form the point pattern to be analysed. For more detailed analysis marks are also considered: the maximum Feret diameters (maximum projection lengths), and the directions of these diameters.

that great values of M are possible for quite different forms of point distribution, both for rather regular point distribution (as in Fig. 96; see also the pair correlation function in Fig. 97) and also for cluster fields with a hard core distance (Fig. 102). But this does not cause problems, in practical investigations because in general the types of point field analysed are known a priori and mostly only fields of the same type are analysed. An example of the application of M as a short-range characteristic is given by Stoyan and Schnabel (1990).

A further parameter is the range, which is equal to r_4: this is the distance for which $g(r) = 1$ for all $r > r_4$. (For empirical pair correlation functions irregular fluctuation of $g(r)$ around 1 is observed.)

Pair correlation functions for some point field models

Poisson field. The pair correlation function is constant in this case:

$$g(r) \equiv 1$$

(p. 217). This form reflects the complete randomness of the point distribution.

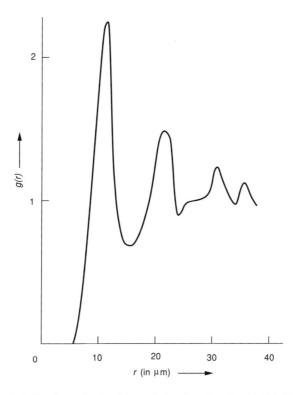

Figure 97 Statistically determined pair correlation function for the Cd-Zn eutectic. The high degree of order is reflected in the form of the pair correlation function.

Lattice point fields. Lattice point fields have a regular point distribution like that of atoms in a crystal lattice. Only certain discrete values are possible as interpoint distances. A pair correlation function can be defined only in a generalized sense. It is non-zero only for discrete interpoint distances and it does not converge to unity as for $r \to \infty$.

Here the case of a quadratic lattice of mesh width a is discussed. Its intensity is

$$\lambda = a^{-2}.$$

The mean number of points, $\lambda K(r)$, in a disc centred at a point of the lattice satisfies

$$\lambda K(r) = \begin{cases} 0 & (r < a), \\ 4 & (a \leq r < \sqrt{2}a), \\ 8 & (\sqrt{2}a \leq r < 2a), \\ 12 & (2a \leq r < \sqrt{5}a), \\ 20 & (\sqrt{5}a \leq r < 2\sqrt{2}a), \\ \vdots & \vdots \end{cases}$$

Analogously to the relation

$$g(r) = \frac{1}{2\pi r} \frac{\mathrm{d}}{\mathrm{d}r} K(r),$$

which is true in the case of a differentiable K-function, the 'function'

$$g(r) = \frac{1}{2\pi r} \sum_{i=1}^{\infty} \delta_{r_i}(r) c_i$$

is now considered, where $r_1 = a$, $r_2 = \sqrt{2}a$, $r_3 = 2a$, $r_4 = \sqrt{5}a$, $r_5 = 2\sqrt{2}a$, ..., $c_1 = 4a^2$, $c_2 = 4a^2$, $c_3 = 4a^2$, $c_4 = 8a^2$, $c_5 = 4a^2$, ... Here $\delta_\varrho(r)$ is the Dirac delta function, which has the properties $\delta_\varrho(r) = 0$ $(r \neq \varrho)$ and $\int_{-\infty}^{\infty} \delta_\varrho(r)\,\mathrm{d}r = 1$.

Randomly disturbed lattices. The points lie in the neighbourhood of the lattice points of a lattice, scattered according to some law. In real structures patterns are observed similar to that of Fig. 96.

For a disturbed quadratic lattice the pair correlation function has maxima at $r = a, \sqrt{2}a, 2a, \ldots$, and minima between them. The sharpness of the maxima depends on the degree of disturbance.

Figure 97 shows the empirical pair correlation function for the pattern of Fig. 96. There are three clear maxima, which belong to the distances to the 1st, 2nd and 3rd neighbours, and two minima, which belong to the gaps between the 1st and 2nd, and the 2nd and 3rd neighbours. For increasing r the curve approaches the value 1.

Random packings. Many point patterns are systems of centres of non-overlapping particles. If the area fraction of the area covered by the particles is large the term 'random packings' is sometimes used. There are several mathematical models of such packings, see e.g. Cowan (1984) or Chapter 16. The pair correlation functions of such point patterns have in general at least two maxima and one minimum, thus reflecting some degree of short-range order.

Figure 98 shows a random packing of discs produced by Cowan (1984). The corresponding pair correlation function is shown in Fig. 99. The pattern of basaltic column cross-sections shown in Fig. 100 may also be interpreted as a packing. The pair correlation function of the polygon centres is given in Fig. 101; see also the detailed discussion in Stoyan and Stoyan (1990a).

Cluster fields. If the points appear in clumps or clusters as shown in Fig. 87 then the pair correlation function is large for small r and increases for increasing r. It is possible that $g(0) = \infty$. The particular form of $g(r)$ depends on the type of distribution of the cluster centres; for a point distribution in the clusters that allows very small interpoint distances $g(r)$ has a form like the dashed curve on Fig. 102. If the interpoint distances in the clusters do not fall below a fixed value r_1 then the other curve of Fig. 102 is obtained. In both cases the range characterizes the spatial size of clusters. (If the cluster points lie in discs of fixed radius R then the range is less than or equal to $2R$.)

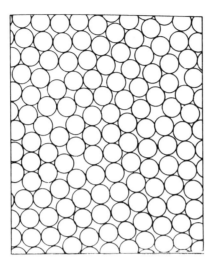

Figure 98   A random packing of discs (Cowan, 1984). The system of disc centres shows a far-reaching order.

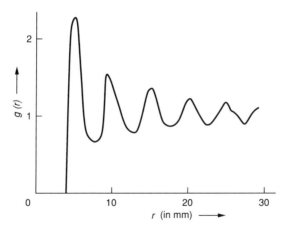

Figure 99   Statistically determined pair correlation function for the pattern of disc centres in Fig. 98. The form of the pair correlation function clearly expresses the high degree of order in the pattern.

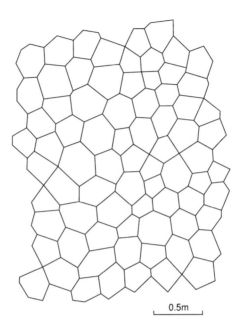

Figure 100   Cross-sections of basaltic columns at Burg Stolpen. The centres of the polygons form a point pattern that is rather regular. The point pattern of the vertices has quite different properties. Here there are points of very small separation, and the pattern has similarities to a cluster field.

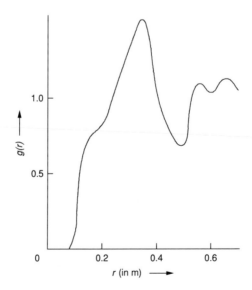

Figure 101   Statistically determined pair correlation function for the pattern of polygon centres in Fig. 100.

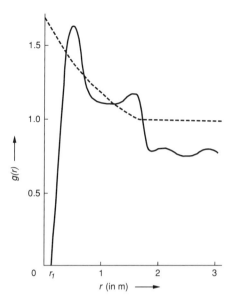

Figure 102 Statistically determined pair correlation function for the vertices of the polygons in Fig. 100. The form of the curve (—) is between that of a cluster field and a hard core field. The second curve (- - -) belongs to a Neyman–Scott cluster field.

If, in contrast, the cluster centres are not randomly scattered (i.e. not according to a Poisson field) then $g(r)$ may take very complicated forms that reflect both the cluster geometry and the distribution of centres.

Points on curves. Somewhat strange forms of the pair correlation function are observed if the points lie on random curve systems. A simple case is that where the points on the single curve are distributed according to a Poisson process (this leads to a particular so-called Cox field). In particular, if the curves are lines then the point distribution can be understood as follows. On each line an origin is chosen independently of the other lines. Then points are scattered on the line in such a manner that their distances from the origin and from another are independent and exponentially distributed. The union of all points is then the point field considered. (Another field of points on lines is that of the intersection points of the lines; but this is more complicated.)

The pair correlation function equals that of the line field. If it is a Poisson line field of line density L_A then

$$g(r) = 1 + \frac{1}{L_A \pi r}$$

(Stoyan and Stoyan, 1986; Stoyan *et al.*, 1987). Clearly there is a pole at $r = 0$.

If the line field is obtained by adding to each line of a Poisson line field a further parallel line of distance d then the pair correlation function has two poles: one at

$r = 0$ and one at $r = d$. It satisfies

$$g(r) = 1 + \frac{1 + h(r)}{\pi r L_A},$$

with

$$h(r) = \begin{cases} 0 & (r \le d), \\ \dfrac{r}{\sqrt{r^2 - d^2}} & (r > d). \end{cases}$$

Here L_A is the line density of the double line field — not that of the original Poisson line field.

A pole also has the pair correlation function of the vertex points of the Poisson–Dirichlet tessellation, as shown by L. Muche, see also (Stoyan and Stoyan, 1990a).

Soft core fields. With respect to the strength of the intersection of the points, so-called soft core fields are intermediate between random packings and Poisson fields. They can be used as models of the centres of non-overlapping discs of variable diameter. Figure 86 shows such a system of non-overlapping discs, and Fig. 103 shows the corresponding pair correlation function. Its form is typical of a soft core point field.

The strength of existing interactions is shown both by the height of the maximum of the pair correlation function and the depth of the possible following minimum. Such a minimum is frequently only very weak or even non-existent; often pair correlation functions are observed as in the second curve of Fig. 103.

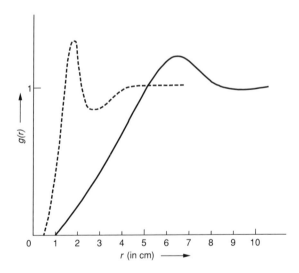

Figure 103 Statistically determined pair correlation functions for two point fields with weak interaction, so-called 'soft core fields': —, pair correlation function for a pattern of sea anemones on a rock (Stoyan, 1987); - - -, pair correlation function for the centres of the silver particles in Fig. 86.

14.5 CHARACTERISTICS OF THIRD AND HIGHER ORDER

Third- and higher-order characteristics enable refinements of the results obtained by second-order characteristics. Since these are more complicated than second-order characteristics, they are not used so frequently. The aim of this section is to sketch the theory and to discuss the third-order and related characteristics found in the literature.

The nth-moment measure $\mu^{(n)}$ is defined by

$$\mu^{(n)}(B_1 \times \cdots \times B_n) = \mathsf{E}N(B_1) \cdots N(B_n)$$

where $B_1, \ldots, B_n$ are Borel sets. Thus nth-order moments of point number can be expressed as

$$\mathsf{E}N(B)^n = \mu^{(n)}(B \times \cdots \times B).$$

For any non-negative measurable function $f(x_1, \ldots, x_n)$

$$\mathsf{E} \sum_{x_1, \ldots, x_n \in N} f(\mathbf{x}_1, \ldots, \mathbf{x}_n) = \int f(x_1, \ldots, x_n) \mu^{(n)}(d(x_1, \ldots, x_n)).$$

The *factorial nth-moment measure* $\alpha^{(n)}$ is that measure which satisfies for any non-negative measurable functions $f(x_1, \ldots, x_n)$ the equation

$$\mathsf{E} \sum_{x_1, \ldots, x_n \in N}^{\neq} f(\mathbf{x}_1, \ldots, \mathbf{x}_n) = \int f(x_1, \ldots, x_n) \alpha^{(n)}(d(x_1, \ldots, x_n)). \qquad (14.30)$$

The inequality sign means that the summation only ranges over n-tuples of pairwise different $\mathbf{x}_i$. This is the difference from $\mu^{(n)}$; n-tuples containing two or more equal points are not included in (14.30).

Assuming that $\alpha^{(n)}$ is suitably continuous, a density function $\varrho^{(n)}(x_1, \ldots, x_n)$ exists:

$$\alpha^{(n)}(B_1 \times \cdots \times B_n) = \int_{B_1} \cdots \int_{B_n} \varrho^{(n)}(x_1, \ldots, x_n) \, dx_1 \cdots dx_n. \qquad (14.31)$$

It is called the nth-product density. In terms of $\varrho^{(n)}$, (14.30) can be rewritten as

$$\mathsf{E} \sum_{x_1, \ldots, x_n \in N}^{\neq} f(\mathbf{x}_1, \ldots, \mathbf{x}_n) = \int f(x_1, \ldots, x_n) \varrho^{(n)}(x_1, \ldots, x_n) \, dx_1 \cdots dx_n. \quad (14.32)$$

Here $\varrho^{(n)}(x_1, \ldots, x_n)$ is a generalization of $\varrho^{(2)}(x_1, x_2)$ on p. 246. Let $C_1, \ldots, C_n$ be infinitesimally small discs with pairwise-different centres $x_1, \ldots, x_n$ and areas $dF_1, \ldots, dF_n$ (because of the smallness of the discs they can be considered as non-intersecting). Then, as on p. 248, the probability that each of the discs contains just one point of the point field N is approximately equal to

$$\varrho^{(n)}(x_1, \ldots, x_n) \, dF_1 \cdots dF_n.$$

In the case of a homogeneous Poisson field of density λ

$$\varrho^{(n)}(x_1, \ldots, x_n) \equiv \lambda^n,$$

and for an inhomogeneous Poisson field with intensity function $\lambda(x)$

$$\varrho^{(n)}(x_1, \ldots, x_n) = \lambda(x_1) \cdots \lambda(x_n).$$

In the following, only third-order product densities are considered. In the absence of homogeneity and isotropy assumptions $\varrho^{(3)}(x_1, x_2, x_3)$ depends on the point triplet x_1, x_2 and x_3. Even for directly congruent triangles of different positions and orientation, the values may be different. If the point field is homogeneous then the product density takes the same values for triplets that can be translated into one another. Then

$$\varrho^{(3)}(x_1, x_2, x_3) = \varrho^{(3)}(0, x_2 - x_1, x_3 - x_1).$$

If one assumes that $h = x_2 - x_1$ and $k = x_3 - x_1$, the product density can be denoted by $\varrho^{(3)}(h, k)$. As in (14.25), the equation

$$\mathrm{E} \sum_{\mathbf{x}_1 \in N} \sum_{\substack{\mathbf{x}_1 \in N \\ (\mathbf{x}_2 \neq \mathbf{x}_1)}} \sum_{\substack{\mathbf{x}_3 \in N \\ (\mathbf{x}_3 \neq \mathbf{x}_1, \mathbf{x}_2)}} f(\mathbf{x}_1, \mathbf{x}_2, \mathbf{x}_3)$$

$$= \int \int \int f(x, x + h, x + k) \varrho^{(3)}(h, k) \, \mathrm{d}h \, \mathrm{d}k \, \mathrm{d}x \qquad (14.33)$$

holds for all non-negative measurable functions $f(x_1, x_2, x_3)$. Finally, if the distribution of the point field is also invariant with respect to rotations and reflections (this is more than isotropy) then the same values are obtained for congruent triangles. Since a congruent triangle is given by the lengths of two sides and the included angle, the second-order product density can be given as a function of a distance r and a vector h in the motion-invariant case (Fig. 104). There

$$r \geq 0, \quad r_h \geq 0 \quad \text{and} \quad 0 \leq \varphi < 2\pi.$$

Thus one arrives at the function $\varrho^{(3)}(r, h)$. Because of the dependence on three real variables, it is still rather complicated, and simplified functions are used that, of course, do not contain all the information of $\varrho^{(3)}(r, h)$.

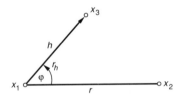

Figure 104 Describing a point triplet (x_1, x_2, x_3) by r and h, or r, r_h and φ.

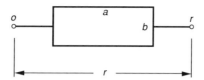

Figure 105 Test rectangle with side lengths a and b, which is used in the definition of $z_B^*(r)$. The origin is denoted by o and a point at a distance r from o by r.

First a function $z_B^*(r)$ related to the pair correlation function is considered:

$$z_B^*(r) = \int \frac{1_B(h)\varrho^{(3)}(r, h)}{\lambda^3 A(B)} \, dh \quad (r \geq 0). \tag{14.34}$$

Here B is a rectangle of side lengths a and b, with central line on the x-axis, as shown in Fig. 105 (it is assumed that $r > a$). The numerator in (14.34) can be interpreted as follows. As in the interpretation of $\varrho^{(2)}(x_1, x_2)$, two infinitesimal discs of areas dF_1 and dF_2 are considered, with distance r between their centres. If there is a point of the point field N in both discs then the number of points in the rectangle that lies between the two points is determined. It is a random variable that is equal to zero (if one of the discs does not contain a point) or to the point number in the rectangle (which can, of course, vanish too). The mean of this random variable is equal to the numerator in (14.34) multiplied by dF_1 and dF_2.

In the case of a homogeneous Poisson point field

$$z_B^*(r) = 1 \quad (r \geq 0). \tag{14.35}$$

Under the mixing assumption,

$$\lim_{r \to \infty} z_B^*(r) = 1. \tag{14.36}$$

If there is local anisotropy in the point field in the sense that some points frequently lie on or near to a line segment then this can be shown by the function $z_B^*(r)$, assuming the rectangle sides to be suitably chosen. Then it takes values greater than one.

Because of the density character of $z_B^*(r)$, its statistical estimation is difficult, especially for small point patterns. Therefore it may be useful not only to consider point pairs of the distance r, but also to average over a certain range of r values. A suitable characteristic of this kind is given by

$$Z_B^*(r_1, r_2) = \pi \int_{r_1}^{r_2} \int \frac{1_B(h)r\varrho^{(3)}(r, h)\, dh\, dr}{\pi\lambda^3(r_2 - r_1)^2 A(B)}. \tag{14.37}$$

The normalization guarantees that in the case of a Poisson point field

$$Z_B^*(r_1, r_2) = 1.$$

The characteristics $Z_B^*(r_1, r_2)$ and $z_B^*(r)$ can be modified in two ways. First, instead of the rectangle, a rhombus can be used, as suggested by Hanisch and Stoyan (1984). (The suggestion of using a rectangle was made by Kendall (1989).) Secondly, the rectangle can be adapted to the distance r by choosing for each r a rectangle B_r that lies symmetrically between the points o and r such that $r - a$ is fixed as in Fig. 105. (A similar case, but for rhombi, has been considered by Hanisch and Stoyan (1984).)

The normalization (division by $\lambda^3 A(B)$) in the definition of $z_B^*(r)$ can be improved. A more informative function is

$$z_B(r) = \int \frac{1_B(h)\varrho^{(3)}(r, h)}{\varrho^2(r)\lambda A(B)} \, dh \quad (r \geq 0), \tag{14.38}$$

where $\varrho^2(r) > 0$.

Without the term $\lambda A(B)$, the right-hand side of (14.38) can be interpreted as a conditional mean of the point number in the set B given that at each of o and r that are positioned relative to B in Fig. 105 there is a point of the point field N. The function $z_B(r)$ satisfies relations analogous to (14.35) and (14.36).

If $z_B(r)$ is greater than unity then B contains more points than expected in the case of a uniform distribution of points. The presence of points at o and r then has an 'exciting' influence on the point number in B. Large values of $z_B(r)$ are thus indicators of a tendency for alignment. In §15.5 the function $z_B(r)$ is statistically determined for a particular example. The values greater than unity obtained there show that the given local anisotropy may be quantified by $z_B(r)$. Similarly to $z_B^*(r)$, $Z_B^*(r_1, r_2)$ can also be modified. The new characteristic is

$$Z_B(r_1, r_2) = \int_{r_1}^{r_2} \int \frac{1_B(h)\varrho^{(3)}(r, h)}{\lambda^3 [K(r_2) - K(r_1)]A(B)} \, dh \, dr, \tag{14.39}$$

assuming $K(r_2) \neq K(r_1)$. It has been used in a slightly different form by Hanisch and Stoyan (1984).

Third-order characteristics are also obtained if the marking used on p. 244 for explaining the K-function is modified. Instead of the marks $n_r(\mathbf{x}_n)$, the squared marks $n_r(\mathbf{x}_n)^2$ are used. In contrast, the marking $\sqrt{n_r(\mathbf{x}_n)}$ used by Getis and Franklin (1987) leads to characteristics outside the field of third-order characteristics.

14.6 SECOND-ORDER CHARACTERISTICS OF MARKED POINT FIELDS

For marked point fields there are further second-order characteristics in addition to the pair correlation function and the K- and L-functions. They describe the correlations between the marks. First the case of 'continuous' real marks is considered (discrete marks are treated on p. 264). Particularly important is the *mark correlation function* $k_f(r)$. Its definition starts with a non-negative test function $f(m', m'')$, depending on the marks m' and m'' of two points x' and x''. Examples of $f(m', m'')$ are

$$f_1(m', m'') = m'm'' \tag{14.40}$$

and

$$f_2(m', m'') = \min\{|m' - m''|, \pi - |m' - m''|\}. \tag{14.41}$$

The first function is used if the marks describe the sizes of objects that are represented by the marked points. Large-values of $f(m', m'')$ are then obtained if both marks are large.

The second function is adapted to the case where the marks are angles between 0 and π. They describe, for example, orientations of particles. If two lines have directions m' and m'' (i.e. angles with the x-axis) then small values of $f_2(m', m'')$ are obtained for nearly parallel lines.

Analogously to $\alpha^{(2)}$, the measure $\alpha_f^{(2)}$ is defined by

$$\alpha_f^{(2)}(B_1 \times B_2) = \sum_{[\mathbf{x}_1; m_1] \in N} \sum_{\substack{[\mathbf{x}_2; m_2] \in N \\ (\mathbf{x}_2 \neq \mathbf{x}_1)}} f(m_1, m_2) 1_{B_1}(\mathbf{x}_1) 1_{B_2}(\mathbf{x}_2). \tag{14.42}$$

As for $\alpha^{(2)}$, the summation is over all pairs $[\mathbf{x}_1; m_1]$, $[\mathbf{x}_2; m_2]$ of marked points of N in B_1 and B_2, where $\mathbf{x}_1 \neq \mathbf{x}_2$.

Assuming continuity, there is a density function for $\alpha_f^{(2)}$, $\varrho_f(x_1, x_2)$, which could be called the f-product density in analogy with the product density introduced on p. 246. An interpretation of $\varrho_f(x_1, x_2)$ is possible as for $\varrho^{(2)}(x_1, x_2)$. Let C_1 and C_2 be two infinitesimally small discs centred at x_1 and x_2 and with areas dF_1 and dF_2. A random variable is considered that vanishes if one of the discs does not contain exactly one point of the field and is otherwise equal to $f(m', m'')$, where m' and m'' are the marks of the points in C_1 and C_2. The corresponding mean is approximately equal to $\varrho_f^{(2)}(x_1, x_2)\, dF_1\, dF_2$. The quotient

$$\kappa_f(x_1, x_2) = \frac{\varrho_f^{(2)}(x_1, x_2)}{\varrho^{(2)}(x_1, x_2)}, \quad \varrho^{(2)}(x_1, x_2) \neq 0,$$

can be interpreted as a conditional mean, namely as the mean of $f(m_1, m_2)$ given that there is a point of the field at both locations x_1 and x_2 with marks m_1 and m_2 respectively.

In the homogeneous case polar coordinates can be used with a function $\kappa_f(r, \varphi)$ analogous to $\varrho^{(2)}(r, \varphi)$. In the isotropic case only distance plays a role, and the functions $\varrho^{(2)}(r)$, $\varrho_f(r)$ and $\kappa_f(r)$ are used.

All of these functions describe the correlations of the marks. Thus the test function $f(m', m'')$ has to be chosen according to the problem and the type of the marks. The above functions are not the only possibilities.

In order to give $\kappa_f(r)$ still more of the character of a correlation function, it is normalized. The result is the mark correlation function

$$k_f(r) = \frac{\kappa_f(r)}{\kappa_f(\infty)} \quad (r \geq 0). \tag{14.43}$$

The quantity $\kappa_f(\infty)$ can be calculated by means of the *mark distribution* $\mathbf{M}$ if the point field (as assumed) has some mixing properties,

$$\kappa_f(\infty) = \int \int f(m', m'')\mathbf{M}(dm')\mathbf{M}(dm''),$$

or, if there is a density function $\mathbf{m}$ for $\mathbf{M}$,

$$\kappa_f(\infty) = \int_{-\infty}^{\infty} \int_{-\infty}^{\infty} f(m', m'')\mathbf{m}(m')\mathbf{m}(m'') \, dm' \, dm''.$$

For the function $f_1(m', m'')$ one obtains

$$\kappa_{f_1}(\infty) = \bar{m}^2 \quad \text{(square of the mean of the marks)}, \tag{14.44}$$

and for $f_2(m', m'')$, assuming isotropy,

$$\kappa_{f_2}(\infty) = \tfrac{1}{4}\pi. \tag{14.45}$$

In the following $k_{f_1}(r)$ and $k_{f_2}(r)$ are always denoted by $k_{mm}(r)$ and $k_d(r)$ respectively.

Figure 124 in Chapter 15 shows the estimated mark correlation function $\hat{k}_{mm}(r)$ for the pattern of silver particles in Fig.86. Here the marks are the diameters. It is obvious that there is a tendency for the particles close together to be small. Figure 107 shows the estimated mark correlation function $\hat{k}_d(r)$ for the line segment field in Fig. 106 (the points here are the segment centres and the marks the angles between the segments and the lower edge, $0 \leq m < \pi$). The local parallelism tendency present in Fig. 106 is well reflected by $\hat{k}_d(r)$. For small values of r, $\hat{k}_d(r)$ is small.

In the case of *discrete* marks the approach is similar. Here each point has a mark in the set of integers $1, \ldots, M$ (characterizing, for example, a qualitative property like the type of tree represented by the point). In principle, as above, functions $f(i, j)$ of pairs of marks could be considered and corresponding mark correlation functions could be used. But more appropriate to the discrete case is the use of indicator functions, which leads to the use of product densities $\varrho_{lm}(x_1, x_2)$. As for the interpretation of $\varrho_{lm}(x_1, x_2)$, $\varrho_{lm}(x_1, x_2) \, dF_1 \, dF_2$ can be interpreted as the probability that C_1 and C_2 each contain a point, namely an l- and an m-point. The quantity

$$p_{lm}(x_1, x_2) = \frac{\varrho_{lm}(x_1, x_2)}{\varrho^{(2)}(x_1, x_2)} \tag{14.46}$$

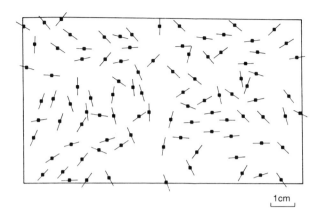

Figure 106   Positions of flies on a leaf. Each fly is shown by a line segment. The pattern can be described by a marked point field, where the centres are the points and the directions (with respect to the lower edge of the figure) the marks. For a detailed discussion see Penttinen and Stoyan (1989).

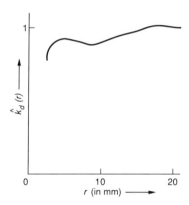

Figure 107   Estimated direction correlation function $\hat{k}_d(r)$ for the flies of Fig. 106. The values of $k_d(r)$ less than unity reflect the tendency towards local parallelism.

can be interpreted as the conditional probability that one of the points x_1 and x_2 is an l-point and the other is an m-point, under the condition that they are points of the field. In the isotropic case the function

$$p_{lm}(r) = \frac{\varrho_{lm}(r)}{\varrho^{(2)}(r)} \quad (r \geq 0; \; l, m = 1, \ldots, M) \tag{14.47}$$

is used. For any r

$$\sum_{l=1}^{M} \sum_{m=1}^{l} p_{lm}(r) = 1.$$

Analogously to the mark correlation function, the normalized functions

$$k_{lm}(r) = \frac{p_{lm}(r)}{p_{lm}(\infty)} \quad (r \geq 0; \; l, m = 1, \ldots, M) \tag{14.48}$$

may also be used. Here the relations

$$p_{lm}(\infty) = \frac{2\lambda_l \lambda_m}{\lambda^2} \quad (l \neq m)$$

and

$$p_{ll}(\infty) = \frac{\lambda_l^2}{\lambda^2} \quad (l, m = 1, \ldots, M)$$

have to be considered, where λ_l is the intensity of the sub-point field of l-points. Finally, the functions $g_{lm}(r)$ can be used; these are defined by

$$\varrho_{lm}(r) = \lambda_1 \lambda_m g_{lm}(r) \quad (r \geq 0; \; l, m = 1, \ldots, M), \tag{14.49}$$

similarly to the pair correlation function.

14.7 NEAREST-NEIGHBOUR CORRELATION

The nearest-neighbour correlation characteristics are parameters $\bar{n}_f$ that are defined like the function $k_f(r)$ by means of a test function $f(m', m'')$. This is done as follows. For each point $\mathbf{x}$ of the point field let its nearest neighbour be denoted $\mathbf{x}^{(n)}$. Let the corresponding marks be $m(\mathbf{x})$ and $m(\mathbf{x}^{(n)})$. Then $f(m(\mathbf{x}), m(\mathbf{x}^{(n)}))$ is a new mark $n_f(\mathbf{x})$ for the point $\mathbf{x}$. The mean of these marks is denoted by $\bar{n}_f$ and used as the nearest-neighbour correlation coefficient.

For the silver particles of Fig. 86 the estimated nearest-neighbour correlation coefficient is $\bar{n}_f = 64.8$ for diameter marks and $f_1(m', m'')$ as in (14.40). Since the mean diameter is 8.4 and its square 72.6, it is clear that particles close together tend to be smaller than isolated ones. For the eutectic structure in Fig. 96 with $f(m', m'')$ as in (14.41) the estimate $\bar{n}_d = 0.262$ is obtained, which corresponds to an angle of 15°. This value indicates very clearly the tendency towards parallelism of particles, since $\bar{n}_d = \frac{1}{4}\pi$ for a completely random orientation.

14.8 DISTANCES TO NEIGHBOURS

In addition to second-order characteristics, quantities related to interpoint distances can also be considered. They give on the one hand further information about the point distribution, and on the other a means by which distributional statements that have been obtained from second-order characteristics can be checked.

Usually the distance from a point to its nearest neighbour is of particular interest, but sometimes the distances to the second, third and kth neighbours are also considered. The distribution function corresponding to the kth neighbour is denoted by $D^{(k)}(r)$; the symbol $D(r)$ is used for $D^{(1)}(r)$. The corresponding density functions are $d(r)$ and $d^{(k)}(r)$, and the means are denoted by m_D and $m_D^{(k)}$.

Mathematically, the $D^{(k)}(r)$ are defined similarly to the K-function. Each point $\mathbf{x}$ of the point field is marked by $\delta^{(k)}(\mathbf{x})$. This is the distance from $\mathbf{x}$ to the kth neighbour. Then $D^{(k)}(r)$ is given by the intensity $\lambda^{(k)}(r)$ of the sub-point field of those points whose mark is less than r:

$$D^{(k)}(r) = \frac{\lambda^{(k)}(r)}{\lambda} \quad (r \geq 0). \tag{14.50}$$

If B is a Borel set of positive area then

$$\lambda D^{(k)}(r) A(B) = \mathsf{E} \sum_{\mathbf{x} \in N} 1_{[o,r]}(d^{(k)}(\mathbf{x})) 1_B(\mathbf{x}). \tag{14.51}$$

For some models of point fields $D(r)$ and (with some difficulty) the $D^{(k)}(r)$ can be given (§§13.2.2 and 16.2).

As in the case of the K-function, $D^{(k)}(r)$ can also be explained by referring to the 'typical' point. That is, $D^{(k)}(r)$ is the distribution function of the distance of the 'typical' point to its kth neighbour.

In general, $D(r)$ is 'less informative' than $K(r)$, though it must be noted that qualitatively different properties are described. It is not difficult to give examples of point fields that have the same $D(r)$ but quite different distributions. Figure 108 shows a simple example.

However, the set of *all* functions $D^{(k)}(r)$ contains more information than the K-function:

$$\lambda K(r) = \sum_{k=1}^{\infty} D^{(k)}(r) \quad (r \geq 0). \tag{14.52}$$

This follows by the well-known formula for the mean of a non-negative random integer X:

$$\mathsf{E}X = \sum_{k=1}^{\infty} \Pr(X \geq k)$$

(in the current case $X = N(b(o,r))$).

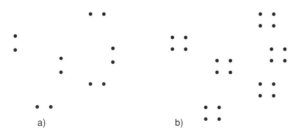

Figure 108 Two point patterns with same nearest-neighbour distances.

For marked point fields, distance distribution functions $D_{ij}(r)$ can also be considered, which give the distribution of the distance of an i-point to its nearest neighbour with the mark j.

14.9   PALM CHARACTERISTICS

Quantities connected with the so-called Palm distribution play an important part in the theory of point fields. Their precise mathematical definition requires rather complicated mathematical tools, which will not be given here. (For an introduction see Stoyan *et al.*, 1987.) What follows aims to give the reader an idea of how these quantities work. The following exposition is in the spirit of Matthes (1963).

Let $f(N)$ be a function that assigns a real number to a point field N, for example

$$f_1(N) = N(b(o, r) \setminus \{o\}) \quad \text{the number of points in the disc } b(o, r),$$
$$\text{where a point at } o \text{ is not counted;}$$

$$f_2(N) = N(B \setminus \{o\}) \quad \text{the number of points in the Borel set } B,$$
$$\text{where a point at } o \text{ is not counted (if } o \in B);$$

$$f_3(N) = \begin{cases} 1 & \text{if the distance of the nearest neighbour of } o \text{ from } o \text{ is less than } r, \\ 0 & \text{otherwise;} \end{cases}$$

$$f_4(N) = \sum_{y \in N} \phi(\|\mathbf{y}\|), \quad \text{where } \phi(r) \text{ is a suitable function and } \|\mathbf{y}\|$$
$$\text{is the distance of } \mathbf{y} \text{ from the origin } o.$$

Now every point $\mathbf{x}$ of the point field N obtains a mark $m_f(\mathbf{x})$ given by

$$m_f(\mathbf{x}) = f(N_{-\mathbf{x}}),$$

with

$$N = \{\mathbf{x}_1, \mathbf{x}_2, \ldots\} \quad \text{and} \quad N_{-\mathbf{x}} = \{\mathbf{x}_1 - \mathbf{x}, \mathbf{x}_2 - \mathbf{x}, \ldots\}.$$

The point field is translated so that the point $\mathbf{x}$ lies at the origin, and then the function value is determined. In the case of the functions above, $m_f(\mathbf{x})$ can be calculated directly, for example

$$m_{f1}(\mathbf{x}) = N(b(\mathbf{x}, r) \setminus \{\mathbf{x}\}).$$

Now let the mean mark $\bar{m}_f$ corresponding to $f(N)$ be finite for a homogeneous N. Because of the reference to the origin, this mean is denoted by $E_0 f(N)$. It is a mean in the sense of the so-called Palm distribution. Such means have already been considered in the §14.4.1. This mean satisfies

$$E_0 f_1(N) = \bar{n}_r = \lambda K(r),$$

$$E_0 f_2(N) = \lambda \Re(B)$$

and

$$E_0 f_3(N) = D(r) \quad (r \geq 0).$$

One often thinks of $E_0 f(N)$ in the following ways.

• $E_0 f(N)$ is the mean of $f(N)$ under the condition that a point of the field lies at o. (For a homogeneous point field the probability of this condition is equal to zero, and it is mathematically difficult to justify this interpretation.)

• $E_0 f(N)$ is the mean of $f(N)$ seen from the 'typical' point of the field N.

14.10 ANISOTROPY CHARACTERISTICS FOR MARKED AND NON-MARKED POINT FIELDS

Many homogeneous point fields are anisotropic; in fact, isotropic fields are really the exception. There are many varied forms of anisotropy. They are particularly multifarious for marked point fields. For example, they include

— anisotropic arrangement of the points;
— anisotropic behaviour of marks if they describe orientations;
— combination of anisotropic point distribution and anisotropic mark behaviour.

In the following some simple anisotropy characteristics are described; see also Stoyan (1991b) and Stoyan and Beneš (1992).

Anisotropic point distribution

A simple characterization uses the 'anisotropic' pair correlation function

$$g(r, \varphi) = \frac{\varrho^{(2)}(r, \varphi)}{\lambda^2} \quad (r \geq 0; \ 0 \leq \varphi < \pi),$$

It is sufficient to consider only values of φ between 0 and π; for φ and $\varphi + \pi$ the same values are obtained. The form of $g(r, \varphi)$ for different φ closely reflects the form of the anisotropy. In applied problems those φ-values at which the differences are greater must be found. Clearly, the precise determination of $g(r, \varphi)$ is difficult for small or medium-size point patterns, since it is a density function of two variables.

Figure 109 shows the function $g(r, \varphi)$ with three angles φ for the pattern of the centres of the particles of Fig. 96. Figure 110 shows $g(r, \varphi)$ for fixed r as a function of φ.

A rougher form of anisotropy analysis is the orientation analysis suggested by Ohser and Stoyan (1981). This uses the directional distribution of line segments connecting point pairs of the point field analysed. There only those pairs are considered which are between r_1 and r_2 apart. The corresponding distribution function is denoted by $O_{r_1 r_2}(\varphi)$ and is equal to the probability that a randomly

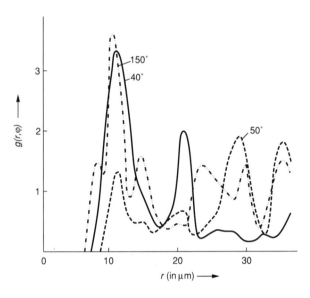

Figure 109 Plot of the anisotropic pair correlation function $g(r, \varphi)$ for the particle centres of Fig. 96. Three values of φ are considered: $\varphi = 40°$, $50°$ and $140°$. The differences demonstrate anisotropy.

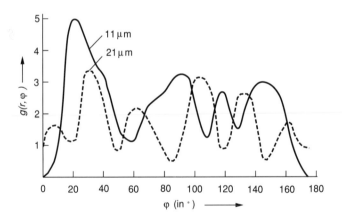

Figure 110 Plot of the anisotropic pair correlation function $g(r, \varphi)$ for the particle centres of Fig. 96. For two values of r (11 μm and 21 μm) the dependence on φ is shown. This clearly demonstrates the main directions of the particles.

chosen line segment forms an angle with the x-axis smaller than φ $(0 \leq \varphi \leq \pi)$. This point pair orientation distribution function $O_{r_1 r_2}(\varphi)$ can be expressed in terms of the reduced second-moment measure $\mathfrak{R}$ by

$$O_{r_1 r_2}(\varphi) = \frac{\mathfrak{R}(S(r_1, r_2, \varphi))}{\mathfrak{R}(b(o, r_2)) - \mathfrak{R}(b(o, r_1))} \quad (0 \leq \varphi < \pi), \tag{14.53}$$

where it is assumed that the denominator is positive. $S(r_1, r_2, \varphi)$ is the sector annulus shown in Fig. 111.

In many cases there is a density function $o_{r_1 r_2}(\varphi)$ for $O_{r_1 r_2}(\varphi)$, with

$$O_{r_1 r_2}(\varphi) = \int_0^\varphi o_{r_1 r_2}(\psi) \, d\psi.$$

This density function is a good characteristic of the anisotropy of a point distribution if the distances r_1 and r_2 are suitably chosen. For example, the values r_1 and r_2 are particularly interesting if they include the r-range in which the pair correlation function has its first maximum.

Figure 112 shows two empirical density functions with $r_1 = 6.25$ μm and $r_2 = 18.75$ μm or 25 μm for the particle centres of Fig. 96. It is not surprising that the function corresponding to the larger r_2 is smoother than that for the smaller r_2. (However, the smoother form here also results from a somewhat greater kernel parameter (§15.4.5).)

In the case of marked point fields the points may be weighted by the marks (Stoyan, 1991b).

Anisotropy of marks

If the marks of the point field are angles then the distribution of these angles alone can be used. This distribution can be interpreted as the rose of directions

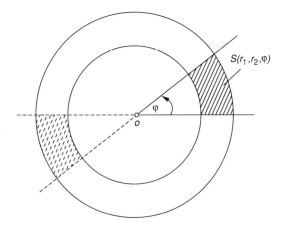

Figure 111 The sector annulus $S(r_1, r_2, \varphi)$ used in the definition of $O_{r_1 r_2}(\varphi)$.

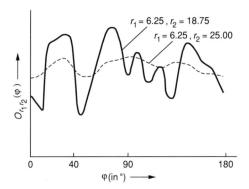

Figure 112 Estimated density functions $o_{r_1 r_2}(\varphi)$ for the point pair orientation distribution function $O_{r_1 r_2}(\varphi)$ of the particle centres in Fig. 96: ——, $r_1 = 6.25$, $r_2 = 18.75$; – – – $r_1 = 6.25$, $r_2 = 25.00$ (lengths in μm). Note the different bandwidths: $h = 5°$ and $20°$ for $r_1 = 18.75$ μm and $r_2 = 25.00$ μm respectively.

(Stoyan *et al.*, 1987, p. 235) of that line segment field which consists of segments of unit length centred at the point and having directions given by the angle marks. Figure 113 shows the estimated directional distribution for the angle marks of the particles on Fig. 96. The mean orientation of the marks around 0.7 (40°) is obvious; angles greater than $\frac{1}{2}\pi$ do not appear.

This approach can be refined if the points have pairs of marks: an angle and a size mark. Then a segment field can be considered, which is constructed as above but with the segment lengths equal to the size marks.

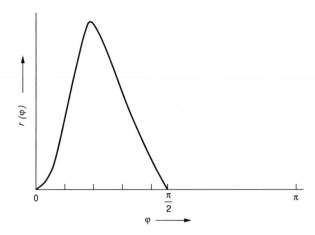

Figure 113 Estimated direction density function for the particles of Fig. 96. It describes the orientations of single particles, i.e. the directions of the lines that belong to the maximum Feret diameters.

Nearest-neighbour direction correlation

Local anisotropies can be expressed by the quantity $\bar{n}_d$ introduced in §14.7. For the pattern of Fig. 96 the estimate of $\bar{n}_d$ is 0.262 (15°). This small value expresses the high degree of parallelism of neighbouring particles. In contrast, for the line segment field of Fig. 106, $\hat{\bar{n}} = 0.7$ (40°), which is close to $\frac{1}{4}\pi$, the value corresponding to complete randomness. It is interesting that the estimated direction mark correlation function $k_d(r)$ for this pattern has values clearly less than unity for small r, which reflects the obvious *local* tendency to parallelism.

CHAPTER 15

Statistics for Homogeneous Point Fields

15.1 INTRODUCTION

This chapter discusses important methods of exploratory data analysis for homogeneous planar point fields. These methods yield fundamental characteristics without particular model assumptions, but of course assume homogeneity and often also isotropy. These characteristics are intensity, second-order characteristics and some further characteristics introduced in Chapter 14. They help to detect and quantify important distributional properties. (Section 14.4.2 has shown in detail what pair correlation functions tell one about point fields.) Based on the exploratory statistics model assumptions can then be made. At the end of the chapter tests and parameter estimations are also discussed.

It is always assumed that the point fields investigated are given in one or k windows $W_1, \ldots W_k$. (When $k > 1$, it is assumed that the point patterns are independent samples corresponding to the same point field.) The following exposition assumes that there is only one window W; for several windows the characteristics are obtained by area weighting.

Some statistical notions will now be explained.

The counterparts of the quantities of Chapter 14 that are statistically obtained by the point pattern in the window W are called *estimators*. The particular result for a particular window is called an *estimate*. When constructing estimators, it is important to assume that the estimator is *unbiased*. In other words, suppose that there is a point field with some characteristics such as the intensity or the K-function $K(r)$ that the statistician aims to determine. But he/she can analyse only one (or more) samples and calculate estimates for these. They will be different for each sample; thus the estimators (e.g. λ and $K(r)$) are random variables. Usually they have a mean (e.g. $E\lambda$ and $EK(r)$). If this mean is equal to the true parameter then the estimator is called unbiased. For example, this is the case if

$$E\hat{\lambda} = \lambda \quad \text{and} \quad E\hat{K}(r) = K(r).$$

In this case the estimator yields the right result on average.

A further important property of estimators is that of *consistency*. If the point field analysed is ergodic then estimators can be constructed that become more accurate with increasing window size. Their variances decrease. ('Increasing window size' means that increasingly larger squares or discs can be placed therein; for details see Stoyan *et al.* (1987, p. 171).) Another form of consistency is given if averaging over several windows gives increasingly accurate values for an increasing number of windows.

A third important property of estimators is their *asymptotic normality* (many frequently used methods of classical statistics are based on asymptotic normality (e.g. the χ^2-goodness-of-fit test)). This property has been intensively studied for estimators in spatial statistics (for example Heinrich, 1988, 1993; and references therein). Of course, asymptotic normality is a quantitative property (corresponding estimates of the speed of convergence have in many cases only theoretical value), and it is quite often difficult to say which window size or point number is necessary to assume asymptotic normality.

In order to ensure unbiasedness, so-called *edge corrections* are frequently necessary. For points near the boundary of the window, the information necessary for correct evaluation is frequently not given. An example is the determination of the nearest neighbour, which may be outside of the window for points close to the boundary. There are several methods of edge correction, which will be discussed in the following sections. Their aim is a possibly complete use of the information given by the point pattern in W. For particularly important cases program sketches are given. Some examples illustrate the application of the estimation methods.

The reader should note that some of the estimators and estimation methods may make sense for inhomogeneous patterns too (p. 193). Some knowledge of the type of inhomogeneity and a good understanding of the estimation method used and its behaviour in the given case is necessary.

15.2   ESTIMATING THE INTENSITY AND THE INTENSITY MEASURE

The beginning of a statistical analysis of point patterns is usually the investigation of the point density.

If it may be assumed that the given data belong to a homogeneous point field then it is sufficient to estimate the intensity λ. The usual estimator is

$$\hat{\lambda} = \frac{\text{number of observed points}}{\text{area of the window(s)}} = \frac{n}{A(W)}. \qquad (15.1)$$

Here and in the following n denotes the number of points in W:

$$n = N(W).$$

There are further estimators of λ for particular point fields, especially Poisson fields. Distance methods are particularly important (p. 219). Särkkä (1992) discusses distance methods for non-Poisson fields, where a certain fraction of the points is marked for the statistical analysis.

Under the assumptions of homogeneity and ergodicity, it makes sense to estimate λ using only one sample. The situation in the inhomogeneous case is quite different. If no distributional assumptions can be made (e.g. that a Poisson field is given; §13.3.4) then several (k) point patterns must be taken from the same window W to estimate the intensity measure. In this case $\Lambda(B)$ (B a subset of W) is estimated by

$$\hat{\Lambda}(B) = \frac{1}{k} \sum_{i=1}^{k} N_i(B), \tag{15.2}$$

where $N_i(B)$ denotes the number of points in the ith pattern in B. If the k point patterns are independent then the usual methods of classical statistics can be used to investigate the properties of $\hat{\Lambda}(B)$.

It is sometimes more interesting to estimate the intensity function $\lambda(x)$. If there is no a priori information on the distribution of the point field then Diggle's estimator should be used, which was already considered in the Poisson case:

$$\hat{\lambda}_h(\mathbf{x}) = \frac{\hat{\Lambda}(W \cap b(\mathbf{x}, h))}{A(W \cap b(\mathbf{x}, h))}. \tag{15.3}$$

The choice of the radius h is a complicated problem. It is reasonable to use different values of h and then to choose that value which gives the 'best' results. $\hat{\lambda}_h(x)$ may be given by isolines, and a 'good' form of $\hat{\lambda}_h(x)$ is one that is sufficiently smooth but still shows important details of the inhomogeneity.

As a form of data analysis, the method described can also be used for a single point pattern. Sometimes an isoline plot may be more instructive than a point pattern. The example in §12.2.2 illustrates this method.

In some statistical problems estimators of λ^2 are needed. Of course, the $\hat{\lambda}$ of (15.1) can be used and squared. But this estimator is in general not unbiased. In the case of a Poisson field

$$\widehat{\lambda^2} = \frac{n(n-1)}{A(W)^2} \tag{15.4}$$

is unbiased (note that $\widehat{\lambda^2}$ is less than $\widehat{\lambda}^2$). This estimator is also used frequently for non-Poisson fields.

15.3 ESTIMATING MARK DISTRIBUTIONS AND RELATED QUANTITIES

Together with the intensity, mark parameters are considered as first-order characteristics. In the case of real-valued marks the mean $\bar{m}$ of the marks or the mark distribution function $M(u)$ are important ($M(u)$ is the probability that the mark of the 'typical' point is less than u; i.e. $M(u) = \mathbf{M}(-\infty, u]$).

The mean mark of a homogeneous marked point field is estimated by

$$\hat{\bar{m}} = \frac{S_N(W)}{n}. \tag{15.5}$$

$S_N(W)$ denotes the sum of the marks of the points in W as in §14.2. In the case of positive marks S_N is the mark sum measure.

Since both $S_N(W)$ and n $(= N(W))$ are random, the distribution of $\hat{\bar{m}}$ may be rather complicated; furthermore, $\hat{\bar{m}}$ is not always unbiased. (However, under suitable assumptions on the marked point field, asymptotical unbiasedness is given for a 'large' window W. In the case of an ergodic field $\hat{\bar{m}}$ is consistent.) In contrast, the quantity $\lambda\bar{m}$ has an unbiased estimator, namely

$$\frac{S_N(W)}{A(W)}.$$

The mark distribution function $M(u)$ can be estimated similarly by

$$\hat{M}(u) = \frac{\text{number of points in } W \text{ with mark} \leq u}{n}, \tag{15.6}$$

or

$$\hat{M}(u) = \frac{N(W \times (-\infty, u])}{n}.$$

Example (silver particles, continued). The system of particles is now considered as a sample of a marked point field, where

point = particle centre,   mark = particle diameter.

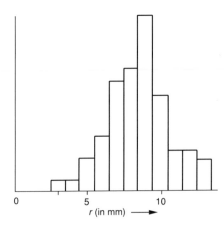

Figure 114   Histogram of the diameters of the silver particles of Fig. 86.

A histogram of the marks, which is an empirical counterpart of the mark distribution density function, is given in Fig. 114. Perhaps it may be approximated by a normal distribution.

15.4   ESTIMATING SECOND-ORDER CHARACTERISTICS

15.4.1   Estimating $\mathfrak{R}(B)$, $K(r)$ and $L(r)$

General

Usually $\lambda^2 K(r)$ and $\lambda^2 \mathfrak{R}(B)$ are first estimated, and estimates of $K(r)$ and $\mathfrak{R}(B)$ are then obtained by dividing through by $\widehat{\lambda^2}$ itself obtained from (15.4). Estimators of $\lambda^2 K(r)$ and $\lambda^2 \mathfrak{R}(B)$ in the anisotropic case are given by (15.11) and (15.12) below. In the isotropic case (15.13) and (15.14) give estimators, with (15.13) being easier to compute but (15.14) being more accurate. If assumptions of isotropy are unwarranted then (15.11) should be used.

The L-function is estimated by

$$\hat{L}(r) = \sqrt{\frac{\hat{K}(r)}{\pi}} \quad (r \geq 0), \tag{15.7}$$

where $\hat{K}(r)$ is an estimator for $K(r)$.

The K- and the L-functions are usually used in the final stages of analysis, in particular for goodness-of-fit tests. For exploratory statistics the pair correlation function is more suitable, since it is easier to interpret.

Explaining and deriving the estimators for $\lambda^2 K(r)$ and $\lambda \mathfrak{R}(B)$

It is always assumed that there are n points $\mathbf{x}_1, \ldots, \mathbf{x}_n$ in the window W; $n = N(W)$. The x- and y-coordinates of $\mathbf{x}_i$ are denoted by ξ_i and η_i respectively.

First let us discuss a naive, almost trivial estimator for $\lambda K(r)$. According to §14.4.1, $\lambda K(r)$ is equal to the mean number of points that lie in a disc of radius r centred at the 'typical' point of the field, which itself is excluded. This interpretation suggests the following method. The points $\mathbf{x}_i$ are considered in turn, and for each the number $N(b(\mathbf{x}_i, r) \backslash \{\mathbf{x}_i\})$ is found. This is the number of points in the disc $b(\mathbf{x}_i, r)$, not counting the centre $\mathbf{x}_i$. The corresponding mean is an estimator $k_1(r)$ for $\lambda K(r)$:

$$k_1(r) = \frac{1}{n} \sum_{i=1}^{n} N(b(\mathbf{x}_i, r) \backslash \{\mathbf{x}_i\}) \quad (r \geq 0). \tag{15.8}$$

Investigation of the distributional properties of this estimator $k_1(r)$ is very difficult, since both the numerator and denominator are random. It is in general biased.

Rather it is natural to use the estimator (15.1) for λ, $\hat{\lambda} = n/A(W)$, and to introduce the estimator

$$k_2(r) = \sum_{i=1}^{n} \frac{N(b(\mathbf{x}_i, r) \backslash \{\mathbf{x}_i\})}{A(W)} \qquad (r \geq 0). \qquad (15.9)$$

This is an unbiased estimator for $\lambda K(r)$ (this can be shown as for $k(r)$ below). Unfortunately this estimator has the disadvantage that it uses information that cannot be completely obtained from the window. It may happen that parts of the discs $b(\mathbf{x}_i, r)$ are outside W (it is assumed that information outside W is not available in principle).

Note that $k_2(r)$ can also be written in the form

$$k_2(r) = \sum_{i=1}^{n} \sum_{\substack{j=1 \\ (j \neq i)}}^{n} \frac{1_{b(\mathbf{x}_i, r)}(\mathbf{x}_j)}{A(W)} = \sum_{i=1}^{n} \sum_{\substack{j=1 \\ (j \neq i)}}^{n} \frac{1_{b(o,r)}(\mathbf{x}_j - \mathbf{x}_i)}{A(W)},$$

where $1_X(x)$ denotes the indicator function of the set X. (The reader should understand that these formulae are equivalent to (15.9). The summation with respect to j ranges over all points of the point field different to $\mathbf{x}_i$.)

If the window W is very large and r relatively small then only a negligible error is made when $N(b(\mathbf{x}_i, r) \backslash \{\mathbf{x}_i\})$ is replaced by $N(W \cap b(\mathbf{x}_i, r) \backslash \{\mathbf{x}_i\})$. Obviously, this method gives a value slightly too small for $\lambda K(r)$, on account of the excluded points outside W. In this situation there is a primitive form of edge correction: so-called 'minus sampling'. To estimate $\lambda K(r)$, points $\mathbf{x}_i$ are used for which the disc $b(\mathbf{x}_i, r)$ lies completely inside W (Fig. 115). This gives the unbiased estimator

$$k^-(r) = \sum_{\mathbf{x}_i \in W \ominus b(o,r)} \frac{N(b(\mathbf{x}_i, r) \backslash \{\mathbf{x}_i\})}{A(W) \ominus b(o, r))} \qquad (r \geq 0). \qquad (15.10)$$

Its disadvantage is that only the inner points (i.e. those in $(W) \ominus b(o, r)$) are disc centres. The information given by the point pattern is obviously only partially used.

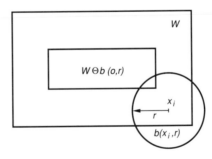

Figure 115   For the point $\mathbf{x}_i$ the disc $b(\mathbf{x}_i, r)$ is not completely in W. Thus it is possible that there are points of a point pattern in the disc which are outside W. If this is not considered, estimation errors result. In $W \ominus b(o, r)$ there are all points x in W for which the whole disc $b(x, r)$ is contained in W.

Of course the r-values are heavily restricted from above; if W is a square of side length a then r in (15.10) has to be less than $\frac{1}{2}a$.

There are more effective methods. The following estimator $k(r)$ of $\lambda^2 K(r)$ does not have the above disadvantages:

$$k(r) = \sum_{i=1}^{n} \sum_{\substack{j=1 \\ (j \neq i)}}^{n} \frac{1_{b(o,r)}(\mathbf{x}_j - \mathbf{x}_i)}{A(W_{\mathbf{x}_j} \cap W_{\mathbf{x}_j})} \quad (r \geq 0). \tag{15.11}$$

Here W_z denotes the window translated by z (Fig. 116). The difference from $k_2(r)$ is that the summation always ranges over the points in W (and thus the estimator is realistic), $A(W_{\mathbf{x}_j} \cap W_{\mathbf{x}_j})$ rather than $A(W)$ being the denominator. Since the area of two shifted windows is in general smaller than that of the original window, the corresponding summand is increased, compensating for the contributions of the points outside of W. It should be noted that

$$A(W_x \cap W_y) = A(W \cap W_{y-x}) = A(W \cap W_{x-y}).$$

In the case of a rectangular window with side lengths a and b $(a > b)$ one requires that $r = b$. Otherwise, the denominator in (15.10) can vanish.

Analogously, $\lambda^2 \mathfrak{K}(B)$ can be estimated by

$$k(B) = \sum_{i=1}^{n} \sum_{\substack{j=1 \\ (j \neq i)}}^{n} \frac{1_B(\mathbf{x}_j - \mathbf{x}_i)}{A(W_{\mathbf{x}_j} \cap W_{\mathbf{x}_j})}. \tag{15.12}$$

Proof of the unbiasedness of $\mathbf{k}(B)$.

$$\mathsf{E}k(B) = \mathsf{E}\sum_{\mathbf{x}} \sum_{\substack{\mathbf{y} \in N \\ \mathbf{y} \neq \mathbf{x}}} 1_W(\mathbf{x}) 1_W(\mathbf{y}) \frac{1_B(\mathbf{y} - \mathbf{x})}{A(W_{\mathbf{y}} \cap W_{\mathbf{x}})}$$

$$= \lambda^2 \int \int 1_W(x) 1(W)(x + h) \frac{1_B(h)}{A(W_{x+h} \cap W_h)} \, dx \, \mathfrak{K}(dh)$$

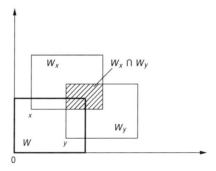

Figure 116 The windows W, W_x and W_y. The latter are W translated by x and y respectively. The shaded area is the intersection $W_x \cap W_y$.

$$= \lambda^2 \int \int 1_W(x) 1(W)(x+h) \frac{1_B(h)}{A(W \cap W_h)} \, dx \, \mathfrak{R}(dh)$$

$$= \lambda^2 \int \left[\int \frac{1_W(x) 1(W)(x+h)}{A(W \cap W_h)} \, dx \right] 1_B \mathfrak{R}(dh)$$

$$= \lambda^2 \int 1_B \mathfrak{R}(dh)$$

$$= \lambda^2 \mathfrak{R}(B).$$

The second equality follows from (14.21) and the third from $A(X) = A(X_z)$ for any X and z, i.e. from the translation invariance of the Lebesgue measure.

Note that in the proof only homogeneity has been used. More accurate estimators are possible assuming isotropy. Two estimators for the homogeneous and isotropic case will be given in due course. But first let us note a particular advantage of the estimator (15.11).

Very often one must estimate the K-function, i.e. the quantity $\mathfrak{R}(b(o, r))$, $r > 0$. Then it may be that the point pattern, which was assumed to be isotropic, has in fact considerable anisotropies. Nevertheless, it belongs to an homogeneous model. Often (15.11) still gives good estimates, while the estimators that are unbiased in the isotropic case may yield unacceptable values (Stoyan, 1991b). This suggests a general use of the estimator $k(r)$.

Ohser's estimator for $\lambda^2 K(r)$ (homogeneous and isotropic case) (Ohser, 1983)

$$k_O(r) = \sum_{i=1}^{n} \sum_{\substack{j=1 \\ (j \neq i)}}^{n} \frac{1_{[0,r]}(\|(\mathbf{x}_j - \mathbf{x}_i)\|)}{\bar{\gamma}_W(\|(\mathbf{x}_j - \mathbf{x}_i)\|)} \quad (r \geq 0). \tag{15.13}$$

Here $\bar{\gamma}_W(r)$ is the isotropized set covariance function of the window W, (§8.4.3). It gives the mean area of the intersection of W with a copy of W shifted by a random vector of length r. For a rectangular window r can be as large as the diagonal. Instead of $1_{[0,r]}(\|(\mathbf{x}_j - \mathbf{x}_i)\|)$, one may also write $1_{b(o,r)}(\mathbf{x}_j - \mathbf{x}_i)$. The proof of the unbiasedness of this estimator is given by Ohser (1983).

Ripley's estimator for $\lambda^2 K(r)$ (homogeneous and isotropic case) (Ripley, 1976; Ohser, 1983)

$$k_R(r) = \sum_{i=1}^{n} \sum_{\substack{j=1 \\ (j \neq i)}}^{n} 1_{[0,r]}(\|\mathbf{x}_j - \mathbf{x}_i\|) \frac{b_{ij}}{A(W^{\|\mathbf{x}_j - \mathbf{x}_i\|})} \tag{15.14}$$

$$(0 \leq r \leq r^* = \sup\{r : A(W^r) > 0\}).$$

Here $W^r = \{x \in W : \partial b(x, r) \cap W \neq \emptyset\}$ is the set of all points x of the window W such that a circular arc of radius r centred at x is not completely outside W. For small r, if the window W is convex, $W = W^r$. Furthermore,

$$b_{ij} = \frac{2\pi}{\alpha_{ij}},$$

where α_{ij} is the sum of all those central angles that belong to arcs of the circle centred at $\mathbf{x}_i$ of radius $\|\mathbf{x}_j - \mathbf{x}_i\|$ and lying in W (Fig. 117).

For rectangular W, r^* is equal to the diagonal length.

A sketch of the proof of unbiasedness of this estimator can be found in Hanisch and Stoyan (1979).

Ripley (1988, p. 35ff) compared the estimated variances of $k_R(r)$ and $k_O(r)$ for the case of a Poisson field. By means of some approximations he found that the variances of both estimators are practically the same; only for large values of $\lambda \pi r^2$, is $k_R(r)$ more accurate. The question still remains as to which estimator is better for other point field models. The estimator $k(r)$ is not so accurate as $k_O(r)$; however, for circular W, $k(r)$ and $k_O(r)$ coincide.

Note that the above estimators only yield $\widehat{\lambda^2 K(r)}$ and $\widehat{\lambda^2 \mathfrak{R}(B)}$. Estimators of $K(r)$ and $\mathfrak{R}(B)$ are obtained by division by $\hat{\lambda}^2$. Here the property of unbiasedness may disappear. This effect may be increased further when estimating the L-function using (15.6). Doguwa and Upton (1989) showed that for small point numbers ($n < 100$) these errors can be considerable. It seems that the following biased estimator of $\lambda K(r)$ behaves better and gives more precise estimates of $K(r)$ and $L(r)$:

$$\widehat{\lambda K}(r) = \frac{1}{n} \sum_{i=1}^{n} \frac{N(W \cap b(\mathbf{x}_i, r) \setminus \{\mathbf{x}_i\}) \pi r^2}{A(W \cap b(\mathbf{x}_i, r))}.$$

This is used to obtain an estimator for $K(r)$ by dividing by $\hat{\lambda}$.

Three outlines for programs

Sketches of BASIC programs follow that yield estimates for $\lambda^2 K(r)$ by (15.11), (15.13) and (15.14). The window is a rectangle of side lengths a and b, where the point coordinates satisfy $0 \le \xi_1 \le a$ and $0 \le \eta_i \le b$. It is assumed that r is sufficiently small that $A(W_{\mathbf{x}_i} - W_{\mathbf{x}_j})$ is always positive and $W^r = W$.

Version using (15.11).

```
1    S = 0
2    FOR i = 2 TO n
3    FOR j = 1 TO i - 1
4    compute ρ = ‖xᵢ - xⱼ‖
```

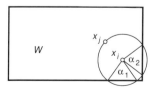

Figure 117 For the points x_i and x_j, $\alpha_{ij} = 2\pi - \alpha_1 - \alpha_2$.

```
5     IF ϱ > r THEN GOTO 30
10    C = (a − |ξᵢ − ξⱼ|) * (b − |ηᵢ − ηⱼ|)
20    S = S + 2/C
30    NEXT j
40    NEXT i
50    k(r) = S
```

In line 10 $A(W_{\mathbf{x}_i} \cap W_{\mathbf{x}_j})$ is calculated.

Version using (15.13). Line 10 must be replaced by

```
10    C = γ̄_W (ϱ).
```

A formula for $\bar{\gamma}_W(\varrho)$ for rectangular W is given on p. 123.

Version using (15.14). Replace lines 10 and 20 in the first program sketch by the following:

```
10    W = 2π : P = r * r
11    E = ξᵢ : IF E < ϱ THEN GOSUB 16
12    E = a − ξᵢ : IF E < ϱ THEN GOSUB 16
13    E = ηᵢ : IF E < ϱ THEN GOSUB 19
14    E = b − ηᵢ : IF E < ϱ THEN GOSUB 19
15    GOTO 25
16    S = SQR (P − E * E) : H = ATN (S/E): W = W − 2 * H :
      IF ηᵢ + S > b THEN W = W + H − ATN ((b − ηᵢ)/E)
17    IF ηᵢ − S < 0 THEN W = W + H − ATN (ηᵢ/E)
18    RETURN
19    S = SQR (P − E * E) : H = ATN(S/E) : W = W − 2 * H :
      IF ξᵢ + S > a THEN W = W + H − ATN ((a − ξᵢ)/E)
20    IF ξᵢ − S < 0 THEN W = W + H − ATN (ξᵢ/E)
21    RETURN
25    S = S + 2 π/W
```

15.4.2 Estimating the pair correlation function

As in the case of estimating $\mathcal{K}(B)$ and $K(r)$, the product density $\varrho^{(2)}(r)$ is first estimated, and an estimate of $g(r)$ is then obtained by dividing by $\hat{\lambda}^2$. It is useful to apply kernel estimators.

For the homogeneous and isotropic case the following estimator of $\varrho^{(2)}(r)$ is recommended:

$$\hat{\varrho}(r) = \frac{1}{2\pi r} \sum_{i=1}^{n} \sum_{\substack{j=1 \\ (j \neq i)}}^{n} \frac{\mathbf{k}_h(r - \|\mathbf{x}_j - \mathbf{x}_i\|)}{A(W_{\mathbf{x}j} \cap W_{\mathbf{x}i})}. \tag{15.15}$$

Here and in the following $\mathbf{k}_h(t)$ denotes the kernel function. The popular

Epanečnikov kernel (Appendix L) is always used

$$
\mathbf{e}_h(t) = \begin{cases} \dfrac{3}{4h}\left(1 - \dfrac{t^2}{h^2}\right) & (-h \le t \le h), \\ 0 & \text{otherwise.} \end{cases}
$$

As for density functions, one may expect that small differences in the kernel function do not have a strong influence on the estimation process, but the choice of the bandwidth is important. Simulations and practical experience suggest that the value h for the Epanečnikov kernel should be

$$
h = c\lambda^{-1/2}, \quad \text{with } c = 0.1\text{-}0.2. \tag{15.16}
$$

Note that small values of h produce 'rough' functions $\hat\varrho(r)$ showing more detail, while larger values of h produce 'smoother' curves.

By comparing (15.15) and (15.11), the idea behind the kernel estimator can easily be understood. Instead of all point pairs of a distance less than or equal to r, only those pairs are considered that are separated by a distance approximately equal to r.

As Fiksel (1988a) has shown, the estimator given by (15.15) has a property similar to unbiasedness:

$$
\mathrm{E}\hat\varrho(r) = \int_{-\infty}^{\infty} \mathbf{k}_h(s)\varrho^{(2)}(r + hs)\,\mathrm{d}s. \tag{15.17}
$$

As $h \to 0$, one obtains $\varrho^{(2)}(r)$ if the product density is continuous at r.

Isotropy is better used with the following estimator, which corresponds to (15.14):

$$
\hat\varrho_R(r) = \frac{1}{2\pi r} \sum_{i=1}^{n} \sum_{\substack{j=1 \\ (j \ne i)}}^{n} \frac{\mathbf{k}_h(r - \|\mathbf{x}_j - \mathbf{x}_i\|)b_{ij}}{A(W^{\|\mathbf{x}_j - \mathbf{x}_i\|})}. \tag{15.18}
$$

It is convenient that the estimator corresponding to (15.13) has a simple form:

$$
\hat\varrho_O(r) = \frac{1}{2\pi r \bar\gamma_W(r)} \sum_{i=1}^{n} \sum_{\substack{j=1 \\ (j \ne i)}}^{n} \mathbf{k}_h(r - \|\mathbf{x}_j - \mathbf{x}_i\|) \tag{15.19}
$$

(Ohser and Tscherny, 1988). $\bar\gamma_W(r)$ is the isotropized set covariance function of W. For fixed r the factor $1/\bar\gamma_W(r)$ — making the edge correction — need be computed only once! If isotropy is assumed then this estimator is the most elegant. In the physical literature similar formulae appeared before Ohser's work; see Hosemann and Bagchi (1962), p. 218.

Furthermore, simulations have shown that there are no great differences between the estimators given by (15.18) and (15.19) and an estimator using Ripley's edge correction (Doguwa, 1990). This contrasts the process of estimating $K(r)$ or $L(r)$.

The paper Stoyan, Bertram and Wendrock (1992) discusses the problem of the accuracy of pair correlation function estimators. For the case of a Poisson point field it gives exact bounds for the estimation variance of $\hat{\varrho}(r)$ given by (15.19). (Here Ripley's ideas are used, which have been mentioned on p. 283.) For general point fields a Poisson approximation suggests the following approximation of the estimation variance $\sigma^2(r)$ of $\hat{g}_0(r)$:

$$\hat{g}_0(r) = \hat{\varrho}_0(r)/\lambda^2$$

$$\sigma^2(r) = \frac{0.6c(r)g(r)}{h\lambda^2} \qquad (r \geq 0) \qquad (15.20)$$

where

$$c(r) = (\pi r \bar{\gamma}_W(r))^{-1}.$$

As simulation experiments have shown, (15.20) yields values which are quite good for Poisson point fields and point fields with short-range order. However, for cluster processes (with heavy clustering) the real estimation variances are much greater than those predicted by (15.20).

The qualitative form of $\sigma(r)$ taken as a function of r is interesting.

$\sigma(r)$ has a pole for $r \to 0$, is then nearly constant and increases then with increasing r. Clearly, the pole results from the 'r' in the denominator in (15.19). This denominator causes a further problem. If $g(r)$ has positive values for $r \to 0$, then for small r estimates which are too high have to be expected. In such cases the recommendation is to use a series of bandwidths h, and to use the results with small h for small r and with great h for larger r.

In the anisotropic case there is an estimator analogous to (15.12) for $\varrho^{(2)}(r, \varphi)$:

$$\hat{\varrho}(r, \varphi) = \frac{1}{2} \sum_{i=1}^{n} \sum_{\substack{j=1 \\ (j \neq i)}}^{n} \frac{\mathbf{K}(\mathbf{x}_j - \mathbf{x}_i, (r, \varphi)) + \mathbf{K}(\mathbf{x}_j - \mathbf{x}_i, (r, \varphi + \pi))}{A(W_{\mathbf{x}_j} \cap W_{\mathbf{x}_i})}. \qquad (15.21)$$

Here $\mathbf{K}$ is a suitable kernel function, which may for example be defined by

$$\mathbf{K}(\mathbf{x}_j - \mathbf{x}_i, (r, \varphi)) = \frac{1}{r} \mathbf{k}_{h_r}(\|\mathbf{x}_i - \mathbf{x}_j\| - r)\mathbf{k}_{h_\varphi}(\alpha(\mathbf{x}_i, \mathbf{x}_j) - \varphi),$$

where $\mathbf{k}_{h_r}(t)$ and $\mathbf{k}_{h_\varphi}(\beta)$ are Epanečnikov kernels with suitable bandwidths h_r and h_φ. Here $\alpha(\mathbf{x}_i, \mathbf{x}_j)$ denotes the angle between the directed line from $\mathbf{x}_i$ to $\mathbf{x}_j$ and the x-axis. As in §14.4.1, (r, φ) denotes the point with polar coordinates r and φ ($r \geq 0; 0 \leq \varphi < 2\pi$). Since

$$\mathfrak{R}(B) = \mathfrak{R}(\breve{B})$$

and

$$\varrho^{(2)}(h) = \varrho^{(2)}(-h),$$

it suffices to consider only the range $0 \leq \varphi < \pi$. The averaging of $h = (r, \varphi)$ and $-h = (r, \varphi + \pi)$ in (15.21) improves accuracy. Figures 109 and 110 show estimated anisotropic pair correlation functions.

Programs for calculating $\hat{\varrho}(r)$, $\hat{\varrho}_O(r)$, $\hat{\varrho}_R(r)$ and $\hat{\varrho}(r, \varphi)$ are very similar to those for $\lambda^2 K(r)$. Line 5 must be replaced by

5   IF NOT $r - h < \varrho < r + h$ THEN GOTO 30

and lines 10–25 must be modified according to the estimator used. For example, in the case (15.15) it suffices to use

10   as on p. 284, line 2
20     $S = S + 1/C * \mathbf{k}_h(r - \varrho)/(\pi r)$

Doguwa (1990) discusses the properties of estimators for the pair correlation function. He considers (15.18) and (15.19), where $g(r)$ is obtained by dividing by $n(n - 1)/A(W)$. Interestingly this gives no great differences in the quality of the estimates, in contrast with the case for $K(r)$ and $L(r)$.

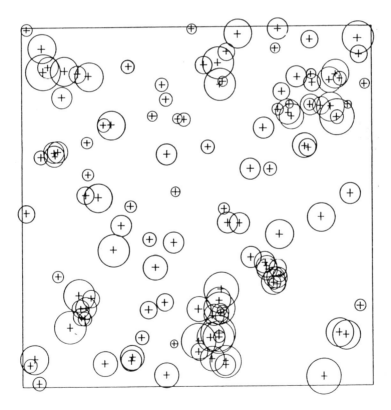

Figure 118   Positions of young pines in a 10m × 10m square in a Finnish forest. The circles characterize the tree heights, where three classes are considered (1, 3 and 5 m).

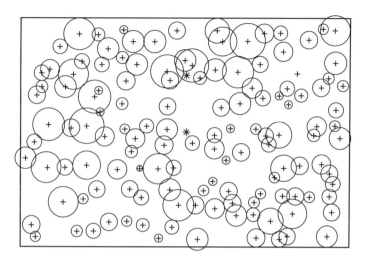

Figure 119   Positions of spruces in a rectangle 56 m×38 m in the Tharandt Forest near Dresden. The circles characterize the stem diameters, where four classes (10, 20, 30 and 40 cm) are considered. The two stars do not belong to the original pattern. Their individual L-functions will be calculated later (p. 291).

Example: *Estimating the pair correlation functions for positions of young pines and older spruces.* Figures 118 and 119 show positions of young pines in a Finnish forest and 60-year-old spruces in the Tharandter Wald near Dresden respectively. For the pines the circles show the tree heights and for the spruces the stem diameters. While the pine forest has grown naturally, the spruce forest has been planted and later cleared by foresters. The result is that the trees in the pine forest have quite irregular positions and even appear in clusters, while the spruces are distributed regularly. If one uses the models of Chapter 16, a cluster field and a Gibbs field with repulsion would probably be appropriate.

Figures 120 and 121 show pair correlation functions obtained using (15.15) for several bandwidths h (in m). The functions clearly show the expected qualitative differences. The influence of the bandwidth h on their form is also shown. In the case of the pines the pair correlation function becomes large as $r \to 0$; perhaps a model for which $g(0) = \infty$ is appropriate. For the very small bandwidth $h = 0.1$ m the low value of $g(0.1)$ results from the minimum distance between the pines. The curve for $h = 0.2$ m (corresponding to (15.16) with a factor 0.2) is perhaps most appropriate to the situation.

The most suitable graph of $g(r)$ seems to be that obtained from (15.16) with $c = 0.2$. There is a positive minimum distance between the trees; namely 1.0 m. Therefore values of $g(r)$ for $r < 1$ are unrealistic. The final estimate of $g(r)$ shown in Fig. 122 has been obtained by modifying the curve for $h = 1$ m of Fig. 121 using the reflection method (Appendix L)

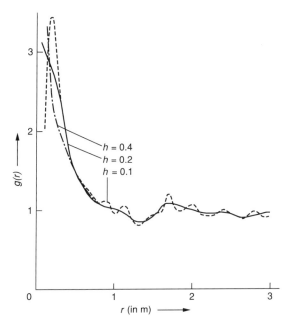

Figure 120   Estimated pair correlation functions for the pattern in Fig. 118 (pines). Three different bandwidths were used: — · — , 0.4 m; — , 0.2 m; - - -, 0.1 m.

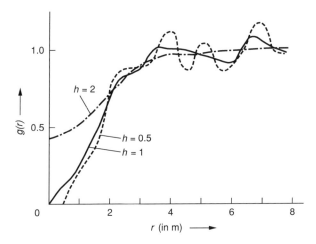

Figure 121   Estimated pair correlation function for the pattern in Fig. 119 (spruces). There were three different bandwidths: — · — , 2 m; — , 1 m; - - -, 0.5 m.

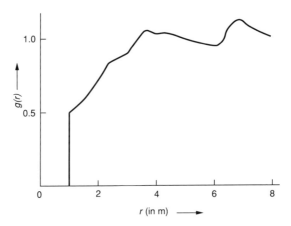

Figure 122 Improved estimate of the pair correlation function for the point pattern of Fig. 121 (spruces). The bandwidth is $h = 1$ m, and the minimum intertree distance (1 m) has been used for the reflection method.

15.4.3 Individual L- and g-functions

It may be that a given point pattern contains single points which play a particular role. These may be points of a different origin from the main body of the points or that have erroneously come into the pattern. In such a case one hopes to observe neighbourhood relationships with the 'normal' points that are different from the usual ones. These differences may be characterized by the following 'individual' L- and g-functions compared with the usual L- and g-functions (Getis and Franklin, 1987).

Let $\mathbf{x}$ be an extra point, and let $\mathbf{x}_1, \ldots, \mathbf{x}_n$ be the 'normal' points of the pattern. Then the individual functions are defined by

$$L_{\mathbf{x}}(r) = \sqrt{\frac{A(W)N(b(\mathbf{x}, r))}{n\pi}} \tag{15.22}$$

and

$$g_{\mathbf{x}}(r) = \frac{A(W) \sum_{i=1}^{n} \mathbf{k}_h(r - \|\mathbf{x} - \mathbf{x}_i\|)}{2\pi rn}. \tag{15.23}$$

Example. In Fig. 119 there are two additional points; they are marked by *. The functions $L_{\mathbf{x}}(r)$ have been computed for both of them (Fig. 123). These functions differ from $L(r)$, particularly for the upper additional point. In its neighbourhood the pattern is rather crowded. In contrast, the central additional point is somewhat isolated from the points of the pattern.

The problem of outlying points is also discussed in Wartenberg (1990). He discusses ways of detecting positional anomalies such as the isolation of an

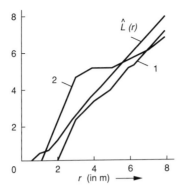

Figure 123   Individual *L*-functions for the additional points of Fig. 119: 1, central point; 2, upper point.

individual data point or point clustering. Nearest-neighbour distances or areas of Dirichlet cells may help here.

15.4.4   Estimating mark correlation functions

The estimation of mark correlation functions and related characteristics is based on that of $\varrho_f(r)$ and $\varrho^{(2)}(r)$. An estimator of $\kappa_f(r)$ is given by

$$\hat{\kappa}_f(r) = \frac{\hat{\varrho}_f(r)}{\hat{\varrho}(r)} \quad (r \geq 0).$$

The formulae (15.15), (15.18) and (15.19) yield estimators $\hat{\rho}(r)$ for $\varrho^{(2)}(r)$. Analogously, $\kappa_f(r)$ can be estimated by

$$\hat{\varrho}_f(r) = \frac{1}{2\pi r} \sum_{i=1}^{n} \sum_{\substack{j=1 \\ (j \neq i)}}^{n} \frac{f(m_i, m_j)\mathbf{k}_h(r - \|\mathbf{x}_i - \mathbf{x}_j\|)}{A(W_{\mathbf{x}j} \cap W_{\mathbf{x}i})}, \tag{15.24}$$

where m_k is the mark of $\mathbf{x}_k(k = i, j)$. This estimator corresponds to (15.15). Analogously, $\varrho_{lm}(r)$ can also be estimated:

$$\hat{\varrho}_{lm}(r) = \frac{1}{2\pi r} \sum_{i=1}^{n} \sum_{\substack{j=1 \\ (j \neq i)}}^{n} \frac{1^{lm}(i, j)\mathbf{k}_h(r - \|\mathbf{x}_i - \mathbf{x}_j\|)}{A(W_{\mathbf{x}j} \cap W_{\mathbf{x}i})}, \tag{15.25}$$

where $1^{lm}(i, j) = 1$ if the points $\mathbf{x}_i$ and $\mathbf{x}_j$ have the marks l and m, or m and l.

Example (silver particles, continued). As before, the points are the particle centres and the marks the particle diameters. Now the mark correlation function $k_{mm}(r)$ is considered. With bandwidth $h = 3.8$ mm the curve in Fig. 124 is obtained.

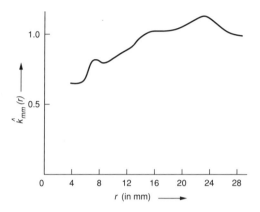

Figure 124 Estimated mark correlation function $\hat{k}_{mm}(r)$ for the silver particle diameters. Values less than unity are obtained for r-values under 15 mm; this indicates the inhibition of the particles.

It quantifies the a priori plausible 'inhibition' of the particle diameters for short distances. This influence extends to a range of 15 mm. (For a detailed discussion, see Stoyan and Wiencek (1991).)

As an example of the use of discrete marks, the same data are analysed again. Particles of diameter less than 9 mm get the mark 1 and all others the mark 2. Figures 125 and 126 show estimates of the functions $p_{lm}(r)$ and $g_{lm}(r)$ ($l, m = 1, 2$). These give similar information on the particle system to $k_{mm}(r)$. The limits of the functions $\hat{p}_{lm}$ as $r \to \infty$ are 0.388 ($l = m = 1$), 0.470 ($l = m = 2$) and 0.142 ($l = m = 2$). These functions also suggest a correlation up to $r = 15$ mm. The relative maximum of $k_{mm}(r)$ at $r = 25$ mm and the maximum of $\hat{p}_{22}(r)$ and the minimum of $\hat{p}_{11}(r)$ for this r-value obviously correspond.

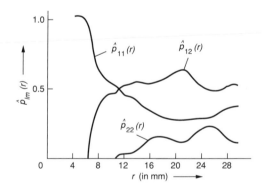

Figure 125 Estimated discrete mark correlation functions $\hat{p}_{lm}(r)$ for the silver particles. 1, small particles; 2, large particles.

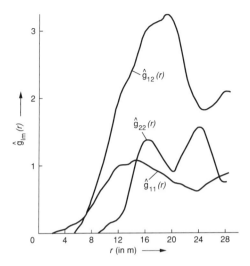

Figure 126   Estimated mark pair correlation functions $g_{lm}(r)$ for the silver particles (cf. Fig. 125).

For further applications of mark correlation functions see Penttinen *et al.* (1992) and Stoyan and Lippmann (1993). There the points are tree positions and the marks are quantities such as trunk diameters and degrees of damage by environmental factors.

15.4.5   Orientation analysis

One possible way of describing anisotropies in point patterns is to estimate and interpret the anisotropic pair correlation function $g(r, \varphi)$. For point patterns of a small number of points this procedure is not possible. Here it is better to average over ranges of distances and to estimate the function $O_{r_1 r_2}(\varphi)$ defined by (14.53). Its estimation is based on (15.11) and (15.12), where B is the time sector annulus $S(r_1, r_2, \varphi)$ introduced in §14.9:

$$\hat{O}_{r_1 r_2}(\varphi) = \frac{\mathbf{k}(S(r_1, r_2, \varphi))}{\mathbf{k}(r_2) - k^{(r)}(r_1)} \quad (0 \le \varphi \le \pi). \tag{15.26}$$

It is perhaps more instructive to use the corresponding density function $O_{r_1 r_2}(\varphi)$. If no normalization is necessary then the following estimator can be used:

$$\hat{o}_{r_1 r_2}(\varphi) = \sum_{i=1}^{n} \sum_{\substack{j=1 \\ (j \neq i)}}^{n} \frac{1_{[r_1, r_2]}(\|\mathbf{x}_i - \mathbf{x}_j\|) \mathbf{k}_h(\varphi - \alpha_{ij})}{A(W_{\mathbf{x}_j} \cap W_{\mathbf{x}_i})} \quad (0 \le \varphi \le \pi), \tag{15.27}$$

where α_{ij} is the angle between the line through $\mathbf{x}_i$ and $\mathbf{x}_j$ and the x-axis. The kernel function $\mathbf{k}_h(\varphi)$ may be chosen to be the Epanečnikov kernel. The calculations must be made modulo π; angles near 0 and π have to be considered as close together. Figure 112 shows two estimates of $o_{r_1 r_2}(\varphi)$ for the particle centres of Fig. 96.

15.5 ESTIMATING THIRD-ORDER CHARACTERISTICS

In the following it is shown how to estimate the quantity $z_B(r)$ defined by (14.38). Further third-order characteristics can be similarly estimated (Hanisch, 1983).

It is

$$z_B(r) = \int \frac{1_B(h)\varrho^{(3)}(r, h)}{\varrho^{(2)}(r)\lambda A(B)}\, dh \quad (r \geq 0).$$

The quantities λ and $\varrho^{(2)}(r)$ in the denominator are estimated by (15.1) and by (15.15), (15.18) or (15.19) respectively. To estimate the integral in the numerator, one uses

$$\hat{z}(r) = \frac{1}{2\pi r} \sum_{i=1}^{n} \sum_{\substack{j=1 \\ (j\neq i)}}^{n} \sum_{\substack{k=1 \\ (k\neq i, j)}}^{n} \frac{k_h(r - \|\mathbf{x}_i - \mathbf{x}_j\|)1_{B_{ij}}(\mathbf{x}_k)}{A(W_{\mathbf{x}_i} \cap W_{\mathbf{x}_j})}, \tag{15.28}$$

where B_{ij} is that rectangle which lies between the points $\mathbf{x}_i$ and $\mathbf{x}_j$ like that in Fig. 105 between o and r.

The estimator $\hat{z}_B(r)$ has an 'unbiasedness' property similar to (15.17). This follows from (14.33).

Example 1. Figure 127 shows the positions of extinctions in a body of steel in planar section. Since extinctions tend to appear at dislocation lines and Fig. 127 shows chains of points, $z_B(r)$ was used to quantify this 'inner alignment tendency'. The values of r used were 10, 20, 30 and 40 mm (length units as in Fig. 127), where the rectangle B depended on the actual r-value: $a = 0.5r$, $b = 0.2r$. The

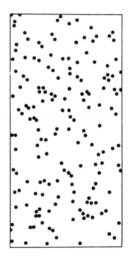

Figure 127 Extinctions in steel shown with high magnification in a planar section. It is known that the extinctions tend to lie on dislocation lines. Therefore it may be reasonable to believe that the pattern has an inner alignment tendency.

Table 20 $z_B(r)$ as a function of r for extinctions in steel.

r	$z_B(r)$
10	0.80
20	0.88
30	1.12
40	1.00

calculation using (15.25) yielded the values as in Table 20. They are close to units; the alignment tendency expected is not shown by the values of $z_B(r)$.

Example 2. Figure 128 shows a point pattern obtained as follows. A point was assigned to the centre and both endpoints of each line segment of a random system of line segments. The mean length of the segments was about 11 mm. Thus one expects large values of $z_B(r)$ for r around 10 mm and for suitable rectangles B. With side lengths as in Example 1, the values in Table 21 were obtained.

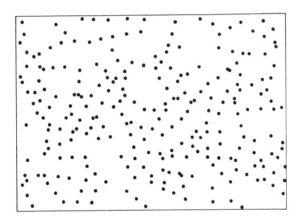

Figure 128 Simulated point pattern with 'inner alignment'. The points lie at the ends and centres of a system of line segments obtained by simulation.

Table 21 $z_B(r)$ as a function of r for the line segment point field.

r	$z_B(r)$
4	0.46
6	0.32
8	0.46
10	1.58
12	1.47
14	1.08

The results are as expected; here $z_B(r)$ indicates 'inner alignment' in the point pattern.

15.6 ESTIMATING NEAREST-NEIGHBOUR DISTANCE DISTRIBUTIONS

A naive method of statistical estimation of $D(r)$ and $D^{(k)}(r)$ is to determine for each point of the pattern in the window the nearest neighbour and to measure the corresponding distance. All these distances form a sample, which can be analysed by the classical methods of statistics. An estimator of $D(r)$ is then the corresponding empirical distribution function, and the corresponding mean m_D can be estimated by the sample mean.

Acceptable results are obtained if the window W is very big. For smaller W the results tend to be too large. The reason for this is simple: for points near the boundary of W the true nearest neighbour is often outside W and thus cannot be considered. Consequently an edge correction is necessary, as for $K(r)$, $L(r)$ and $g(r)$. Several such possibilities are discussed by Stoyan et al. (1987). Here only the edge correction by Hanisch (1984) is treated. It is easily programmed and is not difficult to understand.

Instead of the original problem of estimating $D(r)$ and $D^{(k)}(r)$, the quantities $G(r) = \lambda D(r)$ and $G^{(k)}(r) = \lambda D^{(k)}(r)$ are estimated. This does not lead to difficulties in practice, since nearest-neighbour analysis is usually performed then only if the intensity λ is known accurately. The estimator $\hat{G}(r)$ of $G(r)$ is

$$\hat{G}(r) = \sum_{i=1}^{n} \frac{1_{W \ominus b(o, \delta(\mathbf{x}_i))}(\mathbf{x}_i)}{A(W \ominus b(o, \delta(\mathbf{x}_i)))} 1_{[0,r]}(\delta(\mathbf{x}_i)) \quad (r \geq 0), \tag{15.29}$$

where $\delta(\mathbf{x})$ is the distance from $\mathbf{x}$ to its nearest neighbour.

A sketch of a program for calculating $G(r)$ follows. Given n points with coordinates (ξ_i, η_i) in a rectangular window W of side lengths a and b ($0 \leq \xi_i \leq a$, $0 \leq \eta_i \leq b$), we have

1    $G(r) = 0$
2   FOR $i = 1$ TO n
3    $\beta = \min\{\min\{\xi_i, a - \xi_i\}, \min\{\eta_i, b - \eta_i\}\}$
4   Determine the distance δ_i of x_i to
     its nearest neighbour in the window W.
5   IF $\beta > \delta_i$ AND $\delta_i \leq r$ THEN
     $G(r) = G(r) + 1/(a - 2\delta_i)/(b - 2\delta_i)$
6   NEXT i

Using an estimator $\hat{\lambda}$ for λ or $\hat{G}(\infty)$, for $G(\infty)$, an estimator for $D(r)$ is obtained as

$$\hat{D}(r) = \frac{\hat{G}(r)}{\hat{\lambda}} \quad \text{or} \quad \hat{D}(r) = \frac{\hat{G}(r)}{\hat{G}(\infty)}.$$

As Hanisch (1984) has shown, $\hat{G}(r)$ is unbiased if r is not too large and the denominator in (15.29) does not vanish.

In calculating $\hat{G}(r)$, only those points for which it is certain that their nearest neighbour is in W are considered. Their number divided by $A(W)$ would be a plausible estimator for $G(r)$. However, it would be too small, since many points in W at distances less than r from this nearest neighbour would be ignored. This effect is compensated for by dividing through by $A(W \ominus b(o, \delta(x)))$, which increases the quotient.

It is desirable to use all points x in W with $\delta(\mathbf{x}) < r$ to estimate $G(r)$. Of course, there could also be points among these whose nearest neighbour is not in W (but maybe instead the second neighbour). An estimator of this kind has been suggested

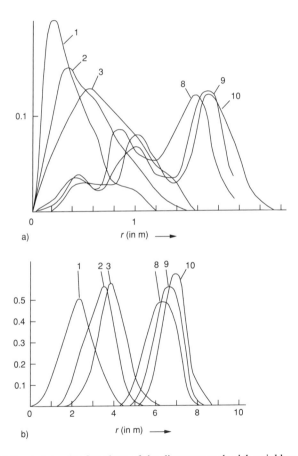

a)

b)

Figure 129 Estimated density functions of the distances to the kth neighbour for the point patterns of Figs. 118 and 119. (a) Pines: the curves are rather chaotic, there are qualitative differences between $k = 1$–4 and $k = 7$–10. (b) Spruces: The curves show a regular form, which corresponds to the order in the point pattern.

by Doguwa and Upton (1990). However, it is not mathematically justified and only approximately unbiased. It is an open problem to construct an unbiased estimator of this kind.

An estimator of $G^{(k)}(r)$ is given by

$$\hat{G}^{(k)}(r) = \sum_{i=1}^{n} \frac{1_{W \ominus (b(o, \delta_k(\mathbf{x}_i)))}(\mathbf{x}_i)}{A(W \ominus b(o, \delta_k(\mathbf{x}_i)))} 1_{[0,r]}(\delta_k(\mathbf{x}_i)) \quad (k = 2, 3, \ldots), \qquad (15.30)$$

where $\delta_k(r)$ is the distance from $\mathbf{x}$ to its kth neighbour. The 'mean' $\lambda m_D^{(k)}$ of $G^{(k)}(r)$ can be estimated similarly by

$$\lambda m_D^{(k)} = \sum_{i=1}^{n} \frac{1_{W \ominus (b(o, \delta_k(\mathbf{x}_i)))}(\mathbf{x}_i)}{A(W \ominus b(o, \delta_k(\mathbf{x}_i)))} \quad (k = 1, 2, \ldots). \qquad (15.31)$$

Finally, the density function $d^{(k)}(r)$ of $D^{(k)}(r)$ is sometimes of interest if it may be assumed that it exists (this is not the case for a lattice point field). A kernel estimator constructed analogously to (15.29) is given by

$$\hat{d}^{(k)}(r) = \sum_{i=1}^{n} \frac{1_{W \ominus (b(o, \delta_k(\mathbf{x}_i)))}(\mathbf{x}_i)}{A(W \ominus b(o, \delta_k(\mathbf{x}_i)))} \mathbf{k}_h(r - \delta_k(\mathbf{x}_i)) \quad (r \geq 0), \qquad (15.32)$$

where $\mathbf{k}_h(z)$ is the kernel function[†].

Example (pines and spruces, continued). In order to make the statistical analysis of the forest data complete, the neighbour distances have also been considered. For both woods, estimates of the density functions $d^{(k)}(r)$ of the distances to the kth neighbours have been determined by the method given above. Figure 129 shows the results, and Table 22 gives the corresponding means $m_D^{(k)}$. The numbers for 'Matern cluster' belong to the cluster field considered in §16.2.

The fact that for the pines the estimated mean distances are smaller than the corresponding theoretical values for a Poisson field of equal intensity for all k shows clearly that this is a point pattern with clustering.

Conversely, for the spruces the estimated mean distances are larger than in the Poisson field case, which is related to the existing short-range order with a tendency to regularity. (Originally the spruces were planted as rows!) The tendency to regularity is also demonstrated in the relatively small variance of the neighbour distance given by the slenderness of the estimated density functions (Fig. 129b). In contrast, the skewness of the estimated density functions in Fig. 129a for the pines is clearly greater than for a Poisson field. A particular characteristic of these density functions is the bimodality for $k \geq 4$. On account of the smaller sample size (126 trees), the densities are not so smooth.

[†] Doguwa (1989) discusses this and similar estimates.

Table 22 Ratios of the mean distances to the kth neighbour $m_D^{(k)}/m_D$.

k	Pines	Spruces	Poisson field (equal intensity)	Matern cluster
1	1 (0.38 m)	1 (2.56 m)	1 (0.45 m; pines) (1.99 m; spruces)	1 (0.34 m)
2	1.59	1.37	1.50	1.64
3	2.02	1.60	1.88	2.09
4	2.41	1.87	2.19	2.55
5	2.76	2.11	2.46	2.94
6	3.04	2.30	2.71	3.34
7	3.41	2.55	2.93	3.63
8	3.53	2.72	4.14	3.93
9	3.68	2.87	3.34	4.23
10	3.95	2.99	3.52	4.59

The italic numbers refer to the labels on the curves in Fig. 129.

15.7 ESTIMATING PALM MEANS

This section shows how to estimate a Palm mean $\bar{m}_f$ or $\mathsf{E}_0 f(N)$. Quantities of this kind were introduced in §14.9 as means of marks. Therefore it is natural to use the estimators described in §15.3.

Consequently, an unbiased estimator of $\lambda \bar{m}_f$ is given by

$$\widehat{\lambda \bar{m}_f} = \sum_{i=1}^{n} \frac{m_f(\mathbf{x}_i)}{A(W)} \tag{15.33}$$

where, as in §14.9, the mark $m_f(\mathbf{x})$ of $\mathbf{x}_i$ is equal to $f(N_{-\mathbf{x}_i})$. Unfortunately, it has similar disadvantages to $k_2(r)$ in (15.9), since it is typical for many functions $f(N)$ that $m_f(x)$ can only then be determined if all points of a certain neighbourhood of $\mathbf{x}$ are given. Because this is not always the case for points near the boundary of the window W, edge effects have to be considered, as in the estimation of $K(r)$ or $D(r)$.

There is no universal method for edge correction. The methods described for second-order characteristics do not always ensure unbiasedness.

If $f(N)$ is such that only points at a distance less than R from o have to be considered (e.g. if, the function $\phi(r)$ in $f_4(N)$ on p. 268 has the property $\phi(r) = 0$ for $r > R$) then for large windows minus sampling can be used. In (15.33) the summation is only over the points in $W \ominus b(o, R)$, and instead of $A(W)$ one divides by $A(W \ominus b(o, R))$. In this way unbiasedness is ensured, but the estimation variance may become rather large because of the potentially small effective window $W \ominus b(o, R)$. An alternative is to use heuristic edge corrections, which follow the example of Ripley's estimator for the K-function (p. 282). An example of the application of this idea is given in §16.3.2.

15.8 GOODNESS-OF-FIT TESTS FOR POINT FIELD MODELS

In point field statistics it is frequently necessary to perform tests, for example testing

- whether or not given model parameters are compatible with the given data for a fixed model; or
- whether a certain model is acceptable.

Their performance is in general rather complicated on account of the complexity and variability of the possible distributions. In Chapter 13 the relatively simple case of a Poisson field has already been considered. It is still possible to solve the test problem in this case with formulae and tables if rough tests are accepted. But even in the Poisson field case it is sometimes suitable to apply Monte Carlo tests. Such tests are for most models (with the possible exception of Cox fields) the most appropriate method. Clearly, they require powerful computers.

In the following the idea of such methods will be described briefly. They were first applied in point field statistics by Ripley (1977) and Besag and Diggle (1977). Note that the hypotheses to be tested in point field statistics are rather numerous, since there are many possible models. Furthermore, the observation conditions can be very different. For example the windows may have many sizes and shapes. Therefore it does not make sense to produce tables with critical values for all possible variations. Rather, it seems to be better to develop a suitable test for each particular test problem, which is characterized by a model, parameter and window. Monte Carlo tests are well suited to this purpose.

In such tests one selects a characteristic Δ that describes the deviations of certain empirical characteristics of a sample from the corresponding theoretical model characteristics of that point field. For example consider the following four characteristics:

$$\Delta^{\lambda} = |\hat{\lambda} - \lambda_M|,$$

where $\hat{\lambda}$ is the estimated intensity and λ_M the intensity of the model field;

$$\Delta_{r_0} = |\hat{L}(r_0) - L_M(r_0)|,$$

where $\hat{L}(r_0)$ is the estimator of the L-function for a given r-value r_0 and $L_M(r_0)$ the theoretical value of the L-function for the model;

$$\Delta^L = \int_0^R [\hat{L}(r) - L_M(r)]^2 f(r) \, dr,$$

where $\hat{L}(r)$ is the estimated L-function, $L_M(r)$ the theoretical L-function for the model, $f(r)$ the weight function and R the upper bound of the r-values (e.g. the half-diagonal length for rectangular W); and

$$\Delta^D = \int_0^R [\hat{D}(r) - D_M(r)]^2 f(r)\, dr,$$

where $\hat{D}(r)$ is the estimated distribution function of the nearest-neighbour distance and $D_M(r)$ the theoretical distribution function of the nearest neighbour distance for the model.

All four Δ-characteristics are chosen so that large positive values suggest rejection of the model hypothesis. The ideal is that the distribution of the characteristic Δ be known under the assumption that the model describes the data. Then it would be possible to construct an exact test. But, unfortunately, the prospects for this are poor, perhaps with the exception of the characteristic Δ^λ (which is in practice not so interesting). Therefore the following approach is in many cases the only possibility for constructing a significance test with probability α of an error of type I.

1. Determine the critical value of the characteristic Δ' for the given point pattern.

2. Generate by simulation k point patterns in W. For k the values $k = 99$ ($\alpha = 0.05$) and $k = 999$ ($\alpha = 0.01$) are suggested in the literature.

3. Determine for these k patterns the values $\Delta_1, \ldots, \Delta_k$ of the characteristic Δ.

4. Order the numbers $\Delta', \Delta_1, \ldots, \Delta_k$ in ascending order. Reject the hypothesis if Δ' belongs to the upper tail of this sequence. More precisely, reject the hypothesis if Δ' is the $(k - k_\alpha + 1)$th, ..., $(k + 1)$th number in the sequence. The critical k_α is given by

$$\alpha = \frac{k_\alpha}{k + 1}. \tag{15.34}$$

A major problem is of course how to arrive at the hypothesis. Only rarely it is given a priori, as in the case of the Poisson field hypothesis. In most cases it is typical that the hypothesis is posed only after exploratory data analysis. After using a suitable model, the model parameters are chosen such that they fit the given data. In particular, it is often possible to force the model intensity λ_M to equal $\hat{\lambda}$. (Thus Δ^λ is often uninteresting for applications.) The test described above does not then have the level α. Rather, one should expect that it is more 'hypothesis-friendly' than actually suited to the α-value chosen. The model parameters have been fitted to the data!

The difficulty described here is well known for other goodness-of-fit tests (for the case of the Kolmogorov–Smirnov test, see Sachs (1984, p. 331) and Lilliefors (1969, 1971)). In the case of point field statistics it is probably very difficult to overcome. The test above would have level α only if those point patterns were

generated that consistently have the same estimated model parameters. In the case
of the Poisson field this is simple. All simulated point patterns have to consist of
exactly n points. Thus the approach of §13.2.5 is justified. For more complicated
models it seems to be difficult to satisfy this requirement. For example, the authors
do not know a method of performing such simulations for Matern cluster fields.
Each pattern has to produce the same estimates of λ_e, R and μ by the methods
described in §16.2. In the case of Gibbs field a first step is to simulate only fields
of n points (Ripley, 1988, p. 48).

For the particularly interesting characteristic Δ_{r_0} the choice of r_0 is problematic.
Furthermore, it is unsatisfactory to consider only one r-value. Therefore when
using L-functions, the following approach is popular. Let k_α be even. For all k
simulated point patterns the L-function is estimated. For all r with $0 \leq r \leq R$ the
$\frac{1}{2}k_\alpha$-smallest and $\left(k - \frac{1}{2}k_\alpha + 1\right)$-largest $L(r)$-values are determined (here R is a
suitable upper limit of the interpoint distances). Thus two functions are obtained:
a lower, $L_{\min}(r)$, and an upper, $L_{\max}(r)$. If the empirical L-function for the pattern
lies between the envelopes $L_{\min}(r)$ and $L_{\max}(r)$.

$$L_{\min}(r) \leq L(r) \leq L_{\max}(r) \quad (0 \leq r \leq R), \tag{15.35}$$

then the hypothesis may be accepted. If the empirical L-function exceeds the limits
set by $L_{\min}(r)$ and $L_{\max}(r)$ then the hypothesis should be rejected. However, the
probability of an error of type I is not exactly α! For each fixed r the probability
of satisfying (15.35) is clearly $1 - \alpha$. But (15.35) has to be satisfied for all r with
$0 \leq r \leq R$. Of course, the events that (15.35) is satisfied for $r = r_1$ and $r = r_2$
$(0 \leq r_1, r_2 \leq R)$ are not stochastically independent, and the probability for their
joint occurrence cannot be computed easily.

Therefore if proceeding as described above then the probability of an error of
type I is not known; it is certainly greater than the chosen α.

Optimistic statisticians hope that the inaccuracies described above compensate
one another. (The first is favourable to the hypothesis, the second unfavourable.)

A rapid method of this kind is to simulate 19 times and choose $L_{\min}(r)$ and
$L_{\max}(r)$ as the minimum and maximum of the 19 $L(r)$-values. This method corresponds to $\alpha = 0.05$. An example is given in §16.2.

At the end of this section a certain hypothesis is treated that does not depend
upon estimated parameters. It says that for a marked point process

$$H_0 : \textit{the marks are independent}$$

Of course, this is an important problem in many applications.

One way of testing it is to use the following Monte Carlo test. First, k marked
point patterns are generated. The points $\mathbf{x}_1, \ldots, \mathbf{x}_n$ are those of the original point
pattern. The marks are determined independently according to the empirical one-
dimensional mark distribution. This can be done by the following algorithm.

1   FOR I = 1 TO n
2   Z = RND(0)

3   J = INT (Z * n) + 1
4   M(I) = ME(J)
5   NEXT I

Here $ME(J)$ denotes the original mark of the jth point $\mathbf{x}_j$ and $M(I)$ the mark of the ith point in the simulated point pattern. The test is carried out similarly to above. A characteristic Δ is chosen that is closely related to correlations in the point pattern, and the empirical Δ-characteristic is compared with the Δ-characteristic that belongs to the simulated point patterns.

While the L-function has been used above, the use of the following function $L_f(r)$ is suggested. Here $L_f(r)$ is connected with a mark function $f(m', m'')$ by

$$L_f(r) = \sqrt{\frac{K_f(r)}{\pi \kappa_f(\infty)}} \quad (r \geq 0). \tag{15.36}$$

Here $\kappa_f(\infty)$ is defined as on p. 264, and $K_f(r)$ is related to $\varrho_f(r)$ by

$$K_f(r) = \frac{2\pi}{\lambda^2} \int_0^r \varrho_f^{(2)}(u) u \, du \quad (r \geq 0)$$

(Penttinen and Stoyan, 1989).

An estimator of $L_f(r)$ is given by

$$\hat{L}_f(r) = \sqrt{\frac{k_f(r)}{\pi \hat{c}_f \hat{\lambda}^2}} \quad (r \geq 0), \tag{15.37}$$

where $k_f(r)$ is obtained as

$$k_f(r) = \sum_{\substack{i=1}}^{n} \sum_{\substack{j=1 \\ (j \neq i)}}^{n} \frac{1_{b(o,r)}(\mathbf{x}_j - \mathbf{x}_i) f(m_i, m_j)}{A(W_{\mathbf{x}_j} \cap W_{\mathbf{x}_i})} \tag{15.38}$$

and

$$\hat{c}_f(r) = \frac{1}{n^2} \sum_{\substack{i=1}}^{n} \sum_{\substack{j=1 \\ (j \neq i)}}^{n} f(m_i, m_j). \tag{15.39}$$

Thus the estimation process is similar to that for the K-function, and it is clear how (15.38) has to be modified using Ripley's or Ohser's edge correction in the isotropic case.

Example (silver particles, continued). By the method described above, 19 marked point patterns have been generated that differ only with respect to the marks. For them and the origin pattern the L_f-functions were estimated for

$$f(m_1, m_2) = m_1 \, m_2.$$

Table 23   Upper and lower envelopes and empirical L_{mm}-functions for the silver particles.

r	$L_{mm,\min}(r)$	$L_{mm}(r)$	$L_{mm,\max}(r)$
7	0	0	0
8	1.3	1.5	2.4
9	1.3	1.5	2.4
10	2.1	2.5	3.4
11	4.0	4.2	5.6
12	5.3	5.3	6.8
13	6.8	6.9	8.4
14	8.7	8.9	10.2
15	10.4	10.3	11.5
16	11.4	11.5	12.6
17	13.5	13.9	14.7
18	15.4	15.7	16.2
19	17.0	17.3	17.8
20	17.8	18.1	18.6
21	18.9	19.2	19.6
22	20.1	20.6	20.6
23	21.0	21.7	21.7
24	22.5	23.2 *	23.0
25	23.1	23.8 *	23.6
26	23.4	24.6 *	24.2
27	25.1	26.0 *	25.6
28	26.3	27.0 *	26.8
29	27.1	28.0 *	27.8
30	27.9	28.7 *	28.6

Table 23 contains values of the empirical L_{mm}-functions and the maximum and minimum envelope $L_{mm,\max}(r)$ and $L_{mm,\min}(r)$. It can be seen that the empirical L_{mm}-function is outside the range determined by $L_{mm,\min}(r)$ and $L_{mm,\max}(r)$ only for very large r. Thus the correlation between the particle diameters is not so strong.

Similar applications with $L_d(r)$ for directional marks are given by Stoyan and Beneš (1991). Harkness and Isham (1983) tested by simulation the hypothesis that two point fields in the same window are independent.

15.9   METHODS FOR ESTIMATING MODEL PARAMETERS

To estimate the parameters of point field models the usual methods of mathematical statistics can be used; for example the maximum-likelihood method, the method of moments or the minimum-contrast method. Until now, the application of the maximum-likelihood method has been possible only in the case of particular point fields, namely Poisson, Cox and Gibbs fields. This is discussed in §§13.3.4, 16.2 and 16.3.

If formulae for model characteristics are given then the method of moments can be applied. Example are given in §16.2. The minimum-contrast method is sometimes used in point field statistics in the following manner.

Assume that for the point field model to be examined the calculation of the L-function is possible (this is not an essential restriction, since an approximation by simulation for given model parameters is always possible). Let $L_\theta(r)$ be the model L-function with parameter θ, where θ may also be a vector. Then that θ which minimizes the integral

$$\int_0^R [\hat{L}(r) - L_\theta(r)]^2 f(r)\, dr$$

is a minimum-contrast estimator of θ (here $f(r)$ is a weight function). Instead of the L-function, the pair correlation function or the distribution function of nearest-neighbour distance can also be used.

In practical applications the integral is replaced by a sum, and suitable optimization methods are used to determine the optimum θ-value. After estimating θ by any method, a model test as suggested in §15.8 should be applied. It is recommended that point field characteristics different from those in the estimation be used. If $\hat{\theta}$ had been determined by means of the L-function then in the test the nearest-neighbour distance distribution function should be used.

CHAPTER 16

Point Field Models

16.1 INTRODUCTION

The theory of point fields offers many mathematical models of such fields. Readers of this book may be interested in Ripley (1981, 1988), Diggle (1983), Stoyan *et al.* (1987) and Daley and Vere-Jones (1988).

Many important models are constructed from Poisson fields, especially *Cox fields*. Details are given by Snyder (1975), Grandell (1976) and Karr (1986). Sometimes these fields are also called doubly stochastic Poisson fields, on account of their construction by a two-step random mechanism. In the first step an intensity function is chosen according to a certain distribution, and in the second a Poisson field having this function as intensity function is formed. The so-called *mixed Poisson field* where $\lambda(x)$ is a constant for each sample is a special case. Samples of this field behave like samples from a homogeneous Poisson field, but the intensity varies from sample to sample. If only one sample is given then it is impossible to distinguish it from a Poisson field. The situation for general Cox fields is similar. If there is only one sample then it is impossible to distinguish it from an inhomogeneous Poisson field. But if there are several samples then the distribution that generates the different $\lambda(x)$ can be estimated (Karr, 1986). For spatial statistics the case where the $\lambda(x)$ are samples of an ergodic homogeneous continuous random field is of particular interest. Then the corresponding Cox field is also ergodic. In applications Cox fields are frequently a first step of generalization of the Poisson field. For them many calculations are still possible with acceptable expense. Therefore they are used extensively in physics. However, it should be noted that homogeneous Cox fields are still 'more variable' than homogeneous Poisson fields; for example, the random point number $N(B)$ in any bounded Borel set B has a greater variance for a homogeneous Cox field than for a homogeneous Poisson field of the same intensity.

Another important class of models comprises the so-called *cluster fields*. There the points appear in clusters or clumps. The idea that there are 'parent points' around which the so-called 'daughter points' lie randomly scattered is quite popular. In many models it is the set of all daughter points that forms the cluster field. A particular class of cluster field models will be treated in §16.2; there the parent points belong to an homogeneous Poisson field. These fields are also 'more variable' than homogeneous Poisson fields. There is no strict border between cluster and Cox fields; rather, there is a large class of point fields that are both cluster and Cox fields.

A third important class of models is given by *hard core fields*. These are mathematical models of random systems of centres of non-overlapping discs of fixed radius R. An example of such a point field is the Poisson hard core field mentioned on pp. 318 and 323. A mathematically rather difficult hard core model that can be simulated easily is the SSI (simple sequential inhibition) field, which is discussed in detail by Diggle (1983). It is generated in a bounded window W, and its samples are obtained as follows. First, according to the uniform distribution in W, a random point is chosen, which is the first point $\mathbf{x}_1$ of the field. It is taken as the centre of a disc of radius $h = 2R$. Then another independent uniform random point is chosen from W. If it lies in the disc $b(\mathbf{x}_1, h)$ then it is rejected and a further random point is chosen, etc. When finally a point outside $b(\mathbf{x}_1, h)$ has been obtained, it is denoted by $\mathbf{x}_2$; it is the second point of the SSI field. The simulation is continued analogously until a planed number n is obtained or until no further point can be placed in W. The result of this procedure is a sample of a hard core field of disc radius R.

The SSI model can be modified in several ways. For example, it is possible to choose the radius R randomly for each disc to use other shapes (e.g. line segments; Penttinen and Stoyan, 1989), which must not mutually intersect.

Sometimes the term 'soft core fields' is used in the case of variable disc radii, in contrast to the terminology of Ogata and Tanemura (1989).

There are further hard and soft core fields, constructed quite differently. The starting point is a homogeneous Poisson field. According to certain rules, some of its points are eliminated so that a hard core field is obtained. Matern (1960,1986) suggested two such thinning rules for which mathematical calculations are possible. The more interesting of them is as follows. All points of the original homogeneous Poisson field independently receive marks that are uniformly distributed on the interval [0, 1]. Points with smaller marks are older and more powerful than those with larger marks. They suppress all younger points, which are closer than h. All points not suppressed form the hard core field. Note that suppressed points can nevertheless suppress other, still weaker, points. Formulae for this point field are given by Matern (1960, 1986) and Stoyan *et al.* (1987).

The model has been generalized by Stoyan and Stoyan (1985); see also Stoyan (1987, 1988) and Cressie (1991). There the radii R may vary as functions of the age marks introduced for the construction. This leads to soft core fields. Mecke *et al.* (1990) give a simulated point pattern for the generalized Matern model, together with the program used. Finally, Penttinen and Stoyan (1989) discuss a model where the discs are replaced by line segments that do not intersect.

Finally, Gibbs point fields should be mentioned here. They will be discussed in detail in §16.3.

16.2   CLUSTER FIELDS: NEYMAN–SCOTT FIELDS

16.2.1   Model description

Neyman–Scott fields form a popular class of cluster fields. They are frequently used in spatial statistics, since they are rather flexible and since some formulae are derived easily.

   The basis of the model is a homogeneous Poisson field of intensity λ_e; its points are called 'parent points'. A cluster of 'daughter points' is scattered around each parent point, and the union of all daughter points forms the Neyman–Scott field. All clusters are mutually independent and generated by the same law. The random numbers of daughter points in the clusters are identically distributed. Let c be a random variable having this distribution. In each of the clusters the points are scattered around the cluster centre (the corresponding parent point) independently and according to the same density function. Imagine the cluster centre translated to the origin o. Then the position of the ith daughter point is given by a random vector X_i, where the X_i are identically independently distributed with the same probability density function $d(x)$. It is frequently assumed that the X_i are isotropic. Then it suffices to find the density function $d(r)$ of the distance of the daughter points from the cluster centre.

Example: *Matern cluster field.* Here the number c of points per cluster has a Poisson distribution with parameter μ. The X_i are uniformly distributed in the disc of radius R centred at o. Thus

$$d(x) = \begin{cases} \dfrac{1}{\pi R^2} & (\|x\| \le R), \\ 0 & \text{otherwise,} \end{cases}$$

or

$$d(r) = \begin{cases} \dfrac{2r}{R^2} & (r \le R), \\ 0 & \text{otherwise.} \end{cases}$$

If the number of points per cluster has a Poisson distribution (as for the Matern cluster field) then the cluster field is also a Cox field (Stoyan *et al.*, 1987, p. 145).

   The *simulation of Neyman–Scott fields* is easy. As is typical for many applications it is assumed that there is a distance R such that the density function $d(r)$ vanishes for $r \ge R$ (distances from the cluster centre to the daughter points larger than R are impossible).

   First a sample of a Poisson field of intensity λ_e is generated in $W \oplus b(o, R)$. (By taking cluster centres outside W, edge effects are avoided, i.e. patterns too thin at the boundary of W.) Then a cluster is generated for each parent point. First a random integer is determined that is equal to the number of daughter points (in the case of a Matern cluster field a Poisson random number of the parameter μ is generated). Then for each daughter point two random numbers are generated, giving polar coordinates.

the distance from the parent point, according to the density function $d(r)$

and

the angle of the ray parent point $\rightarrow$ daughter point with the x-axis, by the uniform distribution on $[0, 2\pi]$.

Figure 87 shows a sample of a Neyman–Scott field with $\lambda_e = 50$ and exactly 5 points per cluster, which are uniformly distributed in discs of radius $R = 0.05$.

16.2.2 Formulae for Neyman–Scott fields

First- and second-order characteristics

By construction, any Neyman–Scott field is homogeneous; if the clusters are isotropic then so is the cluster field. The intensity λ is given by

$$\lambda = \lambda_e \bar{c}, \tag{16.1}$$

where $\bar{c} = \mathbf{E}\mathbf{c}$ is the mean number of points per cluster. In the isotropic case the K-function has the form

$$\lambda K(r) = \lambda \pi r^2 + \frac{1}{\bar{c}} \mathbf{E}\mathbf{c}(\mathbf{c} - 1) F(r) \quad (r \geq 0) \tag{16.2}$$

(Stoyan *et al.*, 1987). Here $F(r)$ is the distribution of the distance of two independent random isotropic points, whose distance from o is distributed according to the density function $d(r)$. The formula (16.2) leads to the pair correlation function

$$g(r) = 1 + \frac{1}{2\pi \lambda_e \bar{c}^2 r} \mathbf{E}\mathbf{c}(\mathbf{c} - 1) f(r) \quad (r \geq 0), \tag{16.3}$$

where $f(r)$ is the density function of $F(r)$. The calculation of $F(r)$ from $d(r)$ is possible, but not so easy, using

$$F(r) = \int_0^\infty \int_0^\infty F(r|r_1, r_2) d(r_1) d(r_2) \, dr_1 \, dr_2,$$

where $F(r|r_1, r_2)$ is the conditional distribution function of the distance of two points with distances r_1 and r_2 from o respectively. It suffices to consider the case $r_1 \leq r_2$:

$$F(r) = 2 \int_0^\infty \int_0^\infty F(r|r_1 \leq r_2) d(r_1) d(r_2) \, dr_1 \, dr_2.$$

The distribution function of the random distance ϱ between the points under the condition that the distances from o are r_1 and r_2 respectively, with $r_1 \leq r_2$, is obtained by geometrical argument as follows.

The random angle φ between the rays from o to the r_1 and r_2 points is uniformly distributed on $[0, 2\pi]$. By symmetry arguments, it suffices to consider only the case $0 \leq \varphi \leq \pi$. The probability that $\varphi \leq r$ can be expressed by a ratio of angles, because

$$\varphi = \arccos \left(\frac{r_1^2 + r_2^2 - \varrho^2}{2 r_1 r_2} \right),$$

and ϱ is increasing as a function of φ. Thus the probability is

$$\frac{1}{\pi} \arccos \left(\frac{r_1^2 + r_2^2 - r^2}{2 r_1 r_2} \right)$$

for $r_2 - r_1 \leq r \leq r_2 + r_1$ and zero otherwise. Thus

$$F(r | r_1 \leq r_2) = \begin{cases} 0 & (r < r_2 - r_1), \\ \frac{1}{\pi} \arccos \left(\dfrac{r_1^2 + r_2^2 - r^2}{2 r_1 r_2} \right) & (r_2 - r_1 \leq r \leq r_1 + r_2), \\ 1 & (r > r_1 + r_2). \end{cases}$$

In the particular case of the Matern cluster field the following formulae hold.

Moment formulae for the Matern cluster field. The intensity λ of the field is

$$\lambda = \lambda_e \mu. \tag{16.4}$$

The pair correlation function satisfies

$$g(r) = 1 + \frac{f(r)}{2 \pi \lambda_e r} \quad (r \geq 0). \tag{16.5}$$

if the formula

$$\mathbf{E}\mathbf{c}(\mathbf{c} - 1) = \mu^2$$

is used for a Poisson-distributed random integer $\mathbf{c}$ of the parameter μ. There,

$$f(r) = \begin{cases} \dfrac{4r}{\pi R^2} \left\{ \arccos \left(\dfrac{r}{2R} \right) - \dfrac{r}{2R} \sqrt{1 - \dfrac{r^2}{4R^2}} \right\} & (0 \leq r \leq 2R), \\ 0 & (r > 2R). \end{cases}$$

In particular,

$$g(0) = 1 + \frac{1}{\pi \lambda_e R^2}.$$

The K-function is given by

$$K(r) = \pi r^2 + \frac{1}{\lambda_e} \times \begin{cases} 2 + \frac{1}{\pi}[(8z^2 - 4)\arccos z - 2\arcsin z \\ \qquad +4z\sqrt{(1-z^2)^3} - 6z\sqrt{1-z^2}] & (r \le 2R), \\ 1 & (r > 2R), \end{cases}$$

where $z = r/2R$.

If the disc $b(o, R)$, in which the daughter points are scattered, is replaced by an arbitrary compact set K, but otherwise the properties of the Matern cluster field are retained, then (16.5) takes a generalized form: $f(r)$ must be replaced by the distance distribution $P(r)$ of K, which is considered in §8.4.3.

Another particular model is the so-called *Thomas field*. Here the number c of points per cluster has a Poisson distribution, but the points follow a symmetric normal distribution of variance σ^2 around the parent points. Its pair correlation function is

$$g(r) = 1 + \frac{\exp\left(-\dfrac{r^2}{4\sigma^2}\right)}{4\pi\lambda_e\sigma^2} \qquad (r \ge 0).$$

Nearest-neighbour distance distribution function. The nearest-neighbour distance distribution function satisfies

$$D(r) = 1 - [1 - H_s(r)]D_c(r) \quad (r \ge 0),$$

where $H_s(r)$ is the spherical contact distribution function of the cluster field and $D_c(r)$ is the probability that in a disc of radius r centred at an arbitrary cluster point there is no other point of the same cluster.

Since the spherical contact distribution function is of interest in its own right in the statistics of Neyman–Scott fields (Baudin, 1981), this is discussed first.

Spherical contact distribution function $H_s(r)$. By definition, $H_s(r)$ is the probability that in the disc $b(o, r)$ there is at least one point of the cluster field. It can be obtained by means of the formulae for the Boolean model. The germs are the parent points, the grains the clusters. The probability that this Boolean model intersects the disc $b(o, r)$ is $H_s(r)$. The general formulae for the Boolean model give

$$H_s(r) = 1 - \exp[-\lambda_e E A(X_0 \oplus b(o, r))] \quad (r \ge 0),$$

where X_0 denotes the random set that has as elements the points of a cluster centred at o.

Thus there is the geometrical problem of determining the mean area of the union X_r of the discs $b(x_i, r)$ with x_i in X_0. It is difficult, since these discs may overlap. A possible solution is as follows.

The desired mean area is determined as an integral over the covering function $p_{X_r}(t)$:

$$EA(X_0 \oplus b(o, r)) = 2\pi \int_0^\infty t p_{X_r}(t) \, dt, \qquad (16.6)$$

where

$$p_{X_r}(t) = \Pr(\mathbf{t} \in X_r), \qquad \|\mathbf{t}\| = t.$$

The calculation of $p_{X_r}(t)$ for the Matern cluster field is relatively easy. Here $p_{X_r}(t)$ is equal to the probability that a homogeneous Poisson field of intensity $\mu/\pi R^2$ has at least one point in the intersection of the discs $b(o, R)$ and $b(\mathbf{t}, r)$. If $A(t, r, R)$ denotes the area of this intersection then

$$p_{X_r}(t) = 1 - \exp\left[-\frac{\mu A(t, r, R)}{\pi R^2}\right] \qquad (t \geq 0).$$

To determine $A(t, r, R)$, the formula in Appendix K can be used. The lower limit of the integral in (16.6) is here $R + r$.

If the distance density function $d(r)$ is given then $p_{X_r}(t)$ can be computed as

$$p_{X_r}(t) = 1 - E[1 - p(t)]^{\mathbf{c}} = 1 - \sum_{k=0}^\infty [1 - p(t)]^k p_k, \qquad p_k = \Pr(\mathbf{c} = k).$$

Here $p(t)$ is the probability that a random disc of radius r, whose centre has a distance from o distributed with density $d(u)$, contains a fixed point $\mathbf{t}$ at distance t from o. Then

$$p(t) = \int_0^{t+r} \frac{b(u)}{2\pi u} d(u) \, du, \qquad (16.7)$$

where $b(u)$ is the length of the circular arc of radius u around o in the disc $b(\mathbf{t}, r)$. This length is

$$b(u) = \begin{cases} 0 & (t \geq u + r, u \geq t + r), \\ 2\pi u & (r \geq u + t), \\ 2u \arccos\left(\dfrac{u^2 + t^2 - r^2}{2ut}\right) & \text{otherwise.} \end{cases}$$

The probability $D_c(r)$ can be obtained similarly:

$$D(r) = \frac{1}{\mathbf{c}} \sum_{k=1}^\infty p_k k (1 - \pi_r)^{k-1} \qquad (r \geq 0).$$

Here π_r is the probability that the disc of radius r, centred at a point of random distance from o with density function $d(r)$, contains an independent random point with the same distance density function. It satisfies

$$\pi_r = \int_0^\infty p(t) d(t) \, dt,$$

where $p(t)$ is the function defined by (16.7). If there is an R with

$$d(t) = 0 \quad (t \geq R)$$

then

$$D_c(r) = \frac{1}{c} p_1 \quad (t \geq 2R).$$

In the case of the Matern cluster field

$$D_c(r) = e^{-\mu \pi_r} \quad (r \leq 2R),$$

with

$$\pi_r = \frac{2}{\pi R^4} \int_0^R t A(t, r, R) \, dt.$$

For all $r \geq 2R$

$$D_c(r) = e^{-\mu}.$$

16.2.3 Statistical methods for Matern cluster fields

Statistical analysis of Neyman–Scott fields can be carried out using of the functions $D(r)$, $H_s(r)$, $g(r)$ and $K(r)$. The approach is the minimum-contrast method (Diggle, 1983; Heinrich, 1992). Model parameters are chosen so that the differences between empirical and theoretical functions are small. The application of the maximum likelihood method leads to great difficulties (Baudin, 1981). In the following the case of the Matern cluster field is discussed mainly by means of an example; the approach for other models may be similar.

For $r \geq 2R$

$$\lambda K(r) = \lambda \pi r^2 + \frac{\mathrm{Ec}(\mathbf{c} - 1)}{\bar{c}},$$

where for the Matern cluster field $\bar{c} = \mu$ and $\mathrm{Ec}(\mathbf{c} - 1) = \mu^2$. Thus the difference

$$\lambda K(r) - \lambda \pi r^2$$

should be independent of r for large r and take the value μ. Therefore there is a theoretical possibility of direct estimation of μ in the case of the Matern cluster field. However, simulation experiments show that this method gives useful results only for very large point patterns (Stoyan, 1991a).

To estimate the parameters of a Matern cluster field, the minimum contrast method is recommended, using $g(r)$, $K(r)$ or $L(r)$. The arguments for this are the simple formulae (in comparison with those for $D(r)$ and $H_s(r)$) and the general experience that statistical analysis for point fields should be done with second-order characteristics rather than with $D(r)$ or $H(r)$. Furthermore, $D(r)$ and $H_s(r)$ are not very sensitive to changes of the model parameters. The biases of the pair correlation function estimators for small r play an important role in the cluster process case: they deform the shape of the function graph in its most interesting part. Therefore, particular care is necessary in the estimation. Following the recommendations on p. 286 three bandwidths h should be used, with c in (15.16) as 0.05, 0.1 and 0.2.

Simulation experiments with this method have yielded acceptable results, (Stoyan, 1991a). Also Barendregt and Rottschäfer (1991) used this method with success. In order to check the quality of their estimators, these authors used a modified bootstrap method, the 'delta-strap-method'. The following example demonstrates the estimation method with the use of $L(r)$.

Example (pines, continued). The estimate $\hat{\lambda}$ of the intensity λ is $126/100$ m^{-2}. Figure 130 shows the empirical pair correlation function obtained using (15.15), which has a form which is typical for a cluster field.

Estimates $\hat{L}(r_i)$ of the L-function yield estimates of R and λ_e. (Thus μ is estimated from λ and λ_e by (16.4).) Let $L(r; \lambda_e, R)$ be the L-function of the Matern cluster field with parameters $\hat{\lambda}$, λ_e and R. For a given R that value of λ_e

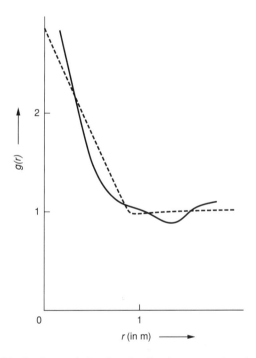

Figure 130 Empirical pair correlation function for the young pines (unbroken line). This function has the typical form of the pair correlation function for a cluster field. For comparison, the broken line shows the pair correlation for a Matern cluster field with the parameters estimated using the given data. It can be seen that the strength of clustering in the young pines is greater than that for the Matern cluster field.

is determined for which the following quantity takes a minimum by the Marquardt procedure:

$$\sum_{i=1}^{n}[\hat{L}(r_i) - L(r_i; \lambda_e, R)]^2;$$

for $r = 0.2i$, $n = 10$, where $R = 0.4, 0.45, \ldots, 0.6$. The R-value from among those five, which yields the smallest value of the sum of squared deviations with the optimum λ_e, is taken as the estimate. For the data this is $R = 0.45$ m. The corresponding λ_e-value is 0.82 m^{-2}. Finally, the value 1.53 is obtained for μ. Figure 130 shows, in addition to the empirical pair correlation function, the theoretical pair correlation function for these parameters The fit is quite good. This is also shown by the L-function. Figure 131 shows the empirical L-function and the envelopes of L-functions obtained by 19 simulations of the Matern cluster field with the estimated parameters. Only for r smaller than 0.4 m does the upper envelope slightly intersects the empirical curve. (As a further model, the Thomas field has also been tried. But the results were worse.)

The results obtained suggest the following model. In the pine stand the trees grow in clusters where the mean number of clusters per m^2 is 0.82. The number of trees per cluster is random, with a Poisson distribution with parameter 1.53. For each cluster the trees are randomly scattered in a disc of radius 0.45 m. (If discs overlap then the point density is increased accordingly.)

The relatively large values of the empirical pair correlation function for small r suggest that perhaps the true degree of clustering is stronger than for the Matern

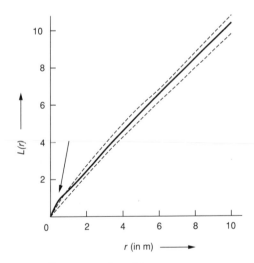

Figure 131 Empirical L-function (unbroken line) for the young pines. The broken lines show the upper and lower envelopes of 19 simulated L-functions. The Matern cluster field was simulated in the same window as the young pine data (cf. Fig. 118). The arrow points to where the upper envelope intersects the estimated L-function.

cluster field, and that the trees are more concentrated around the cluster centres. This is supported by the calculations in §15.7, where the mean distances $m_D^{(k)}$ to the kth neighbour for the pines are compared with those for the Matern cluster field.

Baudin (1983) studied the complicated problem of statistical determination of the cluster centres of a Neyman–Scott field. By definition, they form a Poisson field, and thus are completely randomly distributed. However, if a sample of the cluster field is given then the problem of reconstructing the cluster centres make sense — they will lie in areas of a great point density of the cluster field.

16.3 GIBBS FIELDS

Since the end of the 1970s Gibbs fields (or random Markov fields with interaction) have been intensively used in spatial statistics. They serve as models for many types of point patterns, particularly those with inhibitions, which are more regular than Poisson fields. The distribution of such fields can be characterized by a small number of parameters, which is attractive for applications. On the other hand, it is complicated to calculate the distributional characteristics introduced in Chapter 14. For this and other reasons, simulation methods are important in the theory of Gibbs fields.

16.3.1 Describing the models

Gibbs fields are defined and analysed both as finite point fields in bounded regions and as homogeneous point fields in R^2.

Gibbs fields with a finite point number

Let n points be randomly distributed in a bounded region $B^\dagger$. The joint distribution of the n points is given by a density function $f(x_1, \ldots x_n)$. Since all points are considered to be alike, $f(x_1, \ldots x_n)$ does not depend on the order of the x_i.

The density function has the particular form

$$f(x_1, \ldots x_n) = \frac{1}{Z_n} \exp\left[-\sum_{i=1}^{n} \sum_{j=i+1}^{n} \phi(\|x_i - x_j\|)\right] \quad ((x_1, \ldots x_n) \in B). \quad (16.8)$$

Here $\phi(r)$ is the so-called pair potential. It takes values in the range $-\infty < \phi(r) \le +\infty$, with the convention that $\exp(-\infty) = 0$. The normalizing constant Z_n ('configurational partition function') ensures that $f(x_1, \ldots x_n)$ is indeed a density function. Its determination is in general very difficult.

---

†In the theory of Gibbs fields finite point systems with a random number of points are also studied. There it is possible that with an inappropriate choice of pair potential the distribution may degenerate (Ripley, 1988). For statistical applications the case with a constant number of points usually suffices.

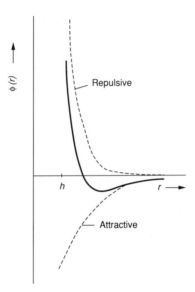

Figure 132 Typical pair potential, as used in physics. It results from the combination of different attracting and repelling forces.

The formula (16.8) shows how n, $\phi(r)$ and B determine the density function. Positive values of $\phi(r)$ leads to 'repulsion', so that interpoint distances r are rarely observed. Conversely, negative values of $\phi(r)$ lead to 'attraction', so that the interpoint distance r appears frequently. Figure 132 shows a typical pair potential. If $\phi(r) = \infty$ for $r < h$ then interpoint distances smaller than h are impossible; h is called the hard core distance. (It is clear that n, B and h can be chosen so that no reasonable density function is possible: large hard core distance h and great n and small region B.) In the case

$$\phi(r) = \begin{cases} \infty & (r \le h), \\ 0 & (r > h), \end{cases}$$

the Gibbs field is called a *Poisson hard core field*.

The form of the density function (16.8) is not arbitrary. It results from the physically motivated demand for fixed total energy

$$\int_B \cdots \int_B U(x_1, \ldots, x_n) f(x_1, \ldots, x_n)\, dx_1 \cdots dx_n,$$

with

$$U(x_1, \ldots, x_n) = \sum_{i=1}^{n} \sum_{j=i+1}^{n} \phi(\|x_i - x_j\|),$$

and maximized entropy

$$-\int_B \cdots \int_B f(x_1, \ldots, x_n) \log f(x_1, \ldots, x_n) \, dx_1 \cdots dx_n.$$

There are several methods for the approximate determination of Z_n; see e.g. the surveys by Diggle *et al.* (1994) and Ripley (1988). Usually large areas with a smooth boundary and a small point density are assumed in these approximations.

The so-called Poisson approximations suggested by Penttinen (1984), which can be used for 'rare' point patterns with weak interactions are particularly simple (Ripley, 1988). It is assumed that some random numbers approximately follow a Poisson distribution.

Example: *Strauss field*

$$\phi(r) = \begin{cases} \beta & (r \le r_0), \\ 0 & (r > r_0) \end{cases} \quad (0 < \beta < \infty).$$

Here

$$f(x_1 \ldots, x_n) = \frac{1}{Z_n} \exp[-\beta t(r_0; x_1, \ldots, x_n)], \tag{16.9}$$

where $t(r_0; x_1, \ldots, x_n)$ is the number of point pairs with a distance less than or equal to r_0; (x_i, x_j) and (x_j, x_i) are identified.

It is assumed that $t(r_0; x_1, \ldots, x_n)$ is approximately Poisson-distributed. For the parameter μ of this Poisson distribution the formula

$$\mu = \frac{n(n-1)}{2} \frac{\pi r_0^2}{A(B)} \tag{16.10}$$

is plausible. The first factor is the number of all pairs, the second is an approximation to the probability that the distance of a randomly chosen pair is less than r_0. The integral

$$Z_n = \int_B \cdots \int_B \exp\left[-\sum_{i=1}^n \sum_{j=i+1}^n \phi(\|x_i - x_j\|)\right] dx_1 \cdots dx_n$$

is, by (16.9), equal to

$$Z_n = A(B)^n \mathsf{E} \exp[-\beta t(r_0; x_1, \ldots, x_n)].$$

This yields, with (16.10),

$$Z_n \approx A(B)^n \exp\left[-\frac{n(n-1)}{2} \frac{\pi r_0^2 \psi}{A(B)}\right], \tag{16.11}$$

where $\psi = 1 - e^{-\beta}$.

Virial expansions are frequently used, as in statistical physics (Ogata and Tanemura, 1981, 1986; Diggle *et al.*, 1994). A simple form is the second cluster approximation

$$Z_n \approx A(B)^n \left[1 - \frac{v_2}{A(B)} \right]^{n(n-1)/2}, \tag{16.12}$$

with

$$v_2 = 2\pi \int_0^\infty (1 - e^{-\phi(r)}) r \, dr.$$

This approximation can be used only if the pair potential is close to zero, when the Gibbs field behaves like a Poisson field. Better approximations are obtained by higher virial expansions (Diggle *et al.*, 1994).

Example *(Strauss field, continued)*. The second cluster approximation (16.12) gives

$$Z_n \approx A(B)^n \left[1 - \frac{\pi r_0^2 \psi}{A(B)} \right]^{n(n-1)/2}.$$

A more precise formula is

$$\begin{aligned}
\log Z_n \approx {} & n \log A(B) - \frac{n(n-1)\pi \psi r_0^2}{2A(B)} - \frac{0.2932516\pi^2 n(n-1)(n-2)\psi^3 r_0^4}{6A(B)^2} \\
& - \frac{n(n-1)(n-2)(n-3)\pi^3}{24A(B)^3} \\
& \times (-0.27432784\psi^6 + 2.18542074\psi^5 - 1.37886114\psi^4) r_0^6
\end{aligned}$$

(Diggle *et al.*, 1994). The first two terms in the expression for $\log Z_n$ yield the same as (16.11).

Homogeneous Gibbs fields

Homogeneous Gibbs fields have infinitely many points, which are distributed in the whole plane. Their formal mathematical definition is more complicated than in the case of a finite point number. For an introduction see Stoyan *et al.* (1987).

Two basic facts are important for the following and for applications. The distribution of an homogeneous and isotropic Gibbs field is given by two characteristics: the *chemical activity* α and the *pair potential* $\phi(r)$. As in the case of a finite point number, $\phi(r)$ determines the character of the point distribution (hard core, cluster etc.), while α determines the intensity. For fixed pair potential, λ increases with decreasing α; α can also take negative values.

The distribution of a homogeneous and isotropic Gibbs field satisfies some continuity properties and the following important mean-value relation:

$$\lambda E_0(T(N \setminus \{o\})) = E \left\{ T(N) \exp \left[-\alpha - \sum_{\mathbf{x} \in N} \phi(\|\mathbf{x}\|) \right] \right\}. \tag{16.13}$$

On the left-hand side there is a Palm mean, as explained in §14.9; $T(N)$ is any measurable function that assigns to a point field a non-negative number.

A heuristic interpretation of (16.13) is given by Stoyan *et al.* (1987, p. 157). For a point pattern X the expression

$$E(\mathbf{x}, X) = \alpha + \sum_{y \in X} \phi(\|\mathbf{x} - \mathbf{y}\|) \quad (\mathbf{x} \in X)$$

is interpreted as the energy needed for adding the point $\mathbf{x}$ to X. The function $E(\mathbf{x}, X)$ is called the *local energy*.

It is a rather difficult problem to calculate the characteristics described in Chapter 14 for Gibbs fields in terms of α and $\phi(r)$. Much work has been done by physicists, who are particularly interested in the pair correlation function in the three-dimensional case. The following is a sketch of methods for the approximate determination of the pair correlation function $g(r)$ and the distribution function $D(r)$ of the nearest-neighbour distance.

The Percus–Yevick approximation. In the statistical mechanics of fluids the Percus–Yevick approximation is frequently used; it gives approximations of the pair correlation functions $g(r)$ if the intensity λ and the pair potential $\phi(r)$ are known. It is given by

$$\phi(r) \approx \frac{g(r)}{g(r) - c(r)} \quad (r \geq 0) \tag{16.14}$$

(Percus, 1964; Hansen and McDonald, 1986; Diggle *et al.*, 1987). $c(r)$ is a further function, the so-called 'direct correlation function', which is given implicitly as the solution of the Ornstein–Zernike equation:

$$c(r) = h(r) - \lambda(c * h)(r) \tag{16.15}$$

with $h(r) = g(r) - 1$ and

$$(c * h)(r) = \int_0^{2\pi} \int_0^{\infty} c(s)h((r^2 + s^2 - 2rs \cos \xi)^{1/2}) \, \mathrm{d}s \, \mathrm{d}\xi.$$

Combining (16.14) and (16.15) yields an integral equation for $g(r)$.

Mase's mean-value approximation. Mase (1990) suggested an approximation for means of the form

$$\mathsf{E} \prod_{\mathbf{x} \in N} f(\mathbf{x}).$$

The function $f(x)$ might be

$$f_1(x) = 1 - 1_B(x)$$

for a Borel set B or

$$f_2(x) = [1 - 1_B(x)]\mathrm{e}^{-\phi(\|x\|)}.$$

In the first case

$$E \prod_{x \in N} f_1(x) = E1_{\{N(B)=0\}} = \Pr(N(B) = 0);$$

thus the mean is the void probability for the set B. In the second case (16.13) gives

$$E \prod_{x \in N} f_2(x) = \lambda e^{\alpha} E_0 1_{\{N(B)=0\}}.$$

In the particular case $B = b(o, r)$ the mean is the same as $D(r)$, the distribution function of the nearest-neighbour distance,

$$D(r) = 1 - E \left\{ \prod_{x \in N} f_2(x) \right\} \frac{e^{-\alpha}}{\lambda} \quad (r \geq 0).$$

Mase (1990) calculated the spherical contact distribution function $H_s(r)$ for a Poisson hard core field. Comparison with simulated results shows that the accuracy of the approximation decreases with increasing α. The numerical and computational expense is large.

Generalizations

Sometimes interactions are considered that generalize those given by (16.8). The pair potential $\phi(r)$ is replaced by more general functions describing the mutual influence of neighbouring points. These point fields are called random Markov point processes. Baddeley and Møller (1989) is an excellent reference.

A special case of such general interactions is given by marks. (If the marks result directly from the point field, like $n_r(x)$ or $\delta^{(k)}(x)$ on pp. 244 and 267 respectively, then the situation is close to the examples considered by Baddeley and Møller (1989).) In this case the pair potential also depends on the marks: instead of

$$\phi(\|x - y\|),$$

the function

$$\phi(\|x - y\|, m(x), m(y))$$

is used, where $m(z)$ is the mark of z. In the case of homogeneous and isotropic fields three characteristics are used to describe the field: α, $\phi(r, m', m'')$ and a 'primary' mark distribution M_1. The formula (16.13) generalizes to

$$\lambda \int E_{o,m}(T(N, m)) M(dm)$$

$$= \int E(T(N, m) \exp \left[-\alpha(m) - \sum_{[x, m(x)] \in N} \phi(\|x\|, m(x), m) \right] M_1(dm). \quad (16.16)$$

Here **M** is the 'usual' mark distribution as introduced in §14.1. $\mathsf{E}_{o,m}$ is a Palm mean-value operator, which gives means under the condition that there is a point with the mark m at o.

Example: *Random systems of non-overlapping discs.* The pair potential has the special form

$$\phi(r, r', r'') = \begin{cases} \infty & \text{if } r \le r' + r'', \\ 0 & \text{otherwise.} \end{cases}$$

The marks are here interpreted as radii; therefore m is replaced by r in the following. If all marks are equal to the fixed number $\frac{1}{2}h$ then a Poisson hard core field is obtained. Replacing '∞' by β gives a point field that is similar to a generalization of the Strauss field discussed in Baddeley and Møller (1989).

If there are density functions then the corresponding density functions $\mathbf{m}(r)$ and $\mathbf{m}_1(r)$ are used instead of the mark distributions **M** and **M**₁. (It is possible to show that the existence of a density function for **M**₁ implies the existence of one for **M**.)

The formula (16.16) gives a useful relation between $\mathbf{m}(r)$ and $\mathbf{m}_1(r)$ obtained by quite a simple test function $T(N, m)$ that is independent of N. For a given x let

$$T(N, m) = \begin{cases} 1 & (r \le x), \\ 0 & (r > x), \end{cases}$$

so that the relation

$$\lambda \int_0^x \mathbf{m}(r)\, \mathrm{d}r = \int_0^x \mathrm{e}^{-\alpha(r)}[1 - H_s(r)]\mathbf{m}_1(r)\, \mathrm{d}r \quad (x \ge 0) \tag{16.17}$$

is obtained (Mase, 1986; and Stoyan, 1989a).

An equivalent form is

$$\lambda m(r) = \mathrm{e}^{-\alpha(r)}[1 - H_s(r)]m_1(r). \tag{16.18}$$

$H_s(r)$ is the spherical contact distribution function of the union set of all discs.

16.3.2 Simulating Gibbs fields

A very popular method of simulating Gibbs field uses so-called spatial birth-and-death processes. (Another possibility is the so-called Metropolis method, see Ripley, 1992.) The simulation begins with a start configuration which is then changed step by step, where points disappear ('die') and new ones are generated ('born'). In principle, any pattern can serve as start configuration. Often Poisson field samples are chosen or, for patterns with greater order, SSI patterns (p. 308). The given pattern to be analysed can also serve as a start configuration. According to Ripley (1987), for the method described below, the influence of the start configuration vanishes after roughly $10n$ steps, where n is the number of points in the pattern. After this point, patterns at a distance of $2n, \ldots, 4n$ steps can be considered as 'independent'.

Fixed point number n in a region B (Ripley, 1979,1987,1990,1992)

Here there are alternating 'deaths' and 'births'. One point from n is chosen at random with uniform probability. This point is then deleted. Then a new point is generated. This happens according to some probabilistic rules. Before explaining them, two simple special cases are described.

Poisson hard core field. Consider the point pattern just after a death. Each of the $n - 1$ points in B is taken as the centre of a disc of radius h. A uniformly random point is chosen in B. If it lies outside all discs then it is the newly born point of the pattern. Otherwise it is rejected and a new random point is generated, and so on.

Systems of non-overlapping discs. The simulation method above is modified. Now each point has an individual disc radius. For the new point a radius is generated according to the density function $\mathbf{m}_1(m)$, and it is rejected if the corresponding disc intersects at least one of the discs (with variable radii) centred at the existing $n - 1$ points. Otherwise it is the newly born point.

General case. After a birth there are n points. The number of the point dying then is determined by

$$k = 1 + \text{int}(un),$$

where u is a uniform random number on $[0,1]$. The position of the new point $\mathbf{x}$ is generated according to the conditional density function

$$f(\mathbf{x}|\mathbf{x}_1, \ldots, \mathbf{x}_{k-1}, \mathbf{x}_{k+1}, \ldots, \mathbf{x}_n).$$

While $f(\mathbf{x}_1, \ldots, \mathbf{x}_n)$ is given by (16.8), it is now

$$f(\mathbf{x}|\mathbf{x}_1, \ldots, \mathbf{x}_{k-1}, \mathbf{x}_{k+1}, \ldots, \mathbf{x}_n) = \frac{1}{c_n} \exp\left[-\sum_{\substack{j=1 \\ (j \neq k)}}^{n} \phi(\|\mathbf{x} - \mathbf{x}_j\|) \right],$$

with a normalizing constant c_n.

Let M be an upper bound of the function

$$\varphi(x) = \exp\left[-\sum_{\substack{j=1 \\ (j \neq k)}}^{n} \phi(\|x - \mathbf{x}_j\|) \right] \quad (x \in B).$$

The generation of $\mathbf{x}$ is carried out by the rejection method. A uniform random point ξ is generated in B and an independent uniform random number ω on $[0, 1]$. One takes

$$\mathbf{x} = \xi$$

if

$$\varphi(\xi) \geq M\omega.$$

Otherwise ξ is rejected and a new point must be generated.

Example: *Simulating a Strauss field.* Here

$$\varphi(\xi) = \exp[-\beta N(x, r_0)],$$

where $N(x, r_0)$ is the number of points in the sequence $x_1, \ldots, x_{k-1}, x_{k+1}, \ldots,$ x_n that are at a distance less than r_0 from x. The upper bound M can be chosen to be 1. The number $N(\xi, r_0)$ has to be determined for the random point ξ, and it is taken as the new point if

$$\exp[-\beta N(\xi, r_0)] \geq \omega.$$

Sample of a homogeneous Gibbs field in a rectangular region B (Stoyan *et al.*, 1987)

The simulation of a sample of a homogeneous Gibbs field is similar to the method used in the case of a fixed point number. The simulation begins with a suitable start configuration, which is changed by birth and deaths. But in this case births and deaths do not strictly alternate, but rather occur randomly. Furthermore, by so-called periodic continuation, patterns are obtained that behave like samples of homogeneous fields. The rectangle B is surrounded by eight further rectangles, as shown in Fig. 133. During the simulation, each outer rectangle contains a copy of

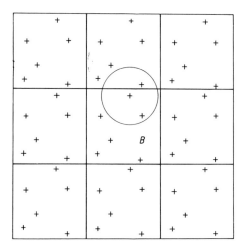

Figure 133 Representation of the periodic continuation of a point field in B. In all rectangles around B there lies a copy of the same point field. For any point in B all points in neighbouring rectangles that intersect the circle are considered.

the point pattern in B. Then for a point near the boundary of B the 'neighbour points' are not only in B but also those in the neighbouring rectangle(s). If B is large enough, there is no danger of undesirable correlations between the points at the edges of B.

Let $E(\mathbf{x}, X)$ be the local energy as in §16.3.1, where $\mathbf{x} \in B$ and X is the actual point pattern in B and the eight neighbour rectangles:

$$E(\mathbf{x}, X) = \alpha + \sum_{y \in X} \phi(\|\mathbf{x} - \mathbf{x}_j\|).$$

Using E, it is decided at each step whether there is a birth or a death. First the quantities

$$\mu_i = \exp[E(\mathbf{x}_i, X \backslash \{\mathbf{x}_i\})] \quad (i = 1, \dots, k)$$

(where k is the actual number of points in B), are calculated;

$$\mu_{k+1} = A(B).$$

Then the quantities

$$p_0 = 0, \quad p_i = \frac{\mu_1 + \cdots + \mu_i}{\mu} \quad (i = 1, \dots, k), \quad \mu = \sum_{i=1}^{k+1} \mu_i$$

are computed. The decision 'death or birth' is made using a uniform random number u on $[0, 1]$. If

$$p_0 + p_1 + \cdots + p_{i-1} < u \le p_0 + p_1 + \cdots + p_i$$

then the ith point dies, while if

$$u > p_0 + \cdots + p_k,$$

a new point is born. The new point $\mathbf{x}$ is chosen as a uniform random point in B.

Note the difference from the case of n points, where each point has an equal chance of death, while here the positions of the points in the pattern play a role. But the position of the new point is independent of the pattern. (This form of simulation is simpler than one where the birth process is analogous to the above, since otherwise the birth probabilities would have a rather complicated form.) In the case of a hard core field it may happen that the newly born point is generated in the hard core region of an already existing point: if so, then it is deleted and the normal process of births and deaths is continued.

16.3.3   Statistical methods for Gibbs fields

In the case of finite Gibbs fields maximum-likelihood and pseudo-likelihood methods are most commonly used, see Ripley (1988, 1992) and Diggle et al. (1994).

The window of observation is denoted by W; it is identified with B in the formulae of §16.3.1.

Approximate maximum-likelihood estimation

The likelihood function of a Gibbs field is given by the density function $f(\mathbf{x}_1, \ldots, \mathbf{x}_n)$:

$$L(\theta; \mathbf{x}_1, \ldots, \mathbf{x}_n) = f(\theta; \mathbf{x}_1, \ldots, \mathbf{x}_n) \quad (\mathbf{x}_1, \ldots, \mathbf{x}_n \in W);$$

the additional parameter θ describes the parameter dependence of the pair potential $\phi(r)$. For given points $\mathbf{x}_1, \ldots, \mathbf{x}_n$ in W the $\hat{\theta}$ that maximizes the expression

$$L(\theta; \mathbf{x}_1, \ldots, \mathbf{x}_n) = \frac{1}{Z_n(\theta)} \exp\left[-\sum_{i=1}^{n} \sum_{j=i+1}^{n} \phi(\theta; \|\mathbf{x}_i - \mathbf{x}_j\|) \right]$$

must be determined. In principle, this is not complicated to do by numerical optimization. However, in fact the problem is not so simple, since $Z_n(\theta)$ is difficult to obtain. Ogata and Tanemura (1981, 1986) suggested the application of the approximations for $Z_n(\theta)$ described in §16.3.1. Diggle *et al.* (1994) have shown that this method gives acceptable estimates if the interaction in the point field is weak; that is, if $\phi(r)$ is close to zero. The asymptotic normality of maximum likelihood estimates is studied in Jensen (1991, 1993); Mase (1992) proved the asymptotic efficiency of these estimates.

In a particular case maximum-likelihood estimation is quite easy. Namely, if a Poisson hard core field is given, i.e.

$$\phi(r) = \begin{cases} \infty & (r \le h), \\ 0 & (r > h), \end{cases}$$

then the estimator for h, the only unknown model parameter, is the minimum interpoint distance in the point pattern.

Sometimes the window W lies in a much larger region, with more points belonging to the same pattern. Ogata and Tanemura (1989) suggested the use of 'edge corrections' in such situations. One possibility is to replace the sum

$$\sum_{\substack{j=1 \\ j \ne i}}^{n} \phi(\theta; \|\mathbf{x}_i - \mathbf{x}_j\|)$$

by

$$\sum_{\substack{j=1 \\ j \ne i}}^{n} \phi(\theta; \|\mathbf{x}_i - \mathbf{x}_j\|) b_{ij}$$

where b_{ij}^{-1} is the length fraction of the circular arc centred at $\mathbf{x}_i$ of radius $\|\mathbf{x}_i - \mathbf{x}_j\|$ in W (p. 282).

Example *(Strauss field, continued).* The parameter β of the Strauss field can be estimated using the approximation formula (16.11) for Z_n. The log-likelihood function is then approximately

$$-\beta t(r_0; \mathbf{x}_1, \ldots, \mathbf{x}_n) - n \log A(B) + \frac{n(n-1)}{2} \frac{\pi r_0^2}{A(B)} (1 - e^{-\beta}).$$

Taking the derivative with respect to β and setting it equal to zero gives

$$t(r_0; \mathbf{x}_1, \ldots, \mathbf{x}_n) = \frac{n(n-1)}{2} \frac{\pi r_0^2}{A(B)} e^{-\beta}.$$

This yields the estimator

$$\hat{\beta} = -\log \left[\frac{2t(r_0; \mathbf{x}_1, \ldots, \mathbf{x}_n) A(B)}{n(n-1)\pi r_0^2} \right]. \tag{16.19}$$

Moyeed and Baddeley (1991) have suggested calculating the maximum-likelihood estimate recursively, by so-called 'stochastic approximation'. This includes a simulation of the Gibbs field for each estimate obtained. The case of the Strauss potential is discussed in detail. Here the maximum-likelihood equation may be reduced to

$$M(\hat{\beta}) = t(r_0; \mathbf{x}_1, \ldots, \mathbf{x}_n), \tag{16.20}$$

where $M(\beta)$ is the mean of $t(r_0; \mathbf{x}_1, \ldots, \mathbf{x}_n)$ for a Strauss field with parameter β. $\hat{\beta}$ can be estimated by the Robbins–Monroe method.

Mase (1992) has also studied the maximum likelihood estimation in the homogeneous case and was able to show that it has asymptotic optimality properties. Furthermore, he proved that the maximum likelihood estimator in the canonical case is as efficient asymptotically as that in the grand canonical case.

Pseudo-likelihood estimation

Besag (1978) suggested the following estimation method. The parameter θ is estimated by maximizing

$$L_p(\theta) = n(\log n - 1) + \sum_{i=1}^{n} \log[\gamma(\mathbf{x}_i | N^i)] - n \log \left[\int_W \gamma(x | N) \, dx \right]. \tag{16.21}$$

Here N denotes the total point pattern in W, and N^i the point pattern in W without the point $\mathbf{x}_i$, $N^i = N \backslash \{\mathbf{x}_i\}$, and

$$\gamma(x | X) = \exp \left[-\sum_{\mathbf{x}_i \in X} \phi(\theta; \|\mathbf{x}_i - x\|) \right].$$

The integral in (16.21) has to be determined numerically. Care is necessary since the integrand is in general a complicated function with many local extrema. In Diggle *et al.* (1994), where W was the unit square, after some experiments the integration was replaced by a summation over a 40×40 lattice.

The theoretical justification is difficult for fields of fixed point number. In the theory of homogeneous Gibbs fields a direct explanation is possible (Diggle *et al.*, 1994). Besag (1978), Ripley (1988), Jensen and Møller (1990) and Diggle *et al.* (1994) use arguments including approximations by lattice fields for the case of finite point fields. The name 'pseudo-likelihood' results from the approximation of the true likelihood function by a simpler term; the simplification results from independence assumptions. The function $L_p(\theta)$ is the corresponding log-pseudo-likelihood function.

Example *(Strauss field, continued).* A pseudo-likelihood estimator of β can be obtained as follows. It is

$$\log[\gamma(\mathbf{x}_i | N^i)] = -\sum_{\substack{j=1 \\ (j \neq i)}}^{n} \phi(\beta; \|\mathbf{x}_i - \mathbf{x}_j\|) = -\beta N(b(\mathbf{x}_i, r_0) \setminus \{\mathbf{x}_i\}).$$

Thus the sum in (16.20) is

$$-2\beta t(r_0; \mathbf{x}_1, \ldots, \mathbf{x}_n),$$

and the integral there is

$$\int_W \exp[-\beta N(b(x, r_0))] \, \mathrm{d}x.$$

Taking the derivative of $L_p(\beta)$ (formed according to (16.20), with $\beta = \theta$), with respect to β and setting it equal to zero yields the equation

$$\begin{aligned}
2t(r_0; \mathbf{x}_1, \ldots, \mathbf{x}_n) &\int_W \exp[-\beta N(b(x; r_0))] \, \mathrm{d}x \\
&= n \int_W N(b(x; r_0) \exp[-\beta N(b(x, r_0))] \, \mathrm{d}x
\end{aligned} \tag{16.22}$$

for β. The integrals have to be determined numerically.

Takacs–Fiksel method

The Takacs–Fiksel method is a method for homogeneous Gibbs fields. Therefore it can only be applied if the analysed point pattern can be considered as a sample of a homogeneous point field. The original idea stems from Takacs (1983, 1986), while Fiksel (1988b) generalized it to the form given here. (The paper by Mase (1984) can be considered as a predecessor.) Tomppo (1986) developed a variant of this method that uses nearest-neighbour distances, and this is suitable for measurements in situ (e.g. in forests, without measurement of tree coordinates).

The basis of the Takacs–Fiksel method is (16.13). The idea consists in

- choosing a series of test functions $T_k(N)$ $(k = 1 \ldots m)$; m should be not less than the number of parameters of the pair potentials;

- calculating estimators $\hat{L}_k(\alpha, \theta)$ and $\hat{R}_k(\alpha, \theta)$ for the left- and right-hand sides of (16.13) for any k;

- choosing the parameters[†] α and θ so that the sum of squared differences

$$S(\alpha, \theta) = \sum_{k=1}^{m} [\hat{L}_k(\alpha, \theta) - \hat{R}_k(\alpha, \theta)]^2$$

is minimized.

Since Takacs (1983), various test functions $T_k(N)$ have been tried. It now seems that two forms are particularly useful:

$$T_k'(N) = N(b(o, r_k)) \exp \left[\alpha + \sum_{\mathbf{x} \in N} \phi(\theta; \|\mathbf{x}\|) \right]$$

and

$$T_k''(N) = N(b(o, r_k)) = \text{ number of points in the disc } b(o, r_k).$$

Both forms have some computational advantages. In the case of $T_k'(N)$, $\hat{R}_k(\alpha, \theta)$ takes the simple form

$$R_k(\alpha, \theta) = \pi \lambda r_k^2.$$

Thus there is no (direct) dependence on α and θ. The estimator is simply

$$\hat{R}_k(\alpha, \theta) = \hat{\lambda} \pi r_k^2. \tag{16.23}$$

An estimator of the left-hand side of (16.13) is

$$L_k^*(\alpha, \theta) = \frac{1}{A(W)} \sum_{i=1}^{n} N(W \cap b(\mathbf{x}_i, r_k) \backslash \{\mathbf{x}_i\}) \exp \left[\alpha + \sum_{\substack{j=1 \\ (j \neq i)}}^{n} \phi(\theta; \|\mathbf{x}_i - \mathbf{x}_j\|) \right].$$

It is not edge-corrected. For an edge correction (which unfortunately does not ensure unbiasedness) it is suggested that one replaces $L_k^*(\alpha, \theta)$ by

$$\hat{L}_k(\alpha, \theta) = \frac{1}{A(W)} \sum_{i=1}^{n} N(W \cap b(\mathbf{x}_i, r_k) \backslash \{\mathbf{x}_i\}) a_i \exp \left[\alpha + \sum_{\substack{j=1 \\ (j \neq i)}}^{n} \phi(\theta; \|\mathbf{x}_i - \mathbf{x}_j\|) b_{ij} \right].$$

$$\tag{16.24}$$

[†]Note that for homogeneous Gibbs fields α is an additional parameter, which in the Takacs–Fiksel method must be estimated alongside θ.

Here

$$a_i = \frac{\pi r_k^2}{A(b(\mathbf{x}_i, r_k) \cap W)},$$

while b_{ij} is defined as on p. 282.

In the case of $T_k''(N)$, $L_k(\alpha, \theta)$ takes the form

$$\lambda^2 K(r_k),$$

since

$$\lambda E_0(N(b(o, r_k) \setminus \{o\})) = \lambda^2 K(r_k),$$

where $K(r)$ is Ripley's K-function. For an unbiased estimation the methods described in §15.4.1 may be used.

An estimator of $R_k(\alpha, \theta)$ can be obtained using a point lattice $\{y_j\}$ $(j = 1, \ldots, l)$ in W:

$$\hat{R}_k(\alpha, 0) = \frac{1}{l} \sum_{j=1}^{l} N(b(y_j, r_k)) c_j \exp\left[-\alpha - \sum_{i=1}^{n} \phi(\|\mathbf{x}_i - y_j\|; \theta) b_{ij} \right],$$

with

$$c_j = \frac{\pi r_k^2}{A(b(y_j, r_k) \cap W)}.$$

It is also possible to use $c_j = 1$ if the lattice points are 'deeper' in the interior of W. (If $\phi(r) = \infty$ for $r \leq r_0$ then the distance of the y_j from the boundary of W should be at least r_0.)

To choose the test function $T_k(N)$, experience shows that with a 'repulsive' pair potential ($\phi(r)$ positive), the form $T_k''(N)$ gives better results. For as 'attractive' pair potential ($\phi(r)$ is negative for some r) $T_k'(N)$ is preferable.

In Diggle *et al.* (1994) the point lattice was quadratic, $l = 196$ (= 14×14). The number of test functions was $m = 10$.

The r_k should be chosen as

$$r_k = k \frac{R}{m} \quad (k = 1, \ldots, m),$$

where R is a relatively large number with $R(\approx 1.2\text{--}1.5) r_{max}$ and r_{max} is the smallest r-value such that

$$\phi(r) = 0 \quad \text{for all } r > r_{max}.$$

In the case of the homogeneous Strauss field the authors do not know of such a simple estimation method as that based on the maximum-likelihood method. The two parameters β (potential parameter) and α (chemical activity) must be estimated, and about 10 radii r_k should be used, where $r_{max} = r_0$.

Comparison of estimation methods

In Diggle *et al.* (1994) the three methods described above were compared. Gibbs fields were simulated in squares of 100 points with weak to strong repulsion and for three different pair potentials and different parameters. It was assumed that the methods for the homogeneous case can also be applied to these patterns. Then the model parameters were re-estimated by the various methods. Thus it was possible to investigate the bias and estimation variances. Two main results were obtained.

(a)   The fairly negative opinion of Ripley (1988) on the Takacs–Fiksel method is not justified. Rather, it may compete with the other two methods. It is not as good only for very weak interactions (for 'almost-Poisson fields'). A practical disadvantage of the application of the Takacs–Fiksel method to point patterns of a density function (16.8) is the need to estimate the chemical activity α, which does not appear in (16.8) at all.

(b)   In the case of stronger interactions between the points the maximum-likelihood method becomes inaccurate. Since the assumptions under which the approximations of Z_n have been obtained do not remain true, this is not surprising. Also, with the pseudo-likelihood method the estimation errors increase with increasing interaction (see also Särkkä, 1993).

Implementation of the Takacs–Fiksel method and the pseudo-likelihood method is possible given a unified plan for arbitrary pair correlation functions. In contrast, in the maximum-likelihood method for each pair potential an adapted and sometimes complicated approximation Z_n must be found.

Cusp point method

For homogeneous Gibbs fields it has sometimes been suggested that one estimate the parameters using approximations for the pair correlation or the K-function (or similar characteristics). For example, Diggle *et al.* (1987) used (16.14) and (16.15). This method needs very precisely estimated pair correlation functions and a careful use of numerical methods. A special case is the cusp point method for the case of a hard core Strauss potential (or 'well potential')

$$\phi(r) = \begin{cases} \infty & (r \leq h), \\ \beta & (h < r \leq r_0), \\ 0 & (r > r_0). \end{cases}$$

Hanisch and Stoyan (1983) showed that the K-function of the corresponding Gibbs field has a cusp point at $r = r_0$ with

$$\frac{\lim\limits_{r \uparrow r_0} K'(r)}{\lim\limits_{r \downarrow r_0} K'(r)} = e^{-\beta}.$$

Thus the pair correlation function has a jump at $r = r_0$. These properties can be used to estimate r_0 and β, where both the K-function and the pair correlation function may be used (Stoyan and Grabarnik, 1991). Clearly, h is estimated by the minimum interpoint distance in the pattern.

As simulations have shown, this leads to useful estimators of r_0 and β. The error in the estimation of β increases with increasing absolute value of β.

When using the K-function, one may proceed as follows.

1. Choose some values $R_1, \ldots, R_k$ as possible candidates for r_0. This should be done using the empirical K-function, and the R_j should lie in an interval that contains the sharpest cusp point of $K(r)$.

2. Determine the minimum $S^{(j)}$ of

$$S_{\sigma, \tau}^{(j)} = \sum_{i=1}^{m} [K(r_i) - K_{R_j}(r_i; \sigma, \tau)]^2 \quad (j = 1, 2, \ldots, k)$$

with respect to the variables σ and τ, where

$$K_{\varrho}(r; \sigma, \tau) = \begin{cases} \sigma(r^2 - \hat{h}^2) & (r \le \varrho), \\ \sigma(\varrho^2 - \hat{h}^2) + \tau(r^2 - \varrho^2) & (r > \varrho). \end{cases}$$

The value of m should be between 10 and 20. The r_i should lie equidistantly in an interval that contains all R_j, with left endpoint greater than h and with the cusp point near the centre. The optimum values of σ and τ are

$$\sigma_j = \frac{A_1 A_2 - c_3 A_3 A_4}{A_2[A_5 + (m-1)c^2] + c^2 A_4^2}, \qquad \tau_j = \frac{A_3 - \sigma c A_4}{A_2},$$

where

$$A_1 = \sum_{i=1}^{l} \hat{K}(r_i)(r_i^2 - h^2) + c \sum_{i=l+1}^{m} \hat{K}(r_i),$$

$$A_2 = \sum_{i=l+1}^{m} (r_i^2 - \varrho^2)^2, \qquad A_3 = \sum_{i=l+1}^{m} \hat{K}(r_i)(r_i^2 - \varrho^2),$$

$$A_4 = \sum_{i=l+1}^{m} (r_i^2 - \varrho^2), \qquad A_5 = \sum_{i=1}^{l} (r_i^2 - \hat{h}^2)^2,$$

and $c = \varrho^2 - \hat{h}^2$ for $r_l \le \varrho$ and $r_{l+1} > \varrho$.

3. Choose as estimator of r_0 that R_j for which the number $S^{(j)}$ is smallest.

4. Estimate β by

$$\hat{\beta} = -\log \frac{\sigma^*}{\tau^*},$$

where σ^* and τ^* correspond to the optimum R-value.

An example of the application of this method can be found in Stoyan and Grabarnik (1991).

Marked Gibbs fields

There are similar statistical methods for marked Gibbs fields as in the case without marks (Harkness and Isham, 1983; Ogata and Tanemura, 1984, 1985; Takacs and Fiksel, 1986; Stoyan, 1989; Grabarnik and Särkkä, 1992; Särkkä, 1993).

Example: *Random systems of non-overlapping discs.* For the model discussed on p. 323 the chemical activity α (considered here as a constant) and the density function $\mathbf{m}_1(r)$ must be estimated. For this (16.18) may be the starting point. The formula (16.18) gives

$$\hat{\mathbf{m}}_1(r) = \frac{\hat{\lambda}\hat{\mathbf{m}}(r)e^{\hat{\alpha}}}{1 - \hat{H}_s(r)} \quad (r \geq 0),$$

where $\hat{\lambda}$ is the estimated intensity, $\hat{\mathbf{m}}(r)$ is an estimator of the radius mark density distribution and $\hat{H}_s(r)$ is an estimator of the spherical contact distribution function of the union set of discs. (An estimator for $\hat{H}_s(r)$ is given by the area fraction of that subset of the window W that is not covered by the discs obtained from the original discs by increasing the radii by r.) The quantity $\hat{\alpha}$ is chosen so that

$$\int_0^\infty \hat{\mathbf{m}}_1(r)\,\mathrm{d}r = 1.$$

An example is studied in Stoyan (1989a).

Appendices

APPENDIX A

Measure and Content

Measures and contents are set functions. That is, they is assign numerical values to sets. The area content is a well-known example:

$$X \to A(X).$$

A somewhat more complicated example depends on a mass distribution given by

$$\mu(X) = \int_X f(x, y) \, dx \, dy,$$

where $f(x, y)$ is a density distribution.

Such set functions may in general be defined on rings or σ-algebras. A *ring* is a collection $\mathcal{R}$ of subsets of a universe $\mathcal{X}$ with the following properties:

(R1) $\emptyset \in \mathcal{R}$ ($\emptyset$ is the empty set);

(R2) if $B, C \in \mathcal{R}$ then $B \setminus C \in \mathcal{R}$;

(R3) if $B, C \in \mathcal{R}$ then $B \cup C \in \mathcal{R}$.

For example the collection of all finite unions of half-open intervals in the real line is a ring.

A *σ-algebra* satisfies more restrictive conditions. (All σ-algebras are rings.) A σ-algebra is a collection $\mathcal{S}$ of subsets of a universe $\mathcal{X}$ where

(S1) $\mathcal{X} \in \mathcal{S}$;

(S2) if $B \in \mathcal{S}$ then $\mathcal{X} \setminus B = B^C \in \mathcal{S}$,

(S3) if $B_1, B_2, B_3, \ldots, \in \mathcal{S}$ then $\bigcup_{n=1}^{\infty} B_n \in \mathcal{S}$.

Then one can further show that

$$\emptyset \in \mathcal{S};$$

$$\text{if } B, C \in \mathcal{S} \text{ then } B \setminus C \in \mathcal{S};$$

$$\text{if } B_1, B_2, B_3, \ldots \in \mathcal{S} \text{ then } \bigcap_{n=1}^{\infty} B_n \in \mathcal{S}.$$

The *Borel σ-algebra* B^d in R^d is a particularly important case. It is the smallest σ-algebra that contains all open subsets of R^d. Roughly speaking, it contains all 'sufficiently well-behaved' sets and also all fractals. All subsets of R^d in this book are Borel. (In fact, it is difficult to give examples of non-Borel sets.)

A set function μ on a ring $\mathcal{R}$ is called a *content* if it has the following properties:

(I1) $\mu(\emptyset) = 0$;

(I2) $\mu(B) \geq 0$ for all $B \in \mathcal{R}$;

(I3) $\mu(B \cup C) = \mu(B) + \mu(C)$ for all $B, C \in \mathcal{R}$, where $B \cap C = \emptyset$.

A *measure* is a set function on a σ-algebra $\mathcal{S}$ with the properties

(M1) $\mu(\emptyset) = 0$;

(M2) $\mu(B) \geq 0$ for all $B \in \mathcal{S}$;

(M3) if $B_1, B_2, \ldots, \in \mathcal{S}$ and $B_l \cap B_k = \emptyset$ for $l \neq k$ then

$$\mu\left(\bigcup_{k=1}^{\infty} B_k\right) = \sum_{k=1}^{\infty} \mu(B_k).$$

The *Lebesgue measure* $\mathcal{L}^d$ is especially important, since it is the d-dimensional volume for all elementary geometrical objects. For countable point sets or for curves in R^d ($d \geq 2$) the Lebesgue measure is 0. In the text $\mathcal{L}^2$ is written as A.

A Borel measure means a measure on $\mathcal{B}^d$.

sup and inf, lim sup and lim inf

Each finite subset A of the integers (or of R^1) has a largest and smallest element max A and min A. With infinite sets this need not be the case.

Consider for example the set $A = \{x_n\}$, $x_n = 1/n$ ($n = 1, 2, \ldots$). It has 1 as the maximum but no minimum, since 0 cannot be written in this form. However, 0 is the so-called *infimum* of A. By the infimum of a subset A in R^1 one means the largest number y (written inf A) with

$$y \leq x \quad \text{for all } x \in A.$$

Analogously there is the *supremum* of A, the smallest z (written sup A) with

$$z \geq x \quad \text{for all } x \in A.$$

In mathematical notation one often sees

$$x = \lim_{n \to \infty} \{x_n\} \quad \text{or} \quad y = \lim_{x \downarrow 0} f(x).$$

In the first case x is the limit of the sequence $\{x_n\}$. The number x has the property that for all small positive reals ϵ there is some $n(\epsilon)$ such that

$$|x - x_n| < \epsilon \quad \text{for all } n \geq n(\epsilon).$$

In the second case one means that for each $\epsilon > 0$ there is a $\delta(\epsilon) > 0$ such that

$$|y - f(x)| < \epsilon \quad \text{for all } x \geq 0 \text{ with } x < \delta(\epsilon).$$

The concepts of lim sup and lim inf are less familiar. It can happen that a sequence $\{x_n\}$ has subsequences with different limits. For example consider

$$x_n = \begin{cases} 1 + 1/n & (n \text{ even}), \\ -1 - 1/n & (n \text{ odd}). \end{cases}$$

This $\{x_n\}$ has two subsequences with values 1 and -1 respectively. The supremum of all limits of convergent subsequences is denoted by lim sup x_n, and likewise the infimum by lim inf x_n.

Analogously there is the notation $\lim_{x \downarrow 0}$ sup $f(x)$. This is the supremum of the set H_f^0, where z is an element of H_f^0 if and only if for all $\epsilon, \delta > 0$ there is an $x \in (0, \delta)$ with $|f(x) - z| < \epsilon$.

Basic Ideas in Topology

At some points in this book ideas such as 'open set' or 'closed hull' are used. These ideas are sketched below — for a thorough treatment see the literature.

All sets considered here are in R^2. For a typical open set consider the open disc $b^o(o, r)$ about the origin with radius r. The circumference is *not* part of the disc.

In general a set O in R^2 is *open* if for all points x in O there is some $\epsilon > 0$ such that the open disc $b^o(x, \epsilon)$ lies completely in O. A rectangle without side edges satisfies this definition. Any line segment or disc including its circumference is not open.

The closed disc $b(o, r)$ about o with radius r is an example of a closed disc; the edge belongs to $b(o, r)$. In general one says that a subset A of R^2 is *closed* if all convergent sequences of elements from A converge to some element of A. Each line or line segment (including endpoints) is a closed set, as is a rectangle with side edges.

There are sets that are neither open nor closed, for example the closed disc $b(o, r)$ without the centre, i.e. $b(o, r) \backslash \{o\}$.

For every subset X in R^2 there is a smallest closed subset $\bar{X}$ containing X. This is the closure or *closed hull* of X. It contains all limits of convergent sequences of elements of X. For example the closed hull of $b^o(o, r)$ is $b(o, r)$.

The *boundary* of a closed set A is the subset ∂A of all those points x for which there is no $r > 0$ such that $b(x, r) \subset A$. The boundary $\partial b(o, r)$ of the disc $b(o, r)$ is the circle around o with radius r.

A set B is called *bounded* if for some finite r

$$B \subset b(o, r).$$

If B is closed and bounded, it is said to be *compact*.

APPENDIX D

Set Operations

In addition to those set theoretic operations that are undoubtedly familiar to the reader, geometric statistics involves a large family of other operations.

In the following A and B denote subsets of R^2 and $b(x, r)$ is the closed disc of radius r about the centre x.

Dilations (multiplication by a real scalar λ)

$$\lambda A = \{\lambda x : x \in A\}.$$

The set λA consists of all points of the form λx, where $\lambda x = (\lambda x', \lambda x'')$, $x = (x', x'')$. For example, if A is the square with corners $(1, 1)$, $(3, 1)$, $(1, 3)$ and $(3, 3)$ then $2A$ is the square with corners $(2, 2)$, $(6, 2)$, $(2, 6)$ and $(6, 6)$. In the special case $\lambda = -1$ one writes $-A$ as $\check{A}$. A is said to be *symmetric* when $A = \check{A}$.

Translation
$$A_x = A + x = \{y + x : y \in A\} \quad (x \in R^2).$$

A_x is the set A shifted by the vector x.

Minkowski addition

$$A \oplus B = \{x + y : x \in A, y \in B\}. \tag{D.1}$$

The set A can be enlarged, displaced or deformed by Minkowski addition by suitable choice of the structuring element B. The special case $B = b(o, r)$ is important. The set $A_r = A \oplus b(o, r)$ is known as the outer parallel set of A. Figure 134 shows A_r for a rectangle and also for a non-convex set.

Equivalently to (D.1), we have

$$A \oplus B = \bigcup_{y \in B} A_y = \bigcup_{x \in A} B_x.$$

The operation $A \rightarrow A \oplus \check{B}$ is often known as dilation. However, in this book the term is reserved for the operation described above.

Minkowski subtraction
$$A \ominus B = \bigcap_{y \in B} A_y. \tag{D.2}$$

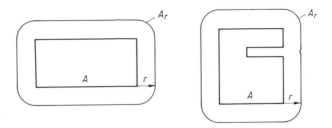

Figure 134   Outer parallel sets for two sets.

In general this operation is unrelated to Minkowski addition. It is not necessarily true that either $(A \ominus B) \oplus \check{B} = A$ or $(A \ominus \check{B}) \oplus B = A$. For the special case $B = b(o, r)$ the set $A^r = A \ominus b(o, r)$ is known as the *inner parallel set* of A. Figure 135 shows A^r for the same rectangle and non-convex set as above. A^r is the set of all those points x of A for which the open disc $b^o(x, r)$ is completely contained by A.

The formula (D.2) can also be written as

$$A \ominus B = \{z : (\check{B})_z \subset A\}.$$

The operation $A \to A \ominus \check{B}$ is often known as *erosion*. One says that A is eroded by B. In this book we are interested above all in the case $B = b(o, r)$, for which $B = \check{B}$.

Since Minkowski addition and subtraction are enlargement and reduction respectively, they are of little use for smoothing and simplification purposes. However, the following operations of *closing* and *opening* are useful tools.

Closing

$$A^B = (A \oplus B) \ominus \check{B}.$$

A is always contained in A^B, though the difference between the two is often not great. The set A^B is smoother and possesses less fine detail than A. Cracks and small notches are removed. Naturally, the choice of B plays a particular role, and the above remarks hold particularly when $B = b(o, r)$. In the special case when $A = b(o, R)$, A^B is equal to A for all radii R.

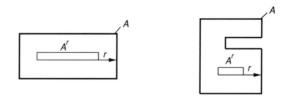

Figure 135   Inner parallel sets for two sets.

Opening

$$A_B = (A \ominus B) \oplus \check{B}.$$

A_B is contained in A, but as above there is usually little difference between the two. Again A_B is smoother and simpler than A. Thin appendages are removed and components disjoint from the main body made more distant.

In the special case $A = b(o, R)$ and $B = b(o, r)$, $A_B = A$ if $R \geq r$, and $A_B = \emptyset$ otherwise.

The Euclidean and Hausdorff Metrics

A metric gives distances between two points of a set. A simple example is for the plane R^2. Here the *Euclidean metric* may be used, which is the usual distance between the two points in the plane. That is,

$$\|x_1 - x_2\| = \sqrt{(x_1' - x_2')^2 + (x_1'' - x_2'')^2},$$

where $x_i = (x_i', x_i'')$ for $i = 1, 2$.

Another example is given by the set $\mathcal{K}$ of all compact subsets of R^2, each subset being viewed as a point of $\mathcal{K}$. There is a metric on $\mathcal{K}$, the so-called Hausdorff metric, given by

$$h(K, G) = \inf\{r > 0 : K \subset G \oplus b(o, r), G \subset K \oplus b(o, r)\}$$

for $K, G \in \mathcal{K}$. Equivalently,

$$h(K, G) = \max\{\sup_{x \in K} d(x, G), \sup_{y \in G} d(y, K)\},$$

where $d(x, K) = \inf_{z \in K} \|x - z\|$.

Corresponding to K, there is the smallest parallel set $K \oplus b(o, r_k)$ containing G; and likewise for G and r_G. The larger of r_K and r_G is the 'distance' $h(K, G)$ between K and G.

Example. Let K be the unit disc $b(o, 1)$ and G the unit square with sides parallel to the coordinate axes and the bottom left corner at the origin. The smallest parallel set for K (for G) that contains G (respectively K) is $K \oplus b(o, \sqrt{2} - 1)$ ($G \oplus b(o, 1)$), which one may easily check. Thus $h(K, G) = \max\{\sqrt{2} - 1, 1\} = 1$.

With the Hausdorff metric, $\mathcal{K}$ is a complete separable metric (i.e. Polish) space.

Boolean Models

The Boolean model[†] is an important model of random sets. It is usually defined as a closed set, but occasionally as an open set as well (p. 33).

It can be defined constructively. Let N be a homogeneous Poisson field of intensity λ and $\{X_n\}$ an independent sequence of random compact sets. The points of N are called *germs* or *germ points*. The sets X_n ($n = 1, 2, \ldots$) are identically distributed and mutually independent; one calls them *grains* or *primary grains*. Moreover, let X_0 be another random compact set with the same distribution as X_n, which one calls the 'prototype grain' or 'typical grain'. In the following it will be assumed that its mean area $\bar{A} = \mathsf{E}A(X_0)$ is finite.

The Boolean model is the union X of the grains X_n, translated by the $\mathbf{x}_n$ in N:

$$X = \bigcup_{\mathbf{x}_n \in N} (X_n + \mathbf{x}_n).$$

The set thus defined is, in the sense of the theory of random sets, homogeneous (or stationary). Clearly this property can be explained as on p. 191 (for a precise definition see the literature). If the X_n are isotropic (e.g. $X_n = b(o, r_n)$ with random r_n) then X is also isotropic.

Figure 136 shows a simulation of a Boolean model with circular grains.

The grains do not necessarily have to be convex (however, the convex case has been extensively discussed). An important special case of the Boolean model with non-convex grains is when X_0 consists of finitely many points. The corresponding Boolean model is the so-called Poisson cluster field; under special assumptions on the distribution, one obtains a Neyman–Scott field (cf. §16.2).

The above definition can be generalized in three ways.

(i)   The grains could be bounded open sets. Then X is a random open set.

[†]The concept of 'Boolean model' derives from Matheron's school in Fontainebleau. They originally considered those random fields $\{Z(x)\}$ whose values vary continuously. For example these may serve as models of geological deposits; so that $Z(x)$ is the thickness of the deposit at x. The Boolean model can be interpreted as a random field, where only the values $Z(x) = 1$ and $Z(x) = 0$ are possible — that is, whether or not x lies in the model. Many mathematicians associate quantities that take only the values 0 or 1 with Boole's name.

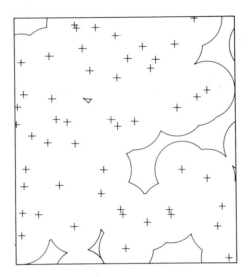

Figure 136   Realization of a Boolean model with disc-shaped grains.

(ii)   It is possible to remove the requirement of homogeneity on N; N is thus
an inhomogeneous Poisson field. The corresponding Boolean model is then
inhomogeneous.

(iii)   Instead of a Poisson field, a general point field could be used.

In the first two cases one calls the resulting set X Boolean as before: in (i) X
is open; in (ii) inhomogeneous. In case (iii) one speaks of the germ–grain model
(Appendix J).

A very important quantity for the homogeneous Boolean model is the area
fraction p_X:

$$p_X = \mathbf{E}A(X \cap Q), \qquad Q = [0, 1] \times [0, 1].$$

One can show that the same value is obtained if Q is replaced by a Borel set B
with $A(B) = 1$. The quantity p_X is also the probability that the origin o (or any
other fixed point) lies in X. It satisfies

$$p_X = 1 - e^{-\lambda \bar{A}},$$

where $\bar{A} = \mathbf{E}A(X_0)$.

In the inhomogeneous case $p_X(x)$ is the probability that the point x is in X.
This can be given as an integral:

$$p_X(x) = 1 - \exp\left[-\int_{R^2} p(x, z)\lambda(z)\,\mathrm{d}z\right] \qquad (x \in R^2).$$

Here $\lambda(z)$ is the intensity function and $p(x, z)$ the probability that the prototype grain X_0, when translated by z contains x:

$$p(x, z) = \Pr(x \in X_0 + z) = \Pr(x - z \in X_0).$$

Finally, for a homogeneous Boolean model one obtains a formula for the spherical contact distribution function $H_s(r)$. This is precisely the probability that the set X intersects with the circle $b(o, r)$ under the condition that $o \notin X$:

$$H_s(r) = 1 - \frac{1}{1 - p_x} \exp[-\lambda E A(X_0 \oplus b(o, r))] \quad (r \geq 0).$$

If X_0 is compact and convex then the mean value $E A(X_0 \oplus b(o, r))$ can be calculated easily with the help of the Steiner formula on p. 108.

The Convex Hull

A subset C of R^2 is said to be *convex* when for each two points x_1, x_2 in C the whole line segment from x_1 to x_2 lies in C, i.e.

$$\lambda x_1 + (1 - \lambda)x_2 \in C \quad \text{for } 0 \leq \lambda \leq 1.$$

The disc $b(o, r)$ is an example of a convex set. The boundary $\partial b(o, r)$ is non-convex.

For each non-convex subset A of R^2 there exists a smallest convex set conv A such that

$$A \subset \text{conv} A.$$

One calls this the *convex hull* of A. The convex hull of $\partial b(o, r)$ is $b(o, r)$. Figures 32 and 53 show the convex hull for another set.

If A is bounded and closed then conv A has these properties as well.

Random Lines and Line Fields

Parametrization of lines

In the following let G be the set of all infinite lines g in the (x, y)-plane. Every such line g may be written as

$$y = a + bx \qquad\qquad (\text{H.1})$$

for some a and b. g is uniquely defined by a and b. In analytical geometry other parametrizations are used. As shown in Fig. 137, g may be characterized by p and ϕ. That is, p is the perpendicular distance of g from the origin o, and ϕ is the angle that the perpendicular makes with the positive x-axis. (In other words, (p, ϕ) are the polar coordinates of the foot of the perpendicular to g from o.)

We have

$$0 \le p < \infty$$

and

$$0 \le \phi < 2\pi.$$

The relationship to the above line representation is given by the so-called Hesse normal form

$$x \cos \phi + y \sin \phi = p.$$

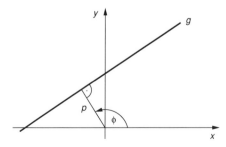

Figure 137   Representation of a line and the corresponding parameters p and ϕ.

Under the (p, ϕ)-parametrization, the set G of all lines in the (x, y)-plane is equivalent to the infinite strip

$$S = \{(p, \phi) : 0 \le p < \infty, 0 \le \phi < 2\pi\}.$$

Each point of the strip corresponds to a line, subsets of S correspond to sets of lines.

Example 1: Let $G_{b(o,r)}$ be the set of all lines that intersect the disc $b(o, r)$. Clearly the corresponding subset of S is the region

$$S_{b(o,r)} = \{(p, \phi) : 0 \le p \le r, \quad 0 \le \phi < 2\pi\}.$$

Example 2: Let G_Q be the set of all lines that cut the square $Q = [-\frac{1}{2}, \frac{1}{2}] \times [-\frac{1}{2}, \frac{1}{2}]$. The corresponding subset of S is

$$S_Q = \{(p, \phi) : p = \tfrac{1}{2}(\sin \phi + \cos \phi); \quad 0 < \phi < 2\pi\}$$

(Fig. 138).

Line measure

In many geometrical problems it is necessary to assign a measure to sets of lines. Usually one uses the Lebesgue measure A on the strip S (which is considered as a subset of R^2). Let γ be the resulting line measure on G. If G' is a (measurable) subset of G then

$$\gamma(G') = A(S'),$$

where S' is the subset of S corresponding to G'. For the examples above

$$\gamma(G_{b(o,r)}) = A(S_{b(o,r)}) = 2\pi r \tag{H.2}$$

and

$$\gamma(G_Q) = A(S_Q) = 4.$$

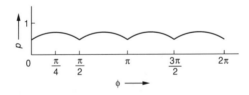

Figure 138   The region between the ϕ-axis and the bow curve is the set S_Q, which corresponds to the lines intersecting the square $Q = [-\frac{1}{2}, \frac{1}{2}] \times [-\frac{1}{2}, \frac{1}{2}]$.

In integral geometry (see Santaló, 1976; Gruber and Wills, 1993) it is shown that for each bounded convex set the set G_K of lines that intersect K satisfies

$$\gamma(G_K) = U(K), \tag{H.3}$$

where $U(K)$ is the circumference of K.

The line measure γ possesses an important invariance property. Let G' be a set of lines and MG' the set of lines that arise from applying a Euclidean transformation M (translation or rotation) to the elements of G', i.e.

$$MG' = \{Mg : g \in G'\}.$$

Then

$$\gamma(G') = \gamma(MG').$$

Random lines

A random line is one with a uniform distribution under γ. Let G' be a set of lines in the (x, y)-plane and S' the corresponding subset of the strip S. Then for every random line in G' there is a corresponding point in S' whose position has the uniform distribution. Thus corresponding probabilities can be calculated in the spirit of the theory of geometrical probability for random lines. For example, the probability that a random line cuts the disc $b(o, r)$, assuming that it cuts the larger disc $b(o, R)$, is

$$\frac{\gamma(G_{b(o,r)})}{\gamma(G_{b(o,R)})}.$$

By (H.3), this gives the value r/R.

Furthermore, let G_K^ℓ be the set of lines that intersect the set K producing a chord of length less than ℓ. Then the distribution function $L(\ell)$ of the random segment is

$$L(\ell) = \text{Pr(segment length} < \ell \mid K \text{ intersects the line)}$$

$$= \frac{\gamma(G_K^\ell)}{\gamma(G_K)}.$$

Random lines are easy to simulate. If one wishes to generate a random line that intersects the discs $b(o, r)$, one takes two independent random variables u and v uniformly distributed on $[0, 1]$ and derives the parameters by

$$p = ru, \quad \phi = 2\pi v.$$

A random line that intersects the square Q can be generated as follows.

```
1   u = RND(0)
2   IF u > 1/SQR(2) THEN 5
```

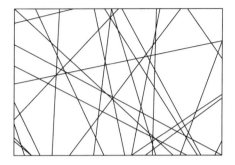

Figure 139   Section of a simulation of a Poisson line field.

```
3    v = RND(0)
4    p = u/SQR(2): φ = 2 * π * v: RETURN
5    v = RND(0): φ = 2 * π * v
6    T = SQR(2) * (SIN (φ) + COS (φ))
7    IF u > T THEN 1
8    p = u/SQR(2): RETURN
```

Line fields

Let there be a point field in R^2. Those points that lie in S form a sub-point field. Each of these points corresponds to a line in the (x, y) plane, and the set of these lines is called a line field.

For each line field there is a corresponding *fibre field* (Stoyan *et al.*, 1987). A fibre field is the random subset of the (x, y)-plane that is the union of all lines in the line field.

Poisson line field

The Poisson line field is a very important example, which is derived from a homogeneous Poisson point field of intensity λ in R^2. The points lying in S correspond to a random set of lines. This is called the Poisson line field with parameter $\varrho = 2\lambda$, and it has a number of important properties.

(a)   It is motion-invariant. That is, if all lines are transformed by the same Euclidean motion, one obtains a new line field with the same distribution as the origin.

(b)   The intersection points with a fixed line form a linear Poisson process with intensity $\varrho = 2\lambda$. The intersecting angles are independent from one another and have distribution function $\frac{1}{2} \sin \alpha$ $(0 \le \alpha \le \pi)$. The number of lines that intersect the plane set K has a Poisson distribution with parameter $\lambda A(S_K)$.

(c)   The mean total length L_A of the line segment of the field in a region of area 1 is $\pi\lambda$. In the theory of fibre fields the quantity L_A is known as the *line density* or *intensity*.

(d)   The probability that more than two lines intersect at some point is exactly 0.

(e)   Suppose one chooses a random line from the field (the meaning of the word 'random' can be defined using the theory of Palm distributions). One then translates the field so that the selected line lies on the x-axis. Then the new line field with x axis removed has the same distribution as the original one.

(f)   The lines of the Poisson line field define a mosaic in the plane, the *Poisson line mosaic*. Its typical cell is known as a *Poisson polygon*.

Simulation of the Poisson line field

Often one has the task of generating a random set of segments that form the lines of a Poisson line field in a set K (e.g. a square). This can be achieved as follows.

1.   Find the set S_K in S and its Lebesgue measure. (If K is convex this is the measure of the boundary of K.)

2.   Generate a Poisson-distributed random number n with parameter $\lambda A(S_K)$.

3.   Generate n independent random lines using the method sketched on p. 357.

The Dirichlet Mosaic and the Delaunay Triangulation

Let $N = \{x_n\}$ be a sequence of points in R^2. One may assign an open set $Z^0(x_i)$ to each point x_i, consisting of all those points in R^2 nearer to x_i than any other point of N. For each x_i, $Z^0(x_i)$ is convex. The $Z^0(x_i)$ are arranged as in a mosaic in R^2. One speaks of the *Dirichlet mosaic relative to N* (often it is also denoted as the *Voronoi* or *Thiessen mosaic*).

The closed hulls $Z(x_i)$ of the $Z^0(x_i)$ are known as the *cells* or *polygons* of the mosaic. They are derived from the $Z^0(x_i)$ by the addition of edges and corners. The x_i are called *generating points*.

Figure 140 shows two such mosaics where (a) one is generated by a finite sequence (so that some cells are unbounded) and (b) the other is on regions from a mosaic generated by an infinite sequence.

Apart from special latticed arrangements, each corner point of the Dirichlet mosaic is the corner point of exactly three cells.

Each Dirichlet mosaic with this property can be assigned a Delaunay triangulation. This consists of those triangles that have corner points in N and whose sides join neighbouring cells. Figure 140 shows some triangles of the triangulation for one of the mosaics.

Many programs exist for generating the Dirichlet mosaic from a given point sequence N.

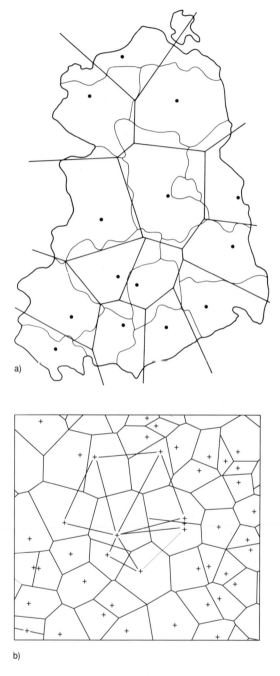

Figure 140   (a) A Dirichlet mosaic, whose points are the centres of the district towns of the former GDR. (b) Section of a Dirichlet mosaic of a simulated unbounded point field. In the centre there are some Delaunay triangles shown.

Germ–Grain Models

Germ-grain models are models for random sets; they are much more general than Boolean models (Appendix F). As before, one obtains the required random set by taking the union of compact sets X_n, which are shifted by the germs $\mathbf{x}_n$:

$$X = \bigcup_{n=1}^{\infty} (X_n + \mathbf{x}_n).$$

The $\mathbf{x}_n$ need not form a Poisson field, nor do the X_n need to be independent from one another or from the $\{\mathbf{x}_n\}$. Under certain assumptions on the distribution, it is possible for the structure to consist of disjoint grains (or particles), as in Figs. 86 and 96.

A homogeneous germ–grain model is one where the pairs $(\mathbf{x}_n, X_n)$ form a homogeneous marked point field; the mark space in this case is $\mathcal{K}$, the set of all compact subsets of R^2. The corresponding mark distribution describes the distribution of the 'typical' grains. It is denoted in this book by P_X.

Some formulae for germ–grain models may be found in Stoyan *et al.* (1987) and Hall (1988).

The Area of Intersection of Two Discs

In various calculations in stochastic geometry one requires the area of intersection of two discs of radius r and R respectively whose centres are separated by a distance t (Fig. 141).

For $r - R < t < r + R$

$$
A(t, r, R) = r^2 \left\{ \arccos\left(\frac{t^2 + r^2 - R^2}{2tr}\right) \right.
$$

$$
\left. -\frac{t^2 + r^2 - R^2}{4t^2 r^2}[4t^2 r^2 - (t^2 + r^2 - R^2)^2]^{1/2} \right\}
$$

$$
+R^2 \left\{ \arccos\left(\frac{t^2 + R^2 - r^2}{2tR}\right) \right.
$$

$$
\left. -\frac{t^2 + R^2 - r^2}{4t^2 R^2}[4t^2 R^2 - (t^2 + R^2 - r^2)^2]^{1/2} \right\}. \tag{K.1}
$$

For $r = R$ this simplifies to

$$
A(t, r) = 2r^2 \arccos\left(\frac{t}{2r}\right) - \tfrac{1}{2}t(4r^2 - t^2)^{1/2}. \tag{K.2}
$$

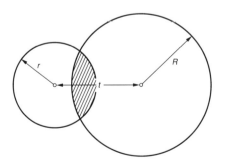

Figure 141   Two mutually intersecting discs, with the intersection shown shaded. This area is denoted by $A(t, r, R)$.

Kernel Estimators for Density Functions

The estimation of (probability) density functions is not simple. The definition of histograms or frequency distributions is well known: the sample values $x_1, \ldots, x_n$ are sorted into classes and the frequencies displayed in a bar chart. Histograms show in a rough way the form of the density functions. In particular, when one is using neither a model nor parameters, the histogram is often inadequate because one requires an estimate of the density function that is first sufficiently smooth and secondly describes the true distributions. Here kernel estimators offer a good way out.

Let $f(x)$ be an unknown density function, which is to be estimated. Furthermore, let $\mathbf{k}(x)$ be another density function, the kernel function. It is usually taken to be symmetric:

$$\mathbf{k}(x) = \mathbf{k}(-x).$$

Then

$$\hat{f}(x) = \frac{1}{n} \sum_{i=1}^{n} \mathbf{k}(x - x_i)$$

is an estimator for $f(x)$. There are many possible kernel functions to choose from. An often used example is the so-called *Epanečnikov kernel*

$$e_h(x) = \begin{cases} \dfrac{3}{4h} \left(1 - \dfrac{x^2}{h^2} \right) & (-h \leq x \leq h), \\ 0 & \text{otherwise.} \end{cases}$$

Note that the choice of this kernel is based on certain optimization considerations. Experimental measurement shows that the particular choice of the form is not as decisive as the bandwidth h. For large h one obtains smooth density functions, which smooth the details of the distribution; for small h the estimated function is rough and may obscure the fundamental structure of the distribution. The correct choice of h is a difficult problem, frequently discussed in the literature. In a large study a preliminary investigation with various bandwidths is recommended.

Consider also the following problem. It is often the case that

$$f(x) = 0 \quad (x < a),$$

where f is the density function to be estimated. In particular, it may be that all data are positive, so that $a = 0$.

Under suitable circumstances a symmetric kernel function $\mathbf{k}(x)$ yields an estimator $\hat{f}$ such that near a

$$\hat{f}(x) > 0,$$

even for $x < a$. A viable means of ensuring a density function that vanishes for $x < a$ is to use the so-called *reflection method*:

$$f(x) = \begin{cases} f(x) + f(2a - x) & (x \geq a), \\ 0 & \text{otherwise.} \end{cases}$$

Kernel estimators are also used in point field statistics. Similar principles to the above hold in this case as well, although the method of choosing optimal bandwidth that was developed for density function estimation is random samples does not apply.

References

Affentranger, F. (1987): The expected volume of the convex hull of n random points on the unit sphere. *Trab. Mat. Inst. Argent. Mat.* **123**, 1-17.

Affentranger, F. (1988): The expected volume of a random polytope in a ball. *J. Micros.* **151**, 277-288.

Affentranger, F. and R.A. Dwyer (1993) The convex hull of the convex hull of random balls. *Adv. Appl. Prob.* **25**, 373-394.

Amit, Y., and P. McCullagh (1994): Extremal convex sets and the three corner theorem. Technical Report 372, Department of Statistics, University of Chicago.

Anderson, D.A. (1981): Maximum likelihood estimation in the non-central chi distribution with unknown scale parameter. *Sankhya* **43**, 58-67.

Araujo, A., and E. Gine (1980): *The Central Limit Theorem for Real and Banach Valued Random Variables.* J. Wiley & Sons, New York, Chichester, Brisbane, Toronto.

Artstein, Z., and R.A. Vitale (1975): A strong law of large numbers for random compact sets. *Ann. Prob.* **5**, 879-882.

Aumann, R.J. (1965): Integrals of set-valued functions. *J. Math. Anal. Appl.* **12**, 1-12.

Avnir, D. (ed.) (1989): *The Fractal Approach to Heterogeneous Chemistry.* J. Wiley & Sons, New York.

Baddeley, A., and J. Møller (1989): Nearest neighbour Markov point processes and random sets. *Int. Statist. Rev.* **57**, 89-21.

Badii, R., and A. Politi (1985): Statistical description of chaoticaltractors: the dimension function. *J. Statist. Phys.* **40**, 725-750.

Bagnold, R.A. (1941): *The Physics of Blown Sands and Desert Dunes.* Methuen, London. Reprinted 1973 by Chapman & Hall, London, New York.

Bak, P. (1986): The devil's staircase. *Phys. Today* **39**, 38-45.

Bandemer, H., M. Albrecht and A. Kraut (1985): On using Fourier series in characterizing particle shapes. *Particle Characterization* **2**, 98-103.

Bandemer, H., A. Kraut and W. Näther (1989): On basic notions of fuzzy set theory and some ideas for their application in image processing. In: Hübler, A., W. Nagel, B.D. Ripley and G. Werner, eds., *Geobild '89.* Akademie-Verlag, Berlin, 153-164.

Bandemer, H., and W. Näther (1992): *Fuzzy Data Analysis.* Kluwer Academic Publ., Dordrecht, Boston, London.

Banerji, K. (1988): Quantitative fractography: a modern perspective. *Metall. Trans.* **19A**, 961-971.

Barendregt, L.G., and M.J. Rottschäfer (1991): A statistical analysis of spatial point patterns. *Statistica Neerlandica* **45**, 345-363.

Barndorff-Nielsen, O.E. (1978): Hyperbolic distributions and distributions on hyperbolae. *Scand. J. Statist.* **5**, 151-157.

Barndorff-Nielsen, O.E. (1986): Sand, wind and statistics: Some recent investigations. *Acta Mech.* **64**, 1-18.

Barndorff-Nielsen, O.E., and C. Christiansen (1988): Erosion, deposition and size distribution of sand. *Proc. R. Soc. Lond.* **A417**, 335-352.

Barndorff-Nielsen, O.E., P. Blaesild, J.L. Jensen and M. Sorensen (1983): The Fascination of Sand. In: Atkinson, A.C., and S.E. Fienberg, eds., *A Celebration of Statistics*. Springer-Verlag, Berlin, Heidelberg, New York, 57–87.

Barnsley, M. (1988): *Fractals Everywhere*. Academic Press, Boston.

Barrett, P.J. (1980): The shape of rock particles, a critical review. *Sedimentology* **27**, 291–303.

Baudin, J. (1981): Likelihood and nearest-neighbour distance properties of multidimensional Poisson cluster processes. *J. Appl. Prob.* **18**, 879–888.

Baudin, M. (1983): Note on the determination of cluster centers from a realization of a multidimensional Poisson cluster process. *J. Appl. Prob.* **20**, 136–143.

Beddow, J.K. (1980): *Particle Science and Technology*. Chemical Publishers, New York.

Beddow, J.K. (1984): *Particle Characterization in Technology*. CRC Press, Boca Raton.

Beddow, J.K., and T. Meloy (1980): *Testing and Characterization of Powder and Fine Particles*. Heyden & Sons, London.

Beddow, J.K., and G.C. Philip (1975): On the use of a Fourier analysis technique for describing the shape of individual particles. *Planseeber. Pulvermetall* **23**, 3–14.

Beddow, J.K., G.C. Philip and A.F. Vetter (1977): On relating some particle profile characteristics to the profile Fourier coefficients. *Powder Technol.* **18**, 19–25.

Beran, J. (1986): *Estimation, Testing and Prediction for Self-Similar and Related Processes*. Dissertation, ETH Zürich.

Beran, J. (1991): *M*-estimators of location for data with slowly decaying serial correlations. *J. Am. Statist. Assoc.* **86**, 704–708.

Beran, J. (1992): Statistical methods for data with long-range dependence. (with discussion) *Statist. Sci.* **7**, 404–421.

Berliner, L.M. (1992): Statistics, probability and chaos. *Statist. Sci.* **7**, 69–122.

Berman, M., and P.J. Diggle (1989). Estimating weighted integrals of the second-order intensity of a spatial point process. *J. R. Statist. Soc.* **B51**, 81–92.

Besag, J. (1978): Some methods of statistical analysis for spatial data. *Bull. Int. Statist. Inst.* **47**, 77–92.

Besag, J., and P.J. Diggle (1977): Simple Monte Carlo tests for spatial pattern. *Appl. Statist.* **26**, 327–333.

Block, A., W.v. Bloh and H.J. Schellnhuber (1990): Efficient box-counting of generalized fractal dimensions. *Phys. Rev.* **A42**, 1869–1874.

Bookstein, F.L. (1978): *The Measurement of Biological Shape and Shape Change*. Lecture Notes in Biomathematics 24, Springer-Verlag, Berlin, Heidelberg, New York.

Bookstein, F.L. (1986): Size and shape spaces for landmark data in two dimensions. *Statist. Sci.* **1**, 181–238.

Bookstein, F.L. (1991): *Morphometric Tools for Landmark Data: Geometry and Biology*. Cambridge University Press, Cambridge, New York, Port Chester, Melbourne, Sydney.

Bookstein, F.L., and P.D. Sampson (1990): Statistical models for geometric components of shape change. *Commun. Statist. Theor. Math.* **19**, 1939–1972.

Bookstein, F.L., R.E. Strauss, J.M. Humphries, B. Chernoff, R.L. Elder and G.R. Smith (1982): A comment upon the uses of Fourier methods in systematics. *Syst. Zool.* **31**, 85–92.

Brillinger, D.R. (1978): Comparative aspects of the study of ordinary time series and point processes. In: Krishnaiah, P.R., ed., *Developments of Statistics* I. Academic Press, New York, 33–133.

Brumberger, H., and J. Goodisman (1983): Voronoi cells: an interesting and potentially useful cell model for interpreting the small-angle scattering of catalysts. *J. Appl. Crystallogr.* **16**, 83–88.

Buchta, Ch. (1984a): Stochastische Approximation konvexer Polygone. *Z. Wahrscheinlichkeitsth. verw. Geb.* **67**, 283–304.

Buchta, Ch. (1984b): Zufallspolygone in konvexen Vielecken. *J. reine angew. Math.* **347**, 212-220.

Buchta, Ch. (1984c): Das Volumen von Zufallspolyedern im Ellipsoid. *Anzeiger Österr. Akad. Wiss., Math.-Naturwiss. Kl.* **121**, 1-4.

Buchta, Ch. (1985): Zufällige Polyeder - eine Übersicht. In: Hlawka, E., ed., *Zahlentheoretische Analysis*. Lecture Notes in Mathematics **1114**, Springer-Verlag, Berlin, Heidelberg, New York, 1-13.

Buchta, Ch. (1987): On nonnegative solutions of random systems of linear inequalities. *Discrete Comput. Geom.* **2**, 85-95.

Buchta, Ch., and J. Müller (1984): Random polytopes in a ball. *J. Appl. Prob.* **21**, 753-762.

Burago, Y. and V. Zalgaller (1988): *Geometric Inequalities*. Springer-Verlag, Berlin, Heidelberg, New York.

Cabo, A.J., and P. Groeneboom (1994): Limit theorems for functionals of convex hulls. *Prob. Theory Rel. Fields*

Carroll, R.J. and F. Lombard (1985): A note on *n* estimators for the binomial distribution. *J. Am. Statist. Assoc.* **80**, 423-426.

Carter, P.H., R. Cawley and R.D. Mauldin (1988): Mathematics of dimension measurement for graphs of functions. In: Seitz, D., L. Sander and B.B. Mandelbrot, eds., *Proc. Materials Research Society Conf. on Fractal Aspects of Materials and Disordered Systems*, 83-186.

Chakerian, G.D., and H. Groemer (1983): Convex bodies of constant width. In: Gruber, P., and J.M. Wills, eds., *Convexity and Its Application*. Birkhäuser, Basel, Boston, Stuttgart, 49-96.

Chatfield, C. (1980): *The Analysis of Time Series: An Introduction*. Chapman & Hall, London, New York.

Chatterjee, S., and M.R. Yilmaz (1992): Chaos, fractals and statistics. *Statist. Sci.* **7**, 49-121.

Clark, M.W. (1981): Quantitative shape analysis: a review. *Math. Geol.* **13**, 304-320.

Cliff, A.D., and J.K. Ord (1981): *Spatial Processes. Models and Applications*. Pion, London.

Coleman, R. (1989): Random sections of a sphere. *Can. J. Statist.* **17**, 27-39.

Cowan, R. (1984): A model for random packing of disks in the neighbourhood of one disk. *SIAM J. Appl. Math.* **44**, 839-853.

Cowan, R. (1987): A bivariate exponential distribution arising in random geometry. *Ann. Inst. Statist. Math.* **39A**, 103-111.

Cowan, R. (1989): Objects arranged randomly in space: an accessible theory. *Adv. Appl. Prob.* **21**, 543-569.

Cox, D., and V. Isham (1980): *Point Processes*. Chapman & Hall, London, New York.

Cox, D., and P.A.W. Lewis (1966): *The Statistical Analysis of Series of Events*. Methuen/J. Wiley & Sons, London, New York.

Crain, I.K., and R.E. Miles (1976): Monte Carlo estimates of the distribution of the random polygons determined by random lines in a plane. *J. Statist. Comput. Simul.* **4**, 293-325.

Cressie, N. (1984): Modelling sets. In: Salinetti, G., ed., *Multifunctions and Integrands*. Lecture Notes in Mathematics **1091**, Springer-Verlag, Berlin, Heidelberg, New York, 138-149.

Cressie, N. (1989): Modeling Growth with Random Sets. *Preprint* 88-24, *Statistical Laboratory, Iowa State University*.

Cressie, N. (1991): *Statistics for Spatial Data*. J. Wiley & Sons, New York.

Cressie, N., and F.L. Hulting (1992): A spatial statistical analysis of tumor growth. *J. Amer. Statist. Assoc.* **87**, 272-283.

Cressie, N., and M. Laslett (1987): Random set theory and problems of modelling. *SIAM Rev.* **29**, 557-574.

Creutzburg, R., A. Mathias & E. Ivanov (1992): Fast algorithm for computing the fractal dimension of binary images. *Physica* **185A**, 56-60.

Crow, E.L., and R.S. Gardner (1959): Confidence intervals for the expectation of a Poisson variable. *Biometrika* **46**, 441–453.

Curl, R.L. (1986): Fractal dimension and the geometries of caves. *Math. Geology* **18**, 765–783.

Cutler, C.D. (1986): The Hausdorff dimension distribution of finite measures in Euclidean space. *Can. J. Math.* **38**, 1459–1484.

Cutler, C.D. (1990): Connecting ergodicity and dimension in dynamical systems. *Ergodic Theory Dyn. Syst.* **10**, 451–462.

Cutler, C.D. (1991): Some results on the behaviour and estimation of the fractal dimensions of distributions on attractors. *J. Statist. Phys.* **62**, 651–708.

Cutler, C.D., and D.A. Dawson (1989): Estimation of dimension for spatially distributed data and related limit theorems. *J. Multivariate Anal.* **28**, 115–148.

Cutler, C.D., and D.A. Dawson (1990): Nearest neighbour analysis of a family of fractal distributions. *Ann. Prob.* **18**, 256–271.

Daley, D.J., and D. Vere-Jones (1988): *An Introduction to the Theory of Point Processes.* Springer-Verlag, New York, Berlin, Heidelberg, London, Paris, Tokyo.

Dauskardt, R.H., F. Habensak and A. Ritchie (1990): On the interpretation of the fractal character of fracture surface. *Acta Metall.* **38**, 143–159.

Davies, K.W., and A.E. Hawkins (1979): Harmonic description of particle shape: their role in discrimination. In: *Proc. PARTECH Symp., Nürnberg*, 582–590.

Davison, M.L. (1988): *Multidimensional Scaling.* J. Wiley & Sons, New York.

Davy, P.J., and F.J. Guild (1988): The distribution of interparticle distance and its application in finite-element modelling of composite materials. *Proc. R. Soc. Lond.* **A418**, 95–112.

Devaney, R.L. (1990): *Chaos, Fractals and Dynamics. Computer Experiments in Mathematics.* Addison-Wesley, Menlo Park, California.

Devroye, L. (1986): *Non-uniform Random Variate Generation.* Springer-Verlag, Berlin, Heidelberg, New York.

Diggle, P.J. (1983): *Stochastical Analysis of Point Processes.* Academic Press, London.

Diggle, P. (1985): A kernel method for smoothing point process data. *Appl. Statist.* **34**, 138–147.

Diggle, P.J., and R.J. Gratton (1984): Monte Carlo methods of inference for implicit stochastic models (with discussion). *J. R. Statist. Soc.* **B46**, 193–227.

Diggle, P.J., D.J. Gates, and A. Stibbard (1987): A non-parametric estimator for pairwise-interaction point processes. *Biometrica* **74**, 763–770.

Diggle, P.J., T. Fiksel, P. Grabarnik, Y. Ogata, D. Stoyan and M. Tanemura (1994): On parameter estimation for pairwise interaction point processes. *Int. Statist. Rev.* **62**.

Dillon, W.R., and M. Goldstein (1984): *Multivariate Analysis. Methods and Applications.* J. Wiley & Sons, New York.

Doguwa, S.I. (1989): A comparative study of the edge-corrected kernel-based nearest neighbour density estimators for point processes. *J. Statist. Comput. Simul.* **33**, 83–100.

Doguwa, S.I. (1990): On edge-corrected kernel-based pair-correlation function estimators for point processes. *Biom. J.* **32**, 95–106.

Doguwa, S.I., and G.J.G. Upton (1989): Edge-corrected estimators for the reduced second moment measure of point processes. *Biom. J.* **31**, 563–576.

Doguwa, S.I. and G.J.G. Upton (1990): On the estimation of the nearest neighbour distribution, $G(t)$; for point processes. *Biom. J.* **32**, 863–876.

Dowdeswell, J.A. (1982): Scanning electron micrographs of quartz sand grains from cold environments examined using Fourier shape analysis. *J. Sed. Petrology* **52**, 1315–1323.

Dryden, L.L., and K.V. Mardia (1990): General shape distributions in a plane. *Adv. Appl. Prob.* **23**, 259–276.

Dryden, I.L., and K.V. Mardia (1992): Size and shape analysis of landmark data. *Biometrika* **79**, 57–68.

Dubois, D., and H. Prade (1980): *Fuzzy Sets and Systems.* Academic Press, New York, London.

Durrett, R. (1988a): *Lecture Notes on Particle Systems and Percolation.* Wadsworth, Pacific Grove, California.

Durrett, R. (1988b): Crabgrass, measles, and gypsy moths: An introduction to modern probability. *Bull. Am. Math. Soc.* **18**, 117–143.

Edelsbrunner, E., D.G. Kirkpatrick and R. Seidel (1983): On the shape of a set of points in the plane. *IEEE Trans. Inf. Theory* **29**, 551–559.

Efron, B., and R.J. Tibshirani (1993): *An Introduction to the Bootstrap.* Chapman & Hall, London, New York.

Efron, S.N. (1965): The convex hull of a Gaussian sample. *Biometrika* **17**, 686–695.

Ehrlich, R., and B. Weinberger (1970): An exact method for characterization of grain shape. *J. Sed. Petrology* **40**, 205–212.

Elliott, D.F., and K.R. Rao (1982): *Fast Transforms.* Academic Press, London, New York.

Engelhardt, W.v. (1979): *Die Bildung von Sedimenten und Sedimentgesteinen. Sediment-Petrologie,* Teil III. E. Schweizerbarthsche Verlags-Buchhandlung, Stuttgart.

Evans, M., N. Hastings and B. Peacock (1993): *Statistical Distributions (Second edition).* J. Wiley & Sons, New York, Chichester, Brisbane, Toronto, Singapore.

Exner, H.E. (1987): Shape a key problem in quantifying microstructures. *Acta Stereol.* **6/III:** 1023–1028.

Exner, H.E. (1988): Quantitative characterization of microstructural geometry of interfaces. In: Pask, E., ed., *Ceramic Microstructures '86.* Plenum, New York, 73–86.

Fairfield-Smith, H. (1983): An empirical law describing heterogeneity in the yields of agricultural crops. *J. Agric. Sci.* **28**, 1–23.

Falconer, K.J. (1985): *The Geometry of Fractal Sets.* Cambridge University Press, Cambridge.

Falconer, K.J. (1986): Random fractals. *Math. Proc. Camb. Phil. Soc.* **100**, 559–582.

Falconer, K.J. (1988): The Hausdorff dimension of self-affine fractals. *Math. Proc. Camb. Phil. Soc.* **103**, 339–350.

Falconer, K.J. (1990): *Fractal Geometry. Mathematical Foundations and Applications.* J. Wiley & Sons, New York.

Farmer, J.D., E. Ott and J.A. Yorke (1983): The dimension of chaotic attractors. *Physica* **7D**, 153–180.

Federer, H. (1969): *Geometric Measure Theory.* Springer-Verlag, Berlin, Heidelberg, New York.

Fiksel, T. (1988a): Edge-corrected density estimators for point processes. *Statistics* **19**, 67–76.

Fiksel, T. (1988b): Estimation of interaction partials of Gibbsian point processes. *Statistics* **19**, 77–86.

Firey, W.J. (1974): Shapes of worn stones. *Mathematika* **21**, 1–11.

Flook, A.G. (1978): The use of dilation logic on the Quantimet to achieve fractal dimension characterization of textured and structured profiles. *Powder Technology* **21**, 295–298.

Flook, A.G. (1982a): Fourier analysis of particle shape. In: Stanley-Wood, N.G. and T. Allen, eds., *Particle Size Analysis 1981.* Wiley/Heyden & Sons, London, 255–262.

Flook, A.G. (1982b): Fractal dimensions: their evaluations and their significance in stereological measurements. *Acta Stereol.* **1**, 79–87.

Flook, A.G. (1987): The quantitative measurement of particle shape. *Acta Stereol.* **6/III:** 1009–1021.

Folk, R.L. (1964): *A Review of Grain-Size Parameters.* Department of Geology, University of Texas.

Fox, R., and M.S. Taqqu (1986): Large-sample properties of parameter estimates for strongly dependent stationary Gaussian time series. *Ann. Statist.* **14**, 517–532.

Fréchet, M. (1948): Les éléments aléatoires de nature quelconque dans un espace distancié. *Ann. Inst. H. Poincaré* **10**, 215–310.

Freund, J.E., and R.E. Walpote (1987): *Mathematical Statistics. (4th Edition)* Prentice Hall, Englewood Cliffs, New Jersey.

Friedman, G.M. (1961): Distinction between dune, beach and river sands from their textural characteristics. *J. Sed. Petrology* **31**, 514–529.

Friedman, G.M. (1967): Dynamic processes and statistical parameters compared for size frequency distributions of beach and river sands. *J. Sed. Petrology* **37**, 327–354.

Frontier, S. (1987): Applications of fractal theory to ecology. In: Legendre, P., and L. Legendre, eds., *Developments in Numerical Ecology.* Springer-Verlag, Berlin, Heidelberg, New York, London, Paris, Tokyo, 335–380.

Füchtbauer, H. and G. Müller (1970): *Sedimente und Sedimentgesteine. Sediment-Petrologie, Teil II.* E. Schweizerbarthsche Verlag-Buchhandlung, Stuttgart.

Full, W.E., and R. Ehrlich (1986): Fundamental problems associated with "eigenshape analysis" and similar "factor" analysis procedures. *Math. Geology* **18**, 451–463.

Gardner, M. (1965): Mathematical games. The "superellipse": a curve that lies between the ellipse and the rectangle. *Sci. Am.* **231**, 222–232.

Gates, J. (1982a): The number of intersections of random chords to a circle-third and fourth moments. *J. Appl. Prob.* **19**, 355–372.

Gates, J. (1982b): Recognition of triangles and quadrilaterals by chord length distribution. *J. Appl. Prob.* **19**, 873–879.

Gates, J. (1987): Some properties of chord length distributions. *J. Appl. Prob.* **24**, 863–874.

Genske, D., H. Herda and Y. Ohnishi (1992): Fractures and fractals. In: Hudson, J.A. (ed.) *Rock Characterization.* (ISRM Symp., EUROROCK'92, Chester). Telford, London, 19–24.

George, E.I. (1987): Sampling random polygons. *J. Appl. Prob.* **24**, 557–573.

Getis, A., and J. Franklin (1987): Second-order neighbourhood analysis of mapped point patterns. *Ecology* **68**, 473–477.

Gille, W. (1988): The chord length distribution of parallelepipeds with their limiting cases. *Exp. Techn. Phys.* **36**, 197–208.

Girling, A.J. (1982): Approximate variances associated with random configurations of hard spheres. *J. Appl. Prob.* **19**, 588–596.

Goodall, C.R. (1991): Procrustes methods in the statistical analysis of shape. *J. R. Statist. Soc.*, **B53**, 285–339.

Goodall, C.R., and A. Bose (1987): Procrustes techniques for the analysis of shape and shape change. In: *Proc. Symp. on the Interface between Computer Science and Statistics, Philadelphia*, 89–92.

Goodall, C.R., and K.V. Mardia (1991): A geometrical derivation of the shape density. *Adv. Appl. Prob.* **23**, 496–514.

Goodall, C.R., and K.V. Mardia (1993): Multivariate aspects of shape theory. *Ann. Statist.* **21**, 848–866.

Goodchild, M.F. (1988): Lakes on fractal surfaces: A null-hypothesis for lake-rich landscapes. *Math. Geol.* **20**, 615–630.

Goodchild, M.F., and D.M. Mark (1987): The fractal nature of geographic phenomena. *Ann. Assoc. Amer. Geogr.* **77**, 265–278.

Gower, J.C. (1975): Generalized procrustes analysis. *Psychometrika* **40**, 33–51.

Grabarnik, P., and A. Särkkä (1992): *On parameter estimation of marked Gibbs point processes.* Preprints from Department of Statistics, University of Jyväskylä, 4.

Graf, H.-P. (1983): *Long-range Correlations and Estimation of the Self-similarity Parameter.* Dissertation, ETH Zürich.

Graf, S. (1987): Statistically self-similar fractals. *Prob. Theory Rel. Fields* **74**, 357–392.

Graf, S., R.D. Mauldin, and S.C. Williams (1988): The exact Hausdorff dimension in random recursive contractions. *Mem. Am. Math. Soc.* **71**, No. 381.

Grandell, J. (1976): *Doubly Stochastic Poisson Processes*. Lecture Notes in Mathematics **529**, Springer-Verlag, Berlin, Heidelberg, New York.

Grassberger, P. (1983): Generalized dimensions of strange attractors. *Phys. Lett.* **97A**, 227–230.

Grassberger, P., and I. Procaccia (1983): Measuring the strangeness of strange attractors. *Physica* **9D**, 189–208.

Grenander, U. (1976): *Pattern Synthesis. Lectures in Pattern Theory,* Vol. 1. Springer-Verlag, New York, Heidelberg, Berlin.

Grenander, U. (1978): *Pattern Analysis. Lectures in Pattern Theory,* Vol. 2. Springer-Verlag, New York, Heidelberg, Berlin.

Grenander, U. (1981): *Regular Structures. Lectures in Pattern Theory,* Vol. 3. Springer-Verlag, New York, Heidelberg, Berlin.

Grenander, U. (1989): Advances in pattern theory. *Ann. Statist.* **17**, 1–30.

Grenander, U., and D.M. Keenan (1987): On the shape of plane images. *Rep. Pattern Anal.* No. 145, *Brown University, Providence, Rhode Island.*

Grenander, U., and D.M. Keenan (1989): A computer experiment in pattern theory. *Commun. Statist. Stoch. Models* **5**, 531–553.

Grenander, U., Y. Chow and D.M. Keenan (1991): *Hands. A Pattern Theoretic Study of Biological Shapes.* Springer-Verlag, New York, Berlin, Heidelberg, London, Paris, Tokyo, Hong Kong, Barcelona.

Groeneboom, P. (1988): Limit theorems for convex hulls. *Prob. Theor. Rel. Fields* **79**, 327–368.

Gruber, P.M., and J.M. Wills (1993): *Handbook of Convex Geometry*. Elsevier Science Publ. ??.

Guckenheimer, J. (1984): Dimension estimates for attractors. *Contemp. Math.* **28**, 357–367.

Günel, E., and D. Chilko (1989): Estimation of parameter n of the binomial distribution. *Commun Statist. Simulation Comp.* **18**, 537–555.

Hadwiger, H. (1958): *Vorlesungen über Inhalt, Oberfläche und Isoperimetrie.* Springer-Verlag, Berlin, Göttingen, Heidelberg.

Hall, P. (1988): *Introduction to the Theory of Coverage Processes.* J. Wiley & Sons, New York, Chichester, Brisbane, Toronto, Singapore.

Hanisch, K.-H. (1983): Reduction of n-th moment measures and the special case of the third moment measure of stationary and isotropic planar point processes. *Statistics* **14**, 421–436.

Hanisch, K.-H. (1984): Some remarks on estimators of the distribution function of nearest neighbour distance in stationary spatial point patterns. *Statistics* **15**, 409–412.

Hanisch, K.-H., and D. Stoyan (1979): Formulas for the second-order analysis of marked point processes. *Math. Operationsf., ser. statist.* **10**, 555–560.

Hanisch, K.-H., and D. Stoyan (1983): Remarks on statistical inference and prediction for a hard-core clustering model. *Statistics* **14**, 559–567.

Hanisch, K.-H., and D. Stoyan (1984): Once more on orientations in point processes. *Elektron. Informationsverarb. Kyb.* **20**, 279–284.

Hansen, J.P., and I.R. McDonald (1986): *Theory of Simple Liquids (2nd ed).* Academic Press, London.

Harkness, R.D., and V. Isham (1983): A bivariate spatial point pattern of ants' nests. *Appl. Statist.* **32**, 293–303.

Hastings, H.M., and S. Sugihara (1993): *Fractals: A User's Guide for the Natural Sciences.* Oxford University Press, Oxford, New York, Tokyo.

Hausner, H.H. (1966): Characterization of the powder particle shape. *Planseeber. Pulvermetall.* **14**, 75–84.

Hawkes, J. (1974): Hausdorff measure, entropy, and the independence of small sets. *Proc. Lond. Math. Soc.* (3) **28**, 700–724.

Hayashi, S. (1985): Self-similar sets as Tarski's fixed points. *Publ. Res. Inst. Math. Sci. Kyoto Univ.* **21**, 1059-1066.

Heesterbeek, J.A.P., J.M.A.M. van Neerven, H.A.J.M. Schellinx and M. Zwaan (1990): The nature of fractal geometry. *CWI Q.* **3**, 137-149.

Heinrich, L. (1988): Asymptotic Gaussianity of some estimators for reduced factorial moment measures and product densities of stationary Poisson cluster processes. *Statistics* **19**, 87-106.

Heinrich, L. (1993): Asymptotic properties of minimum contrast estimators for parameters of Boolean models. *Metrika* **40**, 67-94.

Heinrich, L., and V. Schmidt (1985): Normal convergence of multidimensional shot noise and rates of this convergence. *Adv. Appl. Prob.* **17**, 709-730.

Hermann, H. (1991): A new random surface fractal for application in solid state physics. *phys. stat. sol.* (b) **163**, 329-336.

Hermann, H.J., and Roux, S. (eds.) (1990): *Statistical Models for the Fracture of Disordered Media.* North Holland, Amsterdam.

Hermann, H., H. Wendrock and D. Stoyan (1989): Cell-area distributions of planar Voronoi mosaics. *Metallography* **23**, 189-200.

Hinde, A.L., and R.E. Miles (1980): Monte Carlo estimates of the distribution of the random polygons of the Voronoi tessellation with respect to a Poisson process. *J. Statist. Comp. Simul.* **10**, 205-223.

Holzfuss, J., and G. Mayer-Kress (1986): An approach to error estimation in the application of dimension algorithms. In: Mayer-Kress (1986), 114-122.

Hornbogen, E. (1987): Fractals in microstructure of metals. *Int. Mater. Rev.* **34**, 277-296.

Hosemann, R., and S.N. Bagchi (1962): *Direct Analysis of Diffraction by Matter.* North Holland, Amsterdam.

Huller, D. (1985): *Quantitative Formanalyse von Partikeln.* VDI-Fortschrittsberichte, Reihe 3: Verfahrenstechnik, Nr. 101, VDI-Verlag, Düsseldorf.

Hunt, F. (1990): Error analysis and convergence of capacity dimension algorithms. *SIAM J. Appl. Math.* **50**, 307-321.

Hutchinson, J.E. (1981): Fractals and self-similarity. *Indiana Univ. Math. J.* **30**, 713-747.

Icke, V. and R. van de Weygaert (1987): Fragmenting the universe. I. Statistics of two-dimensional Voronoi foams. *Astron. Astrophys.* **184**, 16-32.

Isham, V. (1984): Multitype Markov point processes: Some approximations. *Proc. R. Soc. Lond.* **A391**, 39-53.

Isham, V. (1993): Statistical aspects of chaos. In: *Chaos and Networks - Statistical and Probabilistic Aspects* (ed. Barndorff-Nielsen, O., J.L. Nielsen und W.S. Kendall), Chapman & Hall, London, New York, pp. 124-200.

Ivanoff, G. (1982): Central limit theorems for point processes. *Stoch. Proc. Appl.* **12**, 171-186.

Jäckel, K.-H. (1986): Monte Carlo tests. *Statist. Software Newsl.* **12**, 35-39.

Jansson, B. (1964): Generation of random bivariate normal deviates and computation of related integrals. *Nordisk Tidskrift Inform. Behandl.* **4**, 205-212.

Jensen, J.L. (1991): A note on asymptotic normality in the thermodynamic limit at low densities. *Adv. Appl. Math.* **12**, 387-399.

Jensen, J.L. (1993): Asymptotic normality of estimates in spatial point processes. *Scand. J. Statist.* **20**, 97-109.

Jensen, J.L. (1993): Chaotic dynamical systems with a view towards statistics: a review. In: *Chaos and Networks - Statistical and Probabilistic Aspects* (ed. O. Barndorff-Nielsen, J.L. Nielsen and W.S. Kendall), Chapman & Hall, London, New York, pp. 201-250.

Jensen, J.L., and J. Møller (1991): Pseudolikelihood for exponential family models of spatial processes. *Ann. Appl. Prob.* **1**, 445-461.

Jewell, N.P., and J.P. Romano (1985): Evaluating inclusion functionals for random convex hulls. *Z. Wahrscheinlichkeitsth. verw. Geb.* **68**, 415-424.

Jourlin, M., and B. Laget (1988): Convexity and Symmetry: Part 1. In: Serra (1988), 343–357.

Kallay, M. (1974): Reconstruction of a plane convex body from the curvature of its boundary. *Isreal J. Math.* **17**, 150–161.

Kallay, M. (1975): The extreme bodies in the set of plane convex bodies with a given width function. *Isreal J. Math.* **22**, 203–207.

Karr, A.F. (1986): *Point Processes and their Statistical Inference.* Marcel Dekker, New York, Zürich.

Kendall, D.G. (1983): The shape of Poisson-Delaunay triangles. In: Demetrescu, M.C., and M. Josifescu, eds., *Studies in Probabilities and Related Topics,* etc., Nagard, Montreal, 321–330.

Kendall, D.G. (1984): Shape manifolds, procrustean metrics, and complex projective spaces. *Bull. Lond. Math. Soc.* **16**, 81–121.

Kendall, D.G. (1985): Exact distributions for shapes of random triangles in convex sets. *Adv. Appl. Prob.* **17**, 308–329.

Kendall, D.G. (1989): A survey of the statistical theory of shape. *Statist. Sci.* **4**, 87–120.

Kendall, D.G. (1991): The Mardia–Dryden shape distribution for triangles: a stochastic calculus approach. *J. Appl. Prob.* **28**, 225–230.

Kerstan, J., K. Matthes, and J. Mecke (1974): *Unbegrenzt teilbare Punktprozesse.* Akademie-Verlag, Berlin.

Koen, C. (1991): Approximate confidence bounds for Ripley's L statistic for random points in a square. *Biom. J.* **33**, 173–178.

Kolmogorov, A.N., and V.M. Tihomirov (1961): ε-entropy and ε-capacity of sets in functional spaces. *Am. Math. Soc. Transl.* **17**, 277–364.

Kôno, N. (1986): Hausdorff dimensions of sample paths for selfsimilar processes. In: Eberlein, E., and M.S. Taqqu, eds., *Dependence in Probability and Statistics.* Birkhäuser, Boston, 109–117.

Kovalevsky, V.A. (1989): Finite topology as applied to image analysis. *Comp. Vision, Graphics, Image Proc.* **46**, 141–161.

Krumbein, W.C. (1941): The effect of abrasion on the size, shape and roundedness of rock fragments. *J. Geol.* **49**, 482–520.

Kruse, R. (1987): On the variance of random sets. *J. Math. Anal. Appl.* **122**, 469–473.

Kruse, R., and K.D. Meyer (1987): *Statistics with Vague Data.* D. Reidel, Dortrecht, Boston.

Künsch, H. (1986): Discrimination between monotonic trends and long-range dependence. *J. Appl. Prob.* **23**, 1025–1030.

Künsch, H. (1987): Statistical aspects of self-similar processes. In: *Proc. 1st World Conf. Bernoulli Society,* Vol. 1. VNU Science Press, 67–76.

Le Caër, G., and J.S. Ho (1990): The Voronoi tessellation generated from eigenvalues of complex random matrices. *J. Phys. A: Math. Gen.* **23**, 3279–3295.

Ledrappier, F. (1992): On the dimension of same graphs. *Contemp. Maths.* **135**, 285–293.

Leichtweiss, K. (1980): *Konvexe Mengen.* Deutscher Verlag der Wissenschaften, Berlin.

Lemaitre, J., A. Gervois, J.P. Troadec, N. Rivier, M. Ammi, L. Oger and D. Bideau (1993): Arrangement of cells in Voronoi tessellations of monosize packings of discs. *Phil. Mag.* **B67**, 347–362.

Lešanovsky, A., and J. Rataj (1990): Determination of compact sets in Euclidean spaces by the volume of their dilatation. *Proc. DIANA III, Bechyně,* Czech Republic, 1990, Math. Institute of ČAV, Prague, pp. 165–177.

Liebovitch, L.S., and T. Toth (1980) A fast algorithm to determine fractal dimension using box counting. *Phys. Lett.* **141A**, 386–390.

Liggett, T.M. (1985): *Interacting Particle Systems.* Springer-Verlag, New York.

Lighthill, M.J. (1958): *Introduction to Fourier Analysis and Generalized Functions.* Cambridge University Press, Cambridge.

Lilliefors, H.W. (1969): On the Kolmogorov-Smirnov test for normality with mean and variance unknown. *J. Am. Statist. Assoc.* **62**, 399–402 (Correction **64**, 1702).

Lilliefors, H.W. (1971): On the Kolmogorov-Smirnov test for the exponential distribution with mean unknown. *J. Am. Statist. Assoc.* **64**, 387–389.

Liu, Lee Tzao, Chen Keh Wei, and R. Attele (1991): Modelling random convex sets. *System Scie. & Math. Scie.* **4**, 111–127.

Lohmann, G.P. (1983): Eigenshape analysis of microfossils: a general morphometric procedure for describing changes in shape. *Math. Geol.* **15**, 659–672.

Lovejoy S., and B.B. Mandelbrot (1985): Fractal properties of rain, and a fractal model. *Tellus* **37A**, 209–232.

Lyashenko, N.N. (1983): Statistics of random compacts in Euclidean space. *Soviet J. Math.* **21**, 76–92.

McKendrick, I.J. (1991): Detecting departures from complete spatial randomness.

Mallows, C.L., and J.M.C. Clark (1970): Linear-intercept distributions do not characterize plane sets. *J. Appl. Prob.* **7**, 240–244 (Correction **8**, (1971), 208–209).

Mandelbrot, B.B. (1969): Long-run linearity, locally Gaussian process, *H*-spectra, and infinite variances. *Int. Econ. Rev.* **10**, 82–113.

Mandelbrot, B.B. (1971): A fast fractional Gaussian noise generator. *Water Resources Res.* **7**, 543–553.

Mandelbrot, B.B. (1972): Renewal sets and random cut-outs. *Z. Wahrscheinlichkeitstheorie verw. Geb.* **22**, 145–157.

Mandelbrot, B.B. (1975): Limit theorems on the self-normalized range for weakly and strongly dependent processes. *Z. Wahrscheinlichkeitstheorie verw. Geb.* **31**, 271–285.

Mandelbrot, B.B. (1977): *Fractals, Form Chance and Dimension.* W.H. Freeman, San Francisco.

Mandelbrot, B.B. (1982): *The Fractal Geometry of Nature,* W.H Freeman, New York.

Mandelbrot, B.B. (1988): Foreword: People and events behind the 'Science of Fractal Images'. In: Peitgen and Saupe (1988), 1–20.

Mandelbrot, B.B., and J.W. van Ness (1968): Fractional Brownian motion, fractional noises and applications. *SIAM Rev.* **10**, 422–437.

Mandelbrot, B.B., and M.S. Taqqu (1979): Robust R/S analysis of long run serial correlations. In. Proc. 42nd Session of the ISI, Manila, *Bulletin of the I.S.I.* **48** (Book 2), 69–100.

Mandelbrot, B.B., and J.R. Wallis (1969): Noah, Joseph, and operational hydrology. *Water Resources Res.* **4**, 909–918.

Mardia, K.V. (1989a): Comment: Some contributions to shape theory. *Statist. Sci.* **4**, 108–111.

Mardia, K.V. (1989b): Shape analysis of triangles through directional techniques. *J. R. Statist. Soc.* **51B**, 449–458.

Mardia, K.V., and I.L. Dryden (1989a): The statistical analysis of shape data. *Biometrika.* **76**, 271–282.

Mardia, K.V., and I.L. Dryden (1989b): Shape distributions for landmark data. *Adv. Appl. Prob.* **21**, 742–755.

Mardia, K.V., and I.L. Dryden (1994): Shape averages and biases. *Adv. Appl. Prob.* **26**.

Mase, S. (1984): Locally asymptotic normality of Gibbs models on a lattice. *Adv. Appl. Prob.* **16**, 585–602.

Mase, S. (1986): On the possible form of size distributions for Gibbsian processes of mutually non-intersecting balls. *J. Appl. Prob.* **23**, 646–659.

Mase, S. (1990): Mean characteristics of Gibbsian point processes. *Ann. Inst. Statist. Math.* **42**, 203–220.

Mase, S. (1992): Uniform LAN condition of planar Gibbian point processes and optimality of maximum likelihood estimators of soft-core potential functions. *Probab. Theory Relat. Fields* **92**, 51–67.

Mason, C.C., and R.L. Folk (1958): Differentiation of beach, dune and aeolian flat environments by size analysis. *J. Sediment. Petrology* **28**, 211-226.

Matern, B. (1960): Spatial Variation. *Medd. Statens Skogforskringsinstitut* **49** (5), 1-144.

Matern, B. (1986): *Spatial Variation*. Lecture Notes in Statistics 36, Springer-Verlag, Berlin, Heidelberg, New York.

Matheron, G. (1975): *Random Sets and Integral Geometry*. J. Wiley & Sons, New York.

Matheron, G. (1989): *Estimating and Choosing*. Springer-Verlag, Berlin Heidelberg, New York.

Matheron, G., and J. Serra (1988): Convexity and symmetry. Part 2. In: Serra (1988), 359-375.

Matthes, K. (1963): Stationäre zufällige Punktfolgen. I. *Jahresbericht d. DMV* **66**, 66-79.

Mattila, P. (1975): Hausdorff dimension, orthogonal projections and intersections with planes. *Ann. Acad. Sci. Fenn. Ser. AI*, **1**, 227-244.

Mattila, P. (1984): Hausdorff dimension and capacities of intersections of sets in n-space. *Acta Math.* **152**, 77-105.

Mauldin, R.D. (1986): On the Hausdorff dimension of graphs and random recursive objects. In: Mayer-Kress (1986), 28-33.

Mauldin, R.D., and S.C. Williams (1986): Random constructions: asymptotic geometric and topological properties. *Trans. Am. Math. Soc.* **295**, 325-346.

Mayer-Kress, G. (ed.) (1986): *Dimensions and Entropies in Chaotic Systems: Quantification of Computer Behaviour*. Springer-Verlag, Berlin, Heidelberg, New York, Tokyo.

Mecke, J. (1987): Extremal properties of some geometric processes. *Acta Appl. Math.* **9**, 61-69.

Mecke, J., R. Schneider, D. Stoyan and W. Weil (1990): *Stochastische Geometrie*. Birkhäuser, Basel, Boston, Berlin.

Meloy, T.P. (1977): A hypothesis for morphological characterization of particle shape and physiochemical properties. *Powder Technol.* **16**, 233-253.

Miles, R.E. (1969): Poisson flats in Euclidean space, I: A finite number of random uniform flats. *Adv. Appl. Prob.* **1**, 211-237.

Miles, R.E. (1970): On the homogeneous planar Poisson point process. *Math. Biosci.* **6**, 85-127.

Miles, R.E. (1972): Multidimensional perspectives in stereology. *J. Microsc.* **95**, 181-185.

Miles, R.E. (1974): On the elimination of edge-effects in planar sampling. In: Harding, E.F., and D.G. Kendall, eds., *Stochastic Geometry* J. Wiley & Sons, Chichester, 228-247.

Miles, R.E. (1985): A comprehensive set of stereological formulae for not-necessarily-convex particles. *J. Microsc.* **138**, 127-136.

Miles, R.E. (1988): An interesting pair of identically distributed random tessellations, with vertices of *Y*-type, which always cross orthogonally. *J. Micros.* **151**, 191-196.

Mohr, D.L. (1981): *Modelling Data as Fractional Gaussian Noise*. Dissertation, Princeton University.

Molchanov, I.S. (1989): On the convergence of the empirical capacity functional of stationary random sets. *Th. Probab. Math. Statist.* **38**, 107-109.

Molchanov, I.S. (1990): Empirical estimation of quantils of random closed sets. *Theor. Prob. Appl.* **35**, 594-600.

Molchanov, I.S. (1991): Random sets. Review of certain results and applications. *Ukr. Mat. Žurn.* **43**, 1587-1599.

Molchanov, I.S. (1993a): *Limit Theorems for Unions of Random Closed Sets*. Lecture Notes in Mathematics, 1561. Springer-Verlag, Berlin, Heidelberg, New York.

Molchanov, I.S. (1993b): Limit theorems for convex hulls of random sets. *Adv. Appl. Prob.*, **25**, 305-414.

Moran, P.A.P. (1946): Additive functions of intervals and Hausdorff measure. *Proc. Camb. Phil. Soc.* **42**, 15-23.

Moreau, J.M., and M. Rubio (1987): The computation of some morphological parameters. *Stereol.* **6/III**: 1029-1034.

Mosimann, J.E. (1970): Size allometry: size and shape variables with characterizations of the lognormal and generalized gamma distributions. *J. Am. Statist. Assoc.* **65**, 930-948.

Moyeed, R.A., and A. Baddeley (1989): Stochastic approximation of the MLE for a spatial point pattern. *Scand. J. Statist.* **18**, 39-50.

Muche, L., and D. Stoyan (1992): Contact and chord length distributions of the Poisson Voronoi tessellation. *J. Appl. Prob.* **29**, 467-471.

Müller, G. (1970): *Methoden der Sedimentuntersuchung. Sediment-Petrologie* I. E. Schweizerbarthsche Verlags-Buchhandlung, Stuttgart.

Müssigmann, U. (1991): Texture analysis using fractal dimensions. In: Encarnago, J.L., and H.-D. Peitgen eds., *Proc. Workshop on Fractal Geometry and Computer Graphics.* Springer-Verlag, Berlin.

Nagel, W. (1993): Orientation-dependent chord length distributions characterize convex polygons. *J. Appl. Prob.* **30**, 730-736.

Näther, W., and M. Albrecht (1990): Linear regression with random fuzzy observation. *Statistics.* **21**, 521-531.

Newell, N.P., and J.P. Romano (1985): Evaluating inclusion functionals for random convex hulls. *Z. Wahrscheinlichkeitsth. verw. Geb.* **68**, 415-424.

Nielsen, H.L. (1985): Shapes of sand grains estimated from grains mass and river size. In: Barndorff-Nielsen, D.E., I.T. Møller, K. Rømer Rasmussen and B.B. Willetts, eds., *Proc. Int. Workshop on Physics of Blown Sand.* Memoirs 8, Department of Theoretical Statistics, University of Aarhus, 677-688.

Nussbaumer, H.J. (1981): *Fast Fourier Transformation and Convolution Algorithms.* Springer-Verlag, Berlin, Heidelberg, New York.

Ogata, Y. (1988): Statistical methods for earthquake occurrences and residual analysis for point processes. *J. Am. Statist. Assoc.* **83**, 9-27.

Ogata, Y., and K. Katsura (1988): Likelihood analysis of spatial inhomogeneity for marked point patterns. *Ann. Inst. Statist. Math.* **40**, 29-39.

Ogata, Y., and K. Katsura (1991): Maximum likelihood estimates of the fractal dimension for random spatial patterns. *Biometrika* **78**, 463-474.

Ogata, Y., and M. Tanemura (1981): Estimation of interaction potentials of spatial point patterns through the maximum likelihood procedure. *Ann. Inst. Statist. Math.* **33B**, 315-338.

Ogata, Y., and M. Tanemura (1984): Likelihood analysis of spatial point patterns. *J. R. Statist. Soc.* **B46**, 496-518.

Ogata, Y., and M. Tanemura (1985): Estimation of interaction potentials of marked spatial point patterns through the maximum likelihood method. *Biometrics* **41**, 421-433.

Ogata, Y., and M. Tanemura (1986): Likelihood estimation of interaction potentials and external fields of inhomogeneous spatial point patterns. In: Francis, I.S., B.F.J. Mauly, and F.C. Lam, eds., *Proc. Pacific Statistical Congress*, North Holland, Amsterdam, 150-154.

Ogata, Y., and M. Tanemura (1989): Likelihood estimation of softcore interaction potentials for Gibbsian point patterns. *Ann. Inst. Statist. Math.* **41**, 583-600.

Ohser, J. (1983): On estimators for the reduced second moment measure of point processes. *Math. Operationsf. Statist., ser. statist.* **14**, 63-71.

Ohser, J., and D. Stoyan (1981): On the second-order and orientation analysis of planar stationary point processes. *Biom. J.* **23**, 523-533.

Ohser, J., and H. Tscherny (1988): *Grundlagen der quantitativen Gefügeanalyse.* Freiberger Forschungshefte B264, Deutscher Verlag für Grundstoffindustrie, Leipzig.

Olkin, I., A.J. Petkau and J.V. Zidek (1981): A comparison of n estimators for the binomial distribution. *J. Am. Statist. Assoc.* **76**, 637-642.

Osborne, A.R., A.D. Kirwan, A. Provenzale and L. Bergamasco (1989): Fractal drifter trajectories in the Kuroshio extension. *Tellus* **41A**, 416-435.

Pahl, M.H., G. Schadel and H. Rumpf (1973): Zusammenstellung von Teilchenformbeschreibungsmethoden. *Aufbereit. Tech.* **14**, 257–264, 672–683, 759–764.

Pape, H., L. Riepe and J.R. Schopper (1987): Theory of self-similar network structures in sedimentary and igneous rocks and their investigation with microscopical and physical methods. *J. Micros.* **148**, 121–147.

Patzschke, N., and U. Zähle (1990): Self-similar random measures. IV-The recursive construction model of Falconer, Graf, and Mauldin and Williams. *Math. Nachr.* **149**, 285–302.

Peitgen, H.-O. and P.H. Richter (1986): *The Beauty of Fractals.* Springer-Verlag, Berlin, Heidelberg, New York, Tokyo.

Peitgen, H.-O. and D. Saupe (eds.) (1988): *The Science of Fractal Images.* Springer-Verlag, Berlin, Heidelberg, New York, Tokyo.

Penttinen, A.K. (1984): *Modelling Interaction in Spatial Point Patterns: Parameter Estimation by the Maximum Likelihood Method.* Jyväskylä Studies in Computer Science, Economics and Statistics 7.

Penttinen, A.K., and D. Stoyan (1989): Statistical analysis for a class of line segment processes. *Scand. J. Statist.* **16**, 153–161.

Penttinen, A., D. Stoyan and H.M. Henttonen (1992): Marked point processes in forest statistics. *Forest Science,* **38**, 4, 806–824.

Percus, J.K. (1964): The pair distribution function in classical statistical mechanics. In: Frisch, H.L., and J.L. Lebowitz, eds., *The Equilibrium Theory of Classical Fluids.* Benjamin, New York.

Pettijohn, F.J. (1975): *Sedimentary Rocks.* 3rd edn. Harper & Brothers, New York.

Piefke, F. (1978): Beziehungen zwischen der Sehnenlängenverteilung und der des Abstandes zweier zufälliger Punkte im Eikörper. *Z. Wahrscheinlichkeitsth. verw. Geb.* **43**, 129–134.

Piefke, F. (1979): The chord length distribution of the ellipse. *Liet. Mat. Rink* **19** (3), 45–54.

Pielou, E.C. (1977): *Mathematical Ecology.* J. Wiley & Sons, New York.

Pirard, E. (1992): Roughness analysis of powders using mathematical morphology. *Acta Stereol.* **11** Suppl. I, 533–538.

Pirard, E. (1994a): Shape processing and analysis using the Calypter. *J. Microscopy*

Pirard, E. (1994b): Analyse morphometrique des poudres: une approche systematique et robuste par la morphologie mathematique. *Revue Metallurgie. Science et Genie des Materiaux.*

Prod'Homme, M., L. Chermant, M. Coster and J.L. Chermant (1991): Étude morphologique des poudres. Applications à titanate de baryum et à l'alumine. *J. Phys. (Paris)* III **1**, 675–688.

Radke, J.D. (1988): On the shape of a set of points. In Toussaint (1988a), 105–136.

Rataj, J. (1994): Characterization of compact sets by their dilatation volume. *Math. Nachr.*

Rayleigh, (1942): The ultimate shape of pebbles, natural and artificial. *Proc. R. Soc. Lond.* **A181**, 107–120.

Reiss, R.-D. (1993): *A Course on Point Processes.* Springer-Verlag, New York, Berlin, Heidelberg.

Renyi, A., and R. Sulanke (1963): Über die konvexe Hülle von n zufälligen gewählten Punkten. *Z. Wahrscheinlichkeitsth. verw. Geb.* **2**, 75–84.

Renyi, A., and R. Sulanke (1964): Über die konvexe Hülle von n zufällig gewählten Punkten II. *Z. Wahrscheinlichkeitsth. verw. Geb.* **3**, 138–147.

Reyment, R.A., R.E. Blackith and N.A. Campbell (1984): *Multivariate Morphometrics*, 2nd edn. Academic Press, London, New York.

Rice, J.A., and B. Silverman (1991): Estimating the mean and covariance structure nonparametrically when the data are curves. *J. R. Statist. Soc.* **53**, 233–243.

Richter, H. (1963): Verallgemeinerung eines in der Statistik benötigten Satzes der Maßtheorie. *Math. Ann.* **150**, 85–90, (Correction **150**, 440–441).

Rigaut, J.P. (1988): Automated segmentation by mathematical morphology and fractal geometry. *J. Micros.* **150**, 21-30.

Rigaut, J.P. (1989): Fractals in biological analysis and vision analysis and vision. In: Losa, G., Merlini, D. and Moresi, R., eds, *Gli oggetti frattali in astrofisica, biologia, fisica e matematica.* Edizioni Cerfim Locarno, 111-145.

Ripley, B.D. (1976): The second-order analysis of stationary point processes. *J. Appl. Prob.* **13**, 255-266.

Ripley, B.D. (1977): Modelling spatial pattern (with discussion). *J. R. Statist. Soc.* **B39**, 172-212.

Ripley, B.D. (1979): Simulating spatial patterns: dependent samples from a multivariate density. *Appl. Stat.* **28**, 109-112.

Ripley, B.D. (1981): *Spatial Statistics.* J. Wiley & Sons, New York, Chichester.

Ripley, B.D. (1987): *Stochastic Simulation.* J. Wiley & Sons, New York, Chichester, Brisbane, Toronto, Singapore.

Ripley, B.D. (1988): *Statistical Inference for Spatial Processes.* Cambridge University Press, Cambridge.

Ripley, B.D. (1990): Gibbsian interaction models. In: Griffiths, D.A., ed., *Spatial Statistics: Past, Present and Future.* Image, New York, 3-28.

Ripley, B.D. (1992): Stochastic models for the distribution of rock types in petroleum reserves. In: Waldau, A., and P. Guttorp, eds., *Statistics in the Environmental and Earth Sciences.* Edward Arnold, London, Melbourne, Auckland, 245-282.

Riss, J. and J. Grohier (1986): Average equivalent shape: definition and identification. *Acta Stereol.* **5**, 37-48.

Rogers, C.A. (1970): *Hausdorff Measures.* Cambridge University Press, Cambridge.

Rogers, C.A. (1976): Problem 4. *Bull. Lond. Math. Soc.* **8**, 29.

Rohlf, F.J. (1986): Relationships among eigenshape analysis, Fourier analysis and analysis of coordinates. *Geology* **18**, 845-854.

Rohlf, F.J., and D. Slice (1990): Methods for comparison of sets of landmarks. *System. Zool.* **39**, 40-59.

Rösler, R., H.A. Schneider and R. Schuberth (1987): Relations between particle shape and Fourier coefficients. *Powder Technol.* **49**, 255-260.

Russ, J.C. (1989): Automatic methods for the curvature of lines, features, and feature alignment in images. *J. Comp.-Ass. Microsc.* **1**, 39-77.

Sachs, L. (1984): *Applied Statistics. A Handbook of Techniques.* Springer-Verlag, New York.

Samarov, A., and M.S. Taqqu (1988): On the efficiency of the sample mean in long-memory noise. *J. Time Series Anal.* **9**, 191-200.

Santaló, L.A. (1976): *Integral Geometry and Geometric Probability.* Addison-Wesley, Reading, Massachusetts.

Särkkä, A. (1992): A note on robust intensity estimator for point processes. *Biom. J.* **34**, 757-767.

Särkkä, A. (1993): *Pseudo-likelihood approach for pair potential estimation of Gibbs processes.* Jyv<skylä Studies in Computer Science, Economics and Statistics **22**.

Saupe, D. (1988): Algorithms for random fractals. In: Peitgen and Saupe (1988), 71-113.

Schmidt, P.W. (1989): Use of scattering to determine the fractal dimension. In: Avnir (1989), 67-79.

Schmidt-Kittler, N. (1984): Pattern analysis of occlusal surfaces in hyprodont herbivores... Palaeontology. *Proc. Koninkl. Akad. van Wetenschappen* **B87**, 453-480.

Schmidt-Kittler, N. (1986): Evaluation of occlusal patterns of hypsodont rodent dentitions by shape parameters. *N. Jb. Geol. Paläont. Abh.* **173**, 75-98.

Schmitt, M. (1992): On two inverse problems in mathematical morphology. In: Dougherty, E., ed., *Mathematical Morphology in Image Processing.* Marcel Dekker, New York, Basel.

Schneider, R. (1988): Random approximation of convex sets. *J. Microsc.* **151**, 211-228.

Schneiderhöhn, P. (1954): Eine vergleichende Studie über Methoden zur quantitativen Bestimmung von Abrundung und Form an Sandkörnern. *Beitr. Miner. Petr.* **4**, 172-191.

Schuberth, R. (1987): *Möglichkeiten und Grenzen der automatischen Bildanalyse zur mathematischen Korngrößen- und Kornformbeschreibung unter besonderer Berücksichtigung der Fourieranalyse.* Dissertation, Bergakademie Freiberg.

Schürger, K. (1983): Ergodic theorems for subadditive superstationary families of convex compact random sets. *Z. Wahrscheinlichkeitstheorie verw. Geb.* **64**, 125-135.

Schuster, H.G. (1984): *Deterministic Chaos – An Introduction.* Physik-Verlag, Weinheim.

Schwarcz, H.P., and K.C. Shane (1969): Measurement of particle shape by Fourier analysis. *Sedimentology* **13**, 213-231.

Scott, G.H. (1980): The value of outline processing in the biometry and systematics of fossils. *Palaeontology* **23**, 757-768.

Serfozo, R.F. (1990): Point processes. In: Heyman, D.P., and M.J. Sobel, eds., *Handbook in Operations Research and Management Science: Stochastic Models.* North Holland, Amsterdam, New York, Oxford, Tokyo, 1-93.

Serra, J.P. (1982): *Image Analysis and Mathematical Morphology.* Academic Press, London.

Serra, J.P. (1988): *Image Analysis and Mathematical Morphology.* Vol. 2: Theoretical Advances. Academic Press, London.

Shaw, M.W. (1990): A test of spatial randomness on small scales, combining information from mapped locations within several quadrats. *Biometrics* **46**, 447-458.

Sibson, R. (1978): Studies in robustness of multidimensioned scaling: procrustes statistics. *J. R. Statist. Soc.* **B40**, 234-238.

Siegel, U., J. Lambrecht and D. (1990): Die Titaninhomogenität in Gußblöcken titanstabilisierter korrosionsbeständiger Stähle. *Neue Hütte* **35**, 81-89.

Silverman, B. (1986): *Density Estimation for Statistics and Data Analysis.* Chapman & Hall, London, New York.

Sinai, Ya.G. (1976): Self-similar probability distributions. *Theor. Prob. Appl.* **21**, 64-80.

Small, C.G. (1988): Techniques of shape analysis on sets of points. *Int. Statist. Rev.* **56**, 243-258.

Small, C.G. (1990): A survey of multidimensional medians. *Int. Statist. Rev.* **58**, 263-277.

Smith, R.L. (1992): Estimating dimension in noisy chaotic time series. *J. R. Statist. Soc.* **B54**, 329-351.

Snyder, D.L. (1975) *Random Point Processes.* J. Wiley & Sons, New York, London.

Solomon, H. (1978): *Geometric Probability.* SIAM, Philadelphia.

Stoyan, D. (1987): Statistical analysis of spatial point processes: a soft-core model and cross-correlations of marks. *Biom. J.* **29**, 971-980.

Stoyan, D. (1988): Thinnings of point processes and their use in the statistical analysis of a settlement pattern of deserted villages. *Statistics* **19**, 45-56.

Stoyan, D. (1989a) Statistical inference for a Gibbs point process of mutually non-intersecting discs. *Biom. J.* **51**, 153-161.

Stoyan, D. (1989b): On means, medians and variances for random compact sets. In: Hübler, A., W. Nagel, B.D. Ripley and G. Werner, eds., *Geobild '89.* Akademie-Verlag, Berlin, 99-104.

Stoyan, D. (1990): Estimation of distances and variances in Bookstein's landmark model. *Biom. J.* **32**, 843-849.

Stoyan, D. (1991a): Statistical estimation of model parameters of planar Neyman–Scott cluster processes. *Metrika* **39**, 67-74.

Stoyan, D. (1991b): Describing the anisotropy of marked planar point processes. *Statistics* **22**, 449-462.

Stoyan, D., and V. Beneš (1992): Anisotropy analysis for particle systems. *J. Microscopy* **164**, 159-168.

Stoyan, D., and M. Frenz (1993): Estimating mean land mark triangles. *Biom. J.* **35**, 643-647.

Stoyan, D., and P. Grabarnik (1991): Statistics for the stationary Strauss model by the cusp method. *Statistics* **22**, 283–289.

Stoyan, D., and G. Lippmann (1993): Models of stochastic geometry - a survey. *ZOR - Methods and Models of Oper. Res.* **38**, 235–260.

Stoyan, D., and H.-D. Schnabel (1990): Description of relations between spatial variability of microstructure and mechanical strength of aluminia ceramics. *Ceramics Int.* **16**, 11–18.

Stoyan, D., and H. Stoyan (1985): On one of Matern's hard-core point process models. *Math. Nachr.* **122**, 205–214.

Stoyan, D., and H. Stoyan (1990a): Exploratory data analysis for planar tessellations: Structural analysis and point processes methods. *Appl. Stoch. Mod. Data Anal.* **6**, 13–25.

Stoyan, D. and H. Stoyan (1990b): Further application of D.G. Kendalls's procrustes analysis. *Biom. J.* **32**, 293–301.

Stoyan, D., and K. Wiencek (1991): Spatial correlations in metal structure and their analysis. *Mater. Char.* **26**, 167–176.

Stoyan, D., W.S. Kendall and J. Mecke (1987): *Stochastic Geometry and Its Applications.* Akademie-Verlag, Berlin / J.Wiley & Sons, Chichester.

Stoyan, H., and D. Stoyan (1986): Simple stochastic models for the analysis of dislocation distributions. *Phys. stat. sol.* (a) **97**, 163–172.

Verteilung der Sehnenlängen an ebenen und räumlichen Figuren. *Math. Nachr.* **23**, 51-74.

Takacs, R. (1983): Estimator for the Pair-potential of a Gibbsian Point Process. *Johannes Kepler Univ. Linz Inst. f. Mathematik, Inst. Ber.* 238.

Takacs, R. (1986): Estimator for the pair-potential of a Gibbsian point process. *Math. Operationsf. Statist., Ser. Statist.* **17**, 429–433.

Takacs, R., and T. Fiksel (1986): Interaction pair-potentials for a system of ants'nests. *Biom. J.* **28**, 1007–1013.

Taqqu, M.S. (1986): A bibliographic guide to self-similar processes and long-range dependence. In: Eberlein, E., and M.S. Taqqu, eds., *Dependence in Probability and Statistics.* Birkhäuser, Boston, 137–162.

Taqqu, M.S. (1987): Toeplitz matrices and estimation of time series with long-range dependence. In: *Proc. 1st World Conf. Bernoulli Society,* Vol.1. VNU Science Press, 75–83.

Taqqu, M.S. (1988): Self-similar processes. In: Kotz, S., and N.L. Johnson, eds., *Encyclopedia of Statistical Sciences,* Vol. 8, J. Wiley & Sons, Chichester.

Taqqu, M.S., and J.B. Levy (1986): Using renewal processes to generate long-range dependence and high variability. In: Eberlein, E., and M.S. Taqqu, eds., *Dependence in Probability and Statistics.* Birkhäuser, Boston, 73–89.

Taylor, S.J. (1986a): The use of packing measure in the analysis of random sets. In: Ito, K., and T. Hida, eds., *Proc. 15th Symp. on Stochastic Processes and their Applications.* Lecture Notes in Mathematics **1203**, Springer-Verlag, Berlin, Heidelberg, New York, pp. 214–222.

Taylor, S.J. (1986b): The measure theory of random fractals. *Math. Proc. Camb. Phil. Soc.* **100**, 383–406.

Taylor, C.C., and S.J. Taylor (1991): Estimating the dimension of a fractal. *J. R. Statist. Soc.* **B53**, 353–364.

Theiler, J. (1990): Estimating fractal dimension. *J. Opt. Soc. Am.* **A7**, 1055–1073.

Tomppo, E. (1986): Models and Methods for Analysing Spatial Patterns of Trees. *Commun. Inst. Forest. Fennicae* **138**, *Helsinki.*

Toussaint, G.T. (ed.) (1988a): *Computational Morphology. A Computational Geometric Approach to the Analysis of Form.* North Holland, Amsterdam, New York, Oxford, Tokyo.

Toussaint, G.T. (1988b): A graph-theoretical primal sketch. In: Toussaint (1988a), 229–260.

Tricot, C. (1973): Sur la notion de densité. *Cahiers du Département d'Économétrie de l'Université de Genève.*

Tricot, C. (1982): Two definitions of fractal dimension. *Math. Proc. Camb. Phil. Soc.* **91**, 57–74.

Tricot, C. (1986): Geometry of the complement of a fractal set. *Phys. Lett.* **114A**, 434–456.

Tricot, C., D. Wehbi, J.-F. Quiniou, and C. Roques-Carmes (1987): Concepts of non-integer dimension applied to image treatment. *Acta Stereol.* **6/III**: 839–844.

Turcotte, D.L. (1992): *Fractals and Chaos in Geology and Geophysics*. Cambridge University Press, Cambridge.

Underwood, E.E. (1980): Stereological analysis of particle characteristics. In: Beddow, J.K., and T.P. Meloy, eds., *Testing and Characteristization of Fine Particles*. Heyden & Sons, Philadelphia, 77–96.

Underwood, E.E. and K. Banerji (1986): Fractals in Fractography. *Mater. Sci. Engng.* **80**, 1–14.

Upton, G. and B. Fingleton (1985): *Spatial Data Analysis by Example*. J. Wiley & Sons, Chichester.

Vervaat, W. (1985): Sample path properties of self-similar processes with stationary increments. *Ann. Prob.* **13**, 1–27.

Vitale, R.A. (1983): Some developments in the theory of random sets. *Bull. Int. Statist. Inst.* **50**, 863–871.

Vitale, R.A. (1984): On Gaussian random sets. In: Ambartzumian, R.V., and W. Weil, eds., *Stochastic Geometry, Geometric Statistics, Stereology*. Teubner-Texte zur Mathematik **65**, 222–224.

Vitale, R.A. (1987a): An alternative formulation of mean value for geometric figures. *J. Microsc.* **151**, 197–204.

Vitale, R.A. (1987b): Expected convex hulls, order statistics, and Banach space probabilities. *Acta Appl. Math.* **9**, 97–102.

Vitale, R.A. (1990): The Brunn–Minkowski inequality for random sets. *J. Multivariat. Anal.* **33**, 286–293.

Vorob'ev, O.Yu. (1984): *Srednemernoje Modelirovanie*. Nauka, Moscow.

Voss, K. (1982): Powers of chords for convex sets. *Biom. J.* **24**, 513–516.

Wadell, H. (1935): Volume, shape and roundness of quartz particles. *J. Geol.* **43**, 250–280.

Waksman, P. (1985): Plane polygons and a conjecture of Blaschke's. *Adv. Appl. Prob.* **17**, 774–793.

Waksman, P. 1986): Hypothesis testing in integral geometry. *Trans. Am. Math. Soc.* **296**, 507–520.

Waksman, P. (1987): Hypothesis testing in integral geometry: guessing the shape of a planar domain. *Contemp. Math.* **63**, 299–305.

Wartenberg, D. (1990): Exploratory spatial analysis: outliers, leverage points, and influence functions. In: Griffith, D.A., ed., *Proc. from a Symposium of Spatial Statistics: Past, Present and Future, Institute of Mathematical Geography, Syracuse University*, 133–162.

Wegmann, R. (1980): Einige Maßzahlen für nichtkonvexe Mengen. *Arch. Math.* **34**, 69–74.

Weil, W. (1982a): An application of the central limit theorem for Banach-space-valued random variables to the theory of random sets. *Z. Wahrscheinlichkeitsth. verw. Geb.* **60**, 203–208.

Weil, W. (1982b): Inner contact probabilities for convex bodies. *Adv. Appl. Prob.* **14**, 582–599.

Weil, W. and J. Wieacker (1993): Stochastic Geometry. In: Gruber, P., and J. Wills (1993), 1391–1438.

Wendel, J.G. (1962): A problem in geometric probability. *Math. Scand.* **11**, 109–111.

Whittaker, J. (1990): *Graphical Methods in Applied Multivariate Statistics*. J. Wiley & Sons, Chichester.

Willets, B.B., and M.A. Rice (1983): Practical representation of characteristic grain shape of sands: a comparison of methods. *Sedimentology* **30**, 557–565.

Willets, B.B., M.A. Rice and S.E. Swaine (1982): Shape effects in aeolian grain transport. *Sedimentology* **29**, 409–417.

Winkelmolen, A.M. (1982): Critical remarks on grain parameters, with special emphasis on shape. *Sedimentology* **29**, 255–265.

Wolff, R.C.L. (1990): A note on the behaviour of the correlation integral in the presence of a time series. *Biometrika* **77**, 689–697.

Wong, P.-Z. and J.-S. Lin (1988): Studying fractal geometry on submicron length scales by small-angle scattering. *Math. Geol.* **20**, 655–665.

Younes, L. (1988a): Estimation and annealing for Gibbsian fields. *Ann. Inst. H. Poincaré* **24**, 269–294.

Younes, L. (1989): Parametric inference for imperfectly observed Gibbsian fields. *Prob. Theory Rel. Fields* **82**, 625-645.

Zähle, U. (1984a): Sets and measures of fractional dimension. *Elektron. Infverarb. Kyb.* **20**, 261–269.

Zähle, U. (1984b): Random fractals generated by random cutouts. *Math. Nachr.* **116**, 27–52.

Zähle, U. (1988): Self-similar random measures. I. Notion, carrying Hausdorff dimension, and hyperbolic distribution. *Prob. Theory Rel. Fields* **80**, 79–100.

Zähle, U. (1990): Self-similar measures II - Generalization to self-affine measures. *Math. Nachr.* **146**, 85–88.

Zähle, U. (1991): Self-similar measures III - Self-similar random process. *Math. Nachr.* **151**, 121–148.

Zahn, C.T., and R.Z. Roskies (1972): Fourier descriptors for plane closed curves. *IEEE Trans. Comp.* **C21**, 269–281.

Zawadzski, A. (1987): Fractal structure and exponential correlation in rain. *J. Geophys. Res.* **92**, 9586–9590.

Ziezold, H. (1977): On expected figures and a strong law of large numbers for random elements in quasi-metric spaces. In: *Trans. 7th Prague Conf.*, Vol. A. Academica, Prague, 591–602.

Ziezold, H. (1989): On expected figures in the plane. In: Hübler, A., W. Nagel, B.D. Ripley and G. Werner, *Geobild '89*, Akademie-Verlag, Berlin, 105–110.

Ziezold, H. (1994): Mean figures and mean shapes applied to biological figure and shape distributions in the plane. *Biom. J.* **36**, 491–500.

Zurmühl, R. (1963): *Praktische Mathematik für Ingenieure und Physiker*. Springer-Verlag, Berlin, Göttingen, Heidelberg.

Index

* Now available in a lower priced paperback edition in the Wiley Classics Library.